Business and Government in the Oil Industry

A Case Study of Sun Oil, 1876–1945

INDUSTRIAL DEVELOPMENT AND THE SOCIAL FABRIC Volume 5

Editor: Glenn Porter, Director, *Regional Economic History Research Center, Eleutherian Mills—Hagley Foundation, Greenville, Delaware*

INDUSTRIAL DEVELOPMENT AND THE SOCIAL FABRIC

An International Series of Historical Monographs

Series Editor: Glenn Porter
Director, Regional Economic History Research Center, Eleutherian Mills—Hagley Foundation, Greenville, Delaware

Volume 1. FORUMS OF ORDER: The Federal Courts and Business in American History
Tony Allan Freyer, *University of Arkansas, Little Rock*

Volume 2. COMPETITION AND REGULATION: The Development of Oligopoly in the Meat Packing Industry
Mary Yeager, *University of California, Los Angeles*

Volume 3. KEEPING THE CORPORATE IMAGE: Public Relations and Business, 1900–1950
Richard S. Tedlow, *Graduate School of Business Administration, Harvard University*

Volume 4. THE GROWTH OF A REFINING REGION
Joseph A. Pratt, *School of Business Administration, University of California, Berkeley*

Volume 5. BUSINESS AND GOVERNMENT IN THE OIL INDUSTRY: A Case Study of Sun Oil, 1876–1945
August W. Giebelhaus, *Georgia Institute of Technology*

Volume 6. PIONEERING A MODERN SMALL BUSINESS: Wakefield Seafoods and the Alaskan Frontier
Mansel G. Blackford, *Ohio State University*

"For Barbara"

Business and Government in the Oil Industry:

A Case Study of Sun Oil, 1876–1945

by AUGUST W. GIEBELHAUS
Georgia Institute of Technology

JAI PRESS INC.

Greenwich, Connecticut

Library of Congress Cataloging in Publication Data

Giebelhaus, August W
Business and government in the oil industry.
(Industrial development and the social fabric;
v. 5)
Bibliography: p.
Includes index.
1. Sun Oil Company—History. 2. Petroleum
industry and trade—United States—History.
3. Industry and state—United States—History.
I. Title. II. Series.
HD9569.S9G53 338.7′6223382′0973 77-7795
ISBN 0-89232-089-3

165 West Putnam Avenue
Greenwich, Connecticut 06830

ISBN NUMBER: 0-89232-089-3
Library of Congress Catalog Card Number: 77-7795
Manufactured in the United States of America

CONTENTS

TABLES

PHOTOGRAPHS

Acknowledgment

Many individuals and institutions have been of great assistance to me in the preparation of this volume. Stephen Salsbury of the University of Sydney, Australia, supervised an earlier version of this study while he was at the University of Delaware, and he has consistently been the source of wise advice and criticism. Richard Ehrlich and Carol E. Hoffecker of the University of Delaware and Lynwood Bryant of the Massachusetts Institute of Technology also read the dissertation in its preliminary versions and provided valuable criticism and suggestions for improvement.

My interest in twentieth-century business-government relations was stimulated in seminars led by professors Salsbury and J.J. Huthmacher at the University of Delaware and Elliot A. Rosen at Rutgers University. Harold F. Williamson, Sr. first introduced me to the study of business history, and the history of the petroleum industry in particular, when I was a member of his seminar at the Eleutherian Mills-Hagley Foundation. Director Richmond D. Williams and his entire staff at the Eleutherian Mills Historical Library have given valuable assistance and support. In particular, I wish to thank John Beverly Riggs and Ruthanna Hindes of the manuscript department, and Betty Bright Low, Susan Danko, and Carol Hallman of the reference department.

Numerous individuals from the Sun Company have extended their cooperation and assistance. Among them are Woody E. Benson and Mildred Poole of the department of records management, Dean S. Chaapel of corporate public relations, and Beverly Knower of the Marcus Hook library. Charles Morris, Jr., Sun's coordinator of photography, provided me with photographs from the public relations archives. Former corporate secretary R. Anderson "Andy" Pew and assistant secretary Donald Ainsworth arranged for me to have access to the board minutes of the Sun Oil Company of New Jersey. Andy Pew, Robert G. Dunlop, Jno. G. "Jack" Pew, and the late Clarence Thayer took time to grant valuable personal interviews. In addition, Mr. Dunlop, Mr. Benson, and Seymour Rutherford of the Sun Company read an earlier version of the manuscript and made several helpful comments aimed at improving the text and correcting inaccuracies. Edward Cage, vice-president of legal services for Sun, also read the manuscript. Although there were no major areas of disagreement between the Sun people and myself, I have incorporated many of their suggestions into either the text or the notes.

Glenn Porter, my editor, has been of enormous help in the process of making the dissertation into a book. His substantive suggestions and many stylistic criticisms have improved the text considerably. Although I wish to acknowledge all of the assistance given to me by the individuals named above, I must emphasize that I alone take responsibility for all matters of accuracy and interpretation.

A Hagley Fellowship from the Eleutherian Mills-Hagley Foundation supported me in the research and writing of the original dissertation. I wish to also thank Patrick Kelly, former head of the department of social sciences at Georgia Tech, and Jon J. Johnston, present acting head, for arranging support during the period that I was revising the dissertation. Mrs. Anita Bryant typed the final version of the manuscript. Finally, I want to express thanks to my wife, Barbara, for her patient understanding and encouragement during the years that I have been engaged in this study.

CHAPTER I

Introduction: Sun Oil, the Petroleum Industry, and Public Policy

The Sun Oil Company has its origins in the early oil and natural gas industry that developed in Pennsylvania, New York, and Ohio during the last three decades of the nineteenth century. This was an industry initially characterized by intense competition and very rapid growth. It was also a chaotic industry that exhibited uneven crude oil production, ease of entry, and a fluctuating price structure for both crude and refined products. Soon, however, in response to the uncertainties of the petroleum business, one corporation emerged dominant: the Standard Oil Company. In order to understand Sun's early development and how the firm perceived its role in the burgeoning oil industry, it is useful to examine briefly the structure of that industry and how it changed in response to the Standard challenge. This chapter will provide a brief overview of the Sun Oil Company and the Pew family that managed it. It concludes with a discussion of the important interrelationship that has grown between the government and the oil industry and suggests how the examination of a particular firm's experience can shed light on the broader industry picture.

A NEW INDUSTRY

Although man has made use of the "rock oil" or petroleum found from natural seepages for centuries, the modern industry is usually seen as having begun with the drilling of Drake's well at Titusville, Pennsylvania, in 1859. American crude oil production rose from 2,000 barrels that year to 4,800,000 in 1869 and 5,250,000 in 1871. Production costs were low and incentives high, a situation that led to conditions in the oil-producing business approaching those of pure competition.[1] In addition to the natural competitive desire to obtain crude quickly, a legal incentive for rapid exploitation derived from the "rule of capture." This doctrine evolved in the courts in response to the problem of many property or lease owners drilling into a common underground pool of oil. The law said that surface owners could take as much oil as they wished from wells on their land, even though they might be draining the pool under a neighbor's property. The Pennsylvania courts looked to the English common law for precedent (*Acton* v. *Blundell*, 1843) and stressed the apparent migratory nature of oil. If game animals migrated from one estate to another, and the owner of the latter estate shot them, the law held that the game belonged to him. By the same reasoning, oil also belonged to the individual who had "captured" it on his land. It therefore became advantageous to drill quicly before someone drained the underground pool. This naturally led to the rapid production of crude oil and accompanying shifts in the price structure.[2]

The early leases granted in the oil regions also influenced the character of field development. Farmer-landowners favored immediate drilling on their land so they could enjoy their royalties. However, the wildcatter who held several leases might prefer to postpone sinking a well until more favorable market conditions prevailed. Consequently, the leases protected the interests of the lessor by requiring that drilling commence by a specific date. Along with the pressures of the rule of capture, these lease arrangements further encouraged rapid exploitation.[3] The pattern of oil field development became one of discovery, rapid competitive drilling, and scarcity, followed by a new discovery. This alternate cycle of glut and scarcity brought concomitant changes in the price structure, a "feast or famine" situation that marked the early phase of United States petroleum history.

Entry into the refining business was equally easy. The distillation equipment then used required relatively small amounts of capital to install and employed a simple technology. As a result of this internal competition and the vicissitudes of crude oil production, refining was also characterized by instability. Initially, the volume of crude might exceed the ability of refiners to process it quickly. Soon, however, those same refiners might discover that they had an excess capacity once the flow of oil slowed. Even if they reached an equilibrium between crude supplies and refining capacity, they might not find a market for their refined products. Again this uncertainty resulted in rapidly fluctuating prices for refined products as well as crude.[4]

Transportation soon emerged as an important factor in the growing industry. Since crude oil strikes usually occurred away from population and export centers, small diameter pipelines or gathering lines were developed to carry crude to railroad tank car facilities. Refinery location was an important factor, since it was most advantageous to be located near more than one rail line. This was of course also a period of intense competition among railroads, and an ability to negotiate preferential freight rates could become a desirable asset. Ultimately, a large refiner could use his market position to secure preferential rates, a practice that became the key to the Standard Oil Company's achievement of a monopolistic position in the industry.[5]

THE EMERGENCE OF STANDARD OIL

The history of Sun and other non-Standard independents must be examined in the light of the history of the Standard Oil Company (New Jersey), the giant enterprise that came to dominate the early petroleum industry and symbolize the evils of monopoly to a generation of Americans. The Standard story is intrinsically tied to the career of John D. Rockefeller, Sr. After beginning his career in the wholesale grocery business, Rockefeller, at the age of twenty-three, decided to try his hand at petroleum refining in Cleveland in 1863. Cleveland had become a major refining center because of its proximity to the Western Pennsylvania producing fields and its good access to both railroad and water transportation. Rockefeller and his partners soon emerged as a major force in the refining industry, and in 1867 formed the Standard Oil Company. Partnerships had been the most common form of organization in the industry, but increasing capital requirements made the corporate form more desirable. Rockefeller had initially obtained working capital by borrowing from local banks and bringing in new partners, but in 1870, he capitalized the Standard Oil Corporation (Ohio) at $1,000,000.[6]

At this point, Standard began the advances later referred to as the "conquest of Cleveland." Rockefeller and other large refiners had been able to obtain preferential railroad rates for some time, but Standard now entered into a contract with the Lake Shore Railroad (a New York Central subsidiary) guaranteeing a price thirty cents per barrel lower than the published rate in exchange for a promised sixty car loads a day. Using this contract as a weapon, Rockefeller pressured competing refiners to sell out to him, and Standard rapidly expanded horizontally. Many smaller units which at first resisted could not compete and were forced out of business. In 1871, Tom Scott of the Pennsylvania Railroad offered Standard and a pool of other large refiners a proposal they could not reject. In exchange for a monopoly of their business, the Pennsylvania would guarantee a significant rebate below the published price. The rebates were secret payments consisting of the difference between the published rates and those actually charged to Standard. In addition,

Scott offered drawbacks, or a commission on all petroleum shipments by non-pool members. The vehicle for this arrangement was a pool, the South Improvement Company.[7]

Critics charged that the pool threatened non-members, and Pennsylvania crude producers and independent refiners joined to defeat the South Improvement scheme by curbing oil shipments to pool members and successfully working to have the state of Pennsylvania revoke its charter. Thus, although this plan received wide attention, it never resulted in the shipment of a barrel of oil. Rockefeller then realized, however, that control of transportation was the key to industry stabilization, and he organized a cartel, the National Refiners Association. A board of fifteen representatives from the main refining centers of New York, Philadelphia, Pittsburgh, Cleveland, and the Northwest Pennsylvania oil regions functioned to purchase crude, allocate refining quotas, fix prices, negotiate railroad freight rates, and distribute profits among members. Opposed to this group was the Petroleum Producers' Association, a similar cartel designed to obtain the best possible crude prices for its members.[8]

The success of these cooperative arrangements depended on the degree of cohesiveness among members. Soon, both producers and refiners saw their efforts thwarted by internal bickering as well as intense competition from outsiders. Standard Oil opted for a system of more formal consolidation. By 1874, Rockefeller had zealously pursued his old strategy of efficient refining operations coupled with transportation rebates, and had begun acquiring large refining interests across the country. In a series of mergers, Rockefeller exchanged stock in Standard Oil for control of several large companies, and by 1878, Standard Oil owned over ninety percent of the total refining capacity in the United States.[9]

In that same year, a competitor, the Tidewater Pipeline Company, began construction of the first major trunk pipeline from the Bradford, Pennsylvania fields to the eastern refining centers. This represented a significant challenge to Standard's stranglehold over transportation, one to which Rockefeller responded. Soon, Standard began work on its own large trunklines to connect its major refineries with the producing regions.[10]

While in the process of major consolidation, Standard Oil developed a new form of business organization, the trust. Ohio law prohibited a corporation from holding stock in another, and also prevented a company chartered in one state from owning property in another. To circumvent the law, Rockefeller and his partners personally held the stock in Standard's various companies as "trustees." They formalized this arrangement with the Trust Agreement of 1882, which exchanged all

Standard Oil stock for trust certificates. This enabled Standard to retain a degree of corporate secrecy since the trustees, as legal agents for the stockholders and not for corporate management *per se*, could testify truthfully to investigators without revealing potentially damaging information.[11]

The trust arrangement continued until Rockefeller, concerned about continuity in the event of the death of major shareholders and continually harrassed by various prosecutions, reorganized Standard as a holding company in 1892. By this time, the holding company had become a more attractive method of industrial combination, following the passage of the New Jersey corporation law in 1889. This new general incorporation law permitted one corporation, such as Standard Oil (New Jersey), to own stock in various other companies. By the turn of the century, the New Jersey Standard Oil corporation owned majority control of all but three of the forty-one affiliated companies in the former trust.[12]

Although the Standard Oil trust dominated refining by 1880 and had expanded significantly into its own transportation and marketing operations, Rockefeller had consciously stayed out of the crude production end of the business. It was unnecessary to drill for his own oil when there was so much available from independent producers at cheap prices. Standard owned only four small producing tracts in Pennsylvania, a marginal share of total Appalachian production. When its managers decided to integrate backward on a large scale, their first thrust did not occur in oil but in gas. By 1886, Standard's subsidiary, the National Transit Company, controlled nine major gas-producing, pipeline, and marketing companies in Pennsylvania, New York, and Ohio. That same year, Rockefeller consolidated his operations with the formation of the National Gas Trust.[13]

National Transit's success encouraged Standard to consider expanding its oil production. A combination of declining supplies from the Pennsylvania fields and the opening of new supplies in Lima, Ohio, prompted this move. When Herman Frasch developed an exclusive Standard process to remove sulphur from the Ohio crude oil in 1890, the trust was in a favorable position to exploit the new supplies of crude. Standard's decision to move into crude oil production was fortuitous. In 1895, a new major force in the industry, the Pure Oil Company, began to challenge Rockefeller in the marketplace. Originally formed by a group of Pennsylvania producers, Pure integrated forward into transportation and refining and signified what would become the new structure of the petroleum industry—oligopolistic competition among large, vertically-integrated units. In this new environment, it was crucial for Standard to have secure crude supplies under its control.[14]

CHANGING INDUSTRY STRUCTURE

By the turn of the century, Standard Oil domination encountered challenges from two fronts. Vertically-integrated firms like Pure proved that they could compete, while at the same time the trust came under increasing attack in state and federal courts. Studies of Standard and the early evolution of public petroleum policy demonstrate how independent oil interests endeavored to use the law to their own advantage. Antitrust attacks and investigations into the trust's use of railroads and, later, trunk pipelines to corner transporation were often initiated by non-Standard interests.[15]

A series of state antitrust prosecutions filed from 1904–1906 had placed the trust on the defensive. But these cases were only a prelude to charges lodged in the United States Circuit Court for the Eastern District of Missouri in November 1906. The government charged the holding company, John D. Rockefeller, six other directors, and various partnerships and corporations within the organization with violations of Sections I and II of the Sherman Act.[16]

After listening to months of testimony, the St. Louis court found Standard Oil (New Jersey), seven individuals, and thirty-seven subsidiaries guilty on November 20, 1909, and ordered that the trust be dissolved. The final decision of the United States Supreme Court on Standard's appeal did not appear until May 15, 1911. Standard lost the appeal, and the only concession made by the high court was to extend the time period for carrying out the break-up to six months. Since the court-ordered plan followed a corporate rather than a functional scheme, the units spun off from the mother corporation varied greatly in size and degree of vertical integration. Some firms were exclusively refiners, others marketers, and still others producers or pipeline operators. In fact, the Standard Oil Company of California remained the only fully-integrated firm of those created out of the old Rockefeller trust.[17]

In analyzing why Standard grew the way it did, some "muckrakers" focused on the apparent avarice of Rockefeller and his associates.[18] Other writers, however, agreed with the rationale Standard itself presented in the brief its lawyers argued before the United States Supreme Court in 1911. This basic argument placed Standard consolidation within the context of thc unstable oil business of the 1870s.[19] From this perspective, Rockefeller's efforts allegedly represented a rational approach designed to solve fundamental economic problems in the most efficient manner. This highlights one of the fundamental issues related to the evolution of business enterprise in the past century: industrial efficiency through consolidation as opposed to enforced competition of many units through the unique American institution of antitrust legislation.

The dissolution of the Standard Trust in 1911 marks a watershed in the development of both industry structure and public oil policy, but historians have demonstrated that the Standard monopoly was already weakened on the eve of the Supreme Court decree. After the Spindletop, Texas discovery, rapid exploitation of new pools in the Gulf Coast, Mid-Continent, and California fields reduced Standard Oil's control of domestic refining from a high of ninety percent in 1880 to between sixty and sixty-five percent in 1911. That same year, Standard's competitors supplied seventy percent of the nation's fuel oil, forty-five percent of its lubricants, and thirty-five percent of the gasoline marketed. (See Appendix, Table 1.) With respect to gasoline, the rapid expansion of the automobile industry and Standard's slow entry into large-scale production enabled independents to capture a particularly large share of the market.[20]

In the years between the dissolution decree and the outbreak of World War I, independents, including Gulf, Texas, Union, Associated, and Sun, competed with the former Standard companies for the markets once dominated by the monopoly. In 1914, however, old Standard companies still controlled sixty-three percent of domestic refining capacity and independents thirty-seven percent. By 1920, the same Standard companies had only approximately thirty-four percent, while the independents, both large and small, controlled sixty-six percent. Competition in the petroleum industry was increasingly becoming oligopolistic (i.e., it was competition on grounds other than price). This trend toward oligopolistic competition appeared in the emergence of competition in terms of service to customers, such as more drive-in gasoline stations and other services, plus a greater emphasis on brand names and the beginning of major advertising and promotional campaigns by the larger firms.[21]

In this oligopolistic environment, there also occurred subtle changes in the ways both oil men and elected officials perceived the proper role of government in the petroleum industry. The independents had earlier united on the issue of antitrust; now the issues were more complex. Monopoly had developed as a response to economic instability. The monopoly was no more, but many of the causes of instability remained.

THE SUN OIL COMPANY

Sun began operation at a time when the wide-open, competitive Pennsylvania oil industry was coming increasingly under the domination of the Standard Oil Company. The firm's "adolescence" followed the Spindletop boom in 1901, and it reached maturity in the new era of oligopolistic competition after 1911. The founder of the firm, Joseph Newton Pew, Sr., initially concentrated on the natural gas business, which was then a peripheral activity related to oil production. He and his partner, Edward Octavius Emerson, expanded into oil after they had acquired capital and experience from their gas operations. After the discovery of oil in Ohio in 1885, they obtained leases in the rich Lima field. Although this marked the beginning of the Sun Oil Company, these oil endeavors were a sideline to the partners' investments in natural gas. This early experience makes Sun one of the oldest independent and continuously operating energy firms in the United States. (The word "independent" is used here to describe a company that operated outside Standard's control.)

Throughout its history, Sun remained relatively small compared to the petroleum giants described as "the Seven Sisters" (Jersey Standard, Texaco, Gulf, Standard of California, Standard of Indiana, Royal Dutch Shell, and British Petroleum).[22] It nevertheless maintained a significant share of the rapidly expanding industry. To compete with Standard Oil, Sun integrated early into the four main branches of the business: crude oil production, refining, transportation, and marketing. Rather than competing directly with Standard in the production of illuminating oil (kerosene), the dominant petroleum product in the nineteenth century, Sun developed specialized markets for gas oil used in the manufacture of illuminating gas, and lubricating oils which it sold at home and abroad. This was a common strategy that smaller firms used when competing with a giant enterprise like Standard Oil, and the firm established a reputation for specialty products which it maintained for many years.

When the huge Spindletop field inaugurated a new era of crude production in Texas in 1901, Sun obtained a large part of this producing business along with other non-Standard companies. The company experienced a growth surge following large sales in World War I and the successful post-war marketing of automobile motor oils and gasoline. In subsequent decades, Sun Oil became an important marketer in the eastern area of the United States. The Temporary National Economic Committee in 1939 ranked Sun fourteenth in size among the twenty top oil companies, and Sun's management described it as a "large independent" or "small major."[23] In the years following the 1911 dissolution of the Standard Oil Trust, the term "independent" came to designate a non-integrated, non-major producer, refiner, or marketer of petroleum. Significantly, when Sun's executives later used this term, they emphasized the older meaning of a competitive, non-Standard firm.

Today, Sun's relative size in the industry remains much the same as it was in 1939, but its influence in the economy has grown along with that of the other oil companies. The 1975 *Fortune* magazine listing of the 500 largest industrial corporations in America ranked the Sun Oil Company thirty-seventh in sales, nineteenth in net income, and twenty-second in stockholder equity. Of the top fifty industrial corporations ranked by sales, sixteen were oil companies and Sun was number fourteen.[24] Sun's rise from sixtieth place (by sales) on the *Fortune* list the previous year to thirty-seventh reflected the unusually large profits enjoyed by oil companies in the crisis year of 1974. However, this high rank also indicates the company's extensive growth in recent years. Changes in organization, management, and business philosophy have accompanied this growth as the older generation of the founding Pew family passed from the active management of the firm.

THE PEWS

Until recently, the Sun Oil Company has been identified closely with the Pew family of Pennsylvania. When Joseph Newton Pew, Sr. died in 1912, the firm's management fell to his sons, John Howard and Joseph Newton Pew, Jr. These men and their cousins, nephews, and close associates dominated the company's fortunes for over half a century. Such a long period of family management was rare in the industry. The experience of Gulf Oil, for example, where the Mellon family soon limited their participation to profits, was more common.[25] The Pews quickly established a reputation for aggressive and independent business practices, and family members were vocal on industry-wide policy issues.

The death of Joseph Newton Pew, Jr. in 1963 and of his brother J. Howard in 1971 signified the end of an era. Although the family still controls approximately forty-five percent of Sun stock, there are no longer any Pews at the top management level.[26]

In 1968, then Board Chairman J. Howard Pew had formally presided over Sun's merger with the Sunray-DX Oil Company of Tulsa, Oklahoma. At the end of his long and active career, he reluctantly acceded to this major expansion and the aquisition of a wider marketing network. A major restructuring of the firm and its management followed the merger. In August 1974, Sun president H. Robert Sharbaugh announced a further policy change with the expansion of overseas operations. One year later, in August 1975, Sun embarked on a radical organizational change still in progress. The new plan decentralized the firm into four separate operating subsidiaries, each with its own management.[27] While much of this recent change reflected particular concerns with current issues, such as the need for foreign crude and the advantages of diversification, they did represent a significant break with the past.

The Pews consciously resisted the firm's becoming too large because they feared the loss of control that it entailed. Under their direction, Sun cultivated a limited marketing area and concentrated on the sale of specialized, high quality products. In gasoline marketing, for example, Sun continued to sell only one brand of unleaded motor fuel at its stations until the fifties. The Pews relied primarily on domestic production of crude oil and were wary of foreign operations. When Sun moved to purchase Middle East crude in 1954, it was the last major integrated firm to do so.

The Pews also maintained tight centralized control of their firm. Although some contemporaries may have viewed Sun's officers as "managing by the seat of their pants," the Pews believed in family ownership and control, and they made their philosophy work. Innovative, competitive, and independent, Sun carved an important niche in the petroleum industry. The two brothers and their cousin, J. Edgar Pew, created a major, fully-integrated unit out of the small company they inherited from J.N. Pew, Sr.

A NEW PUBLIC POLICY

As the Sun Oil Company expanded under their direction, the Pews had to deal with an increasing number of problems. The perennial pattern of feast or famine continued to characterize the production of crude oil in the United States and created uncertainty in all areas of the business. As petroleum products became more important in American life, the public expressed greater concern over a scarcity of reserves. During and after World War I, the military and naval forces began to depend more on oil, and the issue of national security meshed with that of conservation. More and more, the government began to perceive its stake in the industry. However, since the dissolution of the Standard Trust in 1911, the debate over what form the government's relationship to the industry should take had become more complex.

After 1911, a new emphasis began to replace the antitrust phase of government oil policy. Industrial leaders and government officials alike soon saw the advantages of conservation legislation, production regulation, and other cooperative means of stabilizing the petroleum industry. The government moved to protect the public's interest in a valuable national resource, while business executives realized the economic gains possible through legislation to effect stabilization of the industry.

Geologic understanding of oil pools was primitive by today's standards, and the overzealous pumping of underground pools exhausted them so quickly that everyone became the loser.[28] In the nineteenth and early twentieth centuries, however, conservation advocates made little headway in changing accepted practice. Competitive drilling fostered by the "rule of capture" fed a pattern of glut and scarcity.

Photographs of newly opened producing tracts in Texas, Oklahoma, California, and Louisiana, taken in the period 1900–1930, still resembled the "forests of derricks" that earlier dotted the landscape along Oil Creek, Pennsylvania and Spindletop, Texas.[29] Although the principle of regulated production might seem attractive in the long run, who was to decide how much one could produce? In the absence of a willingness to cut back output voluntarily, federal and state governments would eventually have to step in and play a role.

Many petroleum industry histories have discussed such policy issues as conservation legislation, pipeline regulation, oil taxation policies, and government aid to foreign oil exploration. Most of these works, however, including the industry study headed by Harold F. Williamson, Sr., and the company histories written by such scholars as Henrietta Larson, Paul Giddens, Gerald T. White, and George Gibb and Evelyn Knowlton, offer relatively little analysis of the evolutionary relationship between the petroleum industry and government.[30]

Although these accounts contain much valuable information, they emphasize the growth of the industry and its constituent firms; the history of public oil policy is a secondary theme. Moreover, since these studies focus on growth, they treat the development of conservation legislation and government regulatory activity simply as the coming of an enlightened public policy. For example, in considering the industry-sponsored production regulations of the 1930s that finally put to rest the "rule of capture," these authors praise the economic benefits of the programs without questioning the propriety or ethics of such government activity.[31]

More recent studies of United States oil policy have emphasized the growth of consensus and cooperation between government and the petroleum industry in this century. Gerald D. Nash, in the most balanced of these works, argues that events in the oil industry provide a model for the growth of similar cooperative attitudes in the rest of American industry. He maintains that a new era of cooperation following the antitrust phase of government activity in the pre-World War I period abetted the already emerging petroleum oligopoly. Presidents, congressional leaders, and state legislatures became concerned with the exhaustion of reserves and its effect on national security, while industry leaders advocated that government play a role in stabilizing a marketplace characterized by waste, inefficiency, and an uncertain price structure. According to Nash, the views of oil men and policy makers coalesced most closely in periods of strain on the economy. In this century, the response of industry and government toward war and depression provided the occasions for working out the major constituent parts of the growing consensus.[32]

Conservation became a code word for economic stabilization as various industry-sponsored state and federal programs emerged to solve the problem of cyclic glut and scarcity. In addition to conservation legislation, Nash cites related cooperative programs as part of his hypothesis. Chief among these were government acquiescence and participation in plans for the price-fixing of refined products, the exploration and development of foreign crude resources, and favorable tax policies for the petroleum industry.[33]

Nash describes this growing consensus on oil policy as a victory of Theodore Roosevelt's "New Nationalism" over the Wilsonian "New Freedom." To Roosevelt and the "advanced progressivism" embodied in the New Nationalism, bigness itself was not bad. The government would discriminate between "good" and "bad" trusts while maintaining an element of control through regulatory agencies and prosecution of offenders. The New Freedom, on the other hand, emphasized the benefits of free competition its proponents felt would result from the dismemberment of monopoly and encouragement of small competing units.[34]

These divergent views, part of the Roosevelt-Wilson debate in the "Bull Moose" election of 1912, have continued to represent the two poles of the monopoly issue in subsequent administrations. In particular, elements of both the New Nationalism and the New Freedom appeared in the New Deal economic policies of Franklin Roosevelt, which collectively were the most important watershed in the development of public economic policy in this century.[35] Nash argues that government moved further toward comprehensive cooperation and supervision as much of American industry increasingly came to be dominated by large corporations. The petroleum industry emerged in its present form in the midst of this environment, and the resulting compromises on policy reflected both government policy goals and the economic interests of the private sector.[36]

Norman E. Nordhauser's similar 1970 study portrayed the major oil companies in the period 1919–1935 as embracing a three-pronged program of vertical integration, voluntary cooperation, and "political capitalism" to halt the further growth of competition and its attendant economic instability.[37] This third aspect, political capitalism, pervades Nordhauser's work. He borrowed the concept from Gabriel Kolko's controversial 1963 study of the Progressive period, *The Triumph of Conservatism*. There Kolko defined political capitalism as the movement for industry-sponsored legislation to stabilize the marketplace and "rationalize" the economic problems of rampant and uncertain competition.[38] By maintaining that Progressive legislation aided big business instead of curbing its power, Kolko questioned the interpretations of a generation of liberal American historians. In his petroleum industry story, Nordhauser explores Kolko's suggestion that political capitalism extended from the Progressive era of 1900–1914 up through the New Deal.

The Kolko-Nordhauser position, although similar to Nash's, takes a more critical posture. Nordhauser's analysis of the oil industry focuses on the passage of conservation legislation to restrict production. He argues that "industrial science was the handmaiden of profits, that oil men used geological theories and the conservation argument to rationalize their true objective, which was to escape from the pressures of competition."[39] Where Nash accepts the government goals of a strengthened national defense and economic recovery as legitimate grounds for aiding the oil industry, Nordhauser argues that government officials were more often than not tools of the powerful corporations and producers' organizations. He rejects the notion of government acting as a neutral arbiter in the industry.[40] Both analysts, however, describe the same fundamental phenomenon of a growing consensus between industry and government on petroleum policy in the twentieth century.

SUN OIL AS A TEST CASE

The framework suggested by the Nash and Nordhauser studies offers a useful way to examine the development of a particular firm. The history of a company that arose in an era of monopoly and later achieved "major" status in the new environment after 1911 can provide insights into the nature of intra-industry cooperation, government's relationship to industry, and resultant struggles over policy. Further, if there is substance to the thesis of cooperation, it should be possible to document such a relationship by studying the development of that company in the period through 1945.

That year (1945) is a logical stopping point for this story. By the end of World War II, the basic structure of the modern petroleum industry was already well established, and the pattern of regulatory activity that emerged to control the production of domestic crude oil had become an accepted way of life in the oil business. The most significant new development of the post-war era has been the increasing multinational character of the industry as larger firms have sought sources of crude oil in the Middle East, Latin America, Africa, and Asia. That move was at first the result of a search for cheaper supplies of crude, but now the major companies rely on foreign exploration and drilling to meet domestic demand in the face of diminishing reserves at home. Sun was not in the vanguard of foreign exploration efforts, and the firm has only moved heavily into multinational operations in recent years. This study does not, therefore, attempt to deal with this latter phase of Sun and petroleum industry history. Rather, it focuses on the important chapter of that history that concludes with the reinforcement of oligopolistic competition and the attainment of a generally acceptable program of business-government cooperation in oil that was established by the end of World War II.

Thus, this book is in part a study of a single firm, how it survived and prospered among firms larger than itself, and how it did so under the personal style of the Pew family of owner-managers. But even more, it is a case study in the oil industry and public policy. There has been a steady increase in industry-government cooperation, and the private sector has had an enormous input into the formulation of policy. However, the progression has been uneven and the issues complex. An examination of the Sun Oil Company's evaluation of, and response to, the problems of war, over-production, conservation, industrial efficiency, and technological change can illuminate the history of the firm and the industry, as well as that of the climate of public policy surrounding it.

NOTES

1. These summary figures are cited in Alfred D. Chandler, Jr., "The Standard Oil Company—Combination, Consolidation, and Integration," Harvard Business School Case Study #9-362-001: BH 120 (Cambridge: President and Fellows of Harvard College, 1973), p. 3. This concise synthesis by one of America's premier business historians is an extremely useful analysis of the early petroleum industry and Standard Oil's growing role in it. Another short but excellent account is contained in Glenn Porter, *The Rise of Big Business, 1860-1910* (Arlington Heights, Illinois: AHM Corp., 1973), pp. 63-7. More detailed discussion of the emerging oil industry is found in Harold F. Williamson and Arnold R. Daum, *The Age of Illumination, 1859-1899*, Volume I, *The American Petroleum Industry* (Evanston: Northwestern University Press, 1963), especially chapters 6, 7, and 14; and Ralph W. Hidy and Muriel E. Hidy, *Pioneering in Big Business, 1882-1911*, Volume 1, *History of the Standard Oil Company (New Jersey)* (New York: Harper Brothers, 1955), chapter 1.

2. Williamson and Daum, *Illumination*, pp. 375-7; Henrietta M. Larson and Kenneth Wiggins Porter, *History of Humble Oil and Refining Company* (New York: Harper, 1959), pp. 17-8; Chandler, "Standard Oil Company," p. 3.

3. Williamson and Daum, *Illumination*, p. 377.

4. *Ibid.*, pp. 287-93.

5. *Ibid.*, pp. 344-6; Chandler, "Standard Oil," p. 4; Hidy and Hidy, *Pioneering*, pp. 9-11.

6. Hidy and Hidy, *Pioneering*, pp. 13-4; Chandler, "Standard Oil," p. 5-6; for a detailed story of young Rockefeller's first ventures in business and his early partnerships in the oil industry, see Allan Nevins, *Study in Power: John D. Rockefeller, Industrialist and Philanthropist* (New York: Scribner's, 1953), Volume I, Chapters 1 and 2.

7. Williamson and Daum, *Illumination*, pp. 416–29; Chandler, "Standard Oil," pp. 7–9; Porter, *Big Business*, p. 64.
8. Chandler, "Standard Oil," pp. 8–10; Porter, *Big Business*, p. 64; Williamson and Daum, *Illumination*, pp. 356–60.
9. Williamson and Daum, *Illumination*, pp. 416–29; Chandler, "Standard Oil," pp. 10–2.
10. Williamson and Daum, *Illumination*, pp. 430–62; Hidy and Hidy, *Pioneering*, pp. 31–3, 41–2; Chandler, "Standard Oil," p. 13.
11. Hidy and Hidy, *Pioneering*, pp. 40–9; Williamson and Daum, *Illumination*, pp. 466–70.
12. Hidy and Hidy, *Pioneering*, pp. 219–32, 307–13; Williamson and Daum, *Illumination*, p. 715; Porter, *Big Business*, pp. 65–6.
13. Hidy and Hidy, *Pioneering*, pp. 169–75; Williamson and Daum, *Illumination*, pp. 605–7; Chandler, "Standard Oil," pp. 21–2.
14. Chandler, "Standard Oil," pp. 22–3; Hidy and Hidy, *Pioneering*, pp. 176–88.
15. Arthur M. Johnson, "Public Policy and Concentration in the Petroleum Industry, 1870–1911," *Oil's First Century* (Cambridge: Harvard Business School, 1959), pp. 43–56; see also Johnson's larger study, *The Development of American Petroleum Pipelines: A Study in Private Enterprise and Public Policy, 1862–1906* (Ithaca: Cornell University Press, 1957).
16. Hidy and Hidy, *Pioneering*, pp. 671–708; Harold F. Williamson, Ralph L. Andreano, Arnold R. Daum, and Gilbert C. Klose, *The Age of Energy, 1899–1959*, Volume II, *The American Petroleum Industry* (Evanston: Northwestern University Press, 1963), pp. 8–10.
17. Hidy and Hidy, *Pioneering*, pp. 708–11; Williamson et al., *Energy*, pp. 10–2.
18. See Henry Demarest Lloyd, *Wealth Against Commonwealth* (New York: Harper and Brothers, 1894), and Ida Tarbell, *The History of the Standard Oil Company* (New York: Macmillan, 1904). In many ways, Miss Tarbell's book can be read as a positive account of Standard's rise to power. Despite its accounting of alleged unethical practices, there is also a grudging respect for John D. Rockefeller's creative achievement. However, because of the book's broad indictment of corporate power and its contemporary impact, it is almost always described as a part of the muckraking genre.
19. Hidy and Hidy, *Pioneering*; Nevins, *Study in Power*; Williamson and Daum, *Illumination*, Part Five, "Reaction to Instability"; for the original Standard argument, see *U.S.* v *Standard Oil Company of New Jersey et al.*, Supreme Court of the United States, Brief for Appelants, II (Washington, D.C.: U.S. Government Printing Office, 1909).
20. Williamson et al., *Energy*, pp. 4–7; Gerald D. Nash, *United States Oil Policy, 1890–1964* (Pittsburgh: University of Pittsburgh Press, 1968), p. 8; Hidy and Hidy, *Pioneering*, pp. 471–77.
21. Williamson et al., *Energy*, pp. 7, 238–41.
22. See Anthony Sampson, *The Seven Sisters* (New York: Viking Press, 1975).
23. Table 2, API Study of TNEC Questionnaire, Question 30, 31, J. Howard Pew Presidential Papers, Accession #1317, Eleutherian Mills Historical Library, Greenville, Wilmington, Delaware (hereafter cited as J. Howard Pew Papers), Box 45; "Testimony of J. Howard Pew," U.S. Congress, Temporary National Economic Committee, *Hearings Before the Temporary National Economic Committee, Part 14, Petroleum Industry* (Washington, D.C.: U.S. Government Printing Office, 1940), pp. 7163, 7169, 7196, 7197.
24. "The Fortune Directory of the 500 Largest Industrial Corporations," *Fortune* 91, 5 (May 1975), pp. 210–1.
25. Robert Engler, *The Politics of Oil* (Chicago: Macmillan, 1961), p. 47.
26. R. Anderson "Andy" Pew J. Howard Pew's great-nephew, appeared to be a possible successor to family leadership in the firm. (See Michael C. Jensen, "The Pews of Philadelphia," *The New York Times* (October 10, 1971), p. 9.) Pew held the job of Corporate Secretary until June 1977, when he moved to another post within the Sun Company structure and was

passed over for a seat on the Board of Directors. He has been a member of the Board of the Glenmeade Trust Company, the institution that administers the Pew family foundations, and in a recent corporate shake-up, has been elected to the Sun Company Board. It remains unclear at this writing what his future role with the company may be.

27. Memo "Possible Advantages From Merger," (undated) R.G. Dunlop to Sun Oil Company Stockholders, November 6, 1908, Robert G. Dunlop Presidential Papers, Accession #1317, Eleutherian Mills Historical Library, Greenville, Wilmington, Delaware (hereafter cited as Dunlop Papers), Box 20; "Sun Oil Planning to Buy Sunray DX," *The New York Times* (January 20, 1968), p. 37; "The Sun Oil Company," *Moody's Industrial Manual* (New York: Moody's Industrial Services, 1968), pp. 1427-31; "The Sunray-DX Oil Company," (*Moody's* (1968), pp. 1431-5; "Sun's Computer Helps Find Heir," *Business Week* (October 4, 1969), p. 83; Jane Shoemaker, "Sun Oil to Expand Overseas Operations," *The Philadelphia Inquirer* (August 12, 1974), p. 18-C; "Sun Oil Split Into 4 Subsidiaries," *The Philadelphia Inquirer* (August 7, 1975), p. 4-B; *Sun Oil: Building Flexibility for the Future* (St. Davids, Pennsylvania: Sun Oil Company, 1975). Recently the firm has moved to reflect this growing diversification by changing its corporate name from the Sun Oil Company back to the Sun Company, a name first adopted by J.N. Pew, Sr. for his New Jersey Corporation in 1901 and later abandoned (see *Sun Shareholder News* [January 1976]). In this study, the parent firm is referred to as the Sun Oil Company, the name it had used from 1922 to 1976.

28. Williamson and Daum, *Illumination*, pp. 375-6; Erich W. Zimmerman, *Conservation in the Production of Petroleum* (New Haven: Yale University Press, 1957), pp. 51-77.

29. A fascinating comparison can be made between the excellent photographs in both Paul H. Giddens, *Early Days of Oil: A Pictorial History of the Beginnings of the Industry in Pennsylvania* (Princeton: Princeton University Press, 1948), and Walter Rundell, Jr., *Early Texas Oil: A Photographic History, 1866-1936* (College Station: Texas A&M University Press, 1977). The similarities in the pattern of oil field development are striking proof that very little in the way of conservation had been accomplished in thirty years.

30. Williamson and Daum, *Illumination*; Williamson et al., *Energy*; Larson and Porter, *Humble*; Paul H. Giddens, *Standard Oil Company (Indiana): Oil Pioneer of the Middle West* (New York: Appleton-Century-Crofts, 1955); Gerald T. White, *Formative Years in the Far West (Standard Oil of California* (New York: Appleton-Century-Crofts, 1962); George S. Gibb, and Evelyn H. Knowlton, *The Resurgent Years, 1911-1927*, Volume II, *History of the Standard Oil Company (New Jersey)* (New York: Harper Brothers, 1956); Henrietta M. Larson, Evelyn H. Knowlton, and Charles S. Popple, *New Horizons*, Volume III, *History of the Standard Oil Company (New Jersey)* (New York: Harper and Row, 1971).

31. An exception to this is the work of Arthur M. Johnson (see *The Development of American Petroleum Pipelines: A Study in Private Enterprise and Public Policy 1862-1906* [Ithaca: Cornell University Press, 1957], and *Petroleum Pipelines and Public Policy, 1906-1959* [Cambridge: Harvard University Press, 1907]). Johnson examines the important interrelationships between the development of regulatory activity by state and federal government and the growth of major pipeline systems.

32. *United States Oil Policy, 1890-1914*, preface, pp. 238-51.

33. *Ibid.*, pp. 249-50.

34. *Ibid.*, preface, p. vi; pp. 243-4.

35. See especially Ellis Hawley, *The New Deal and the Problem of Monopoly* (Princeton: Princeton University Press, 1963). These ideas are also discussed in standard New Deal histories, including Arthur M. Schlessinger, Jr., *The Coming of the New Deal* (Boston: Little Brown, 1960) and William E. Leuchtenburg, *Franklin D. Roosevelt and the New Deal, 1932-1940* (New York: Harper and Row, 1963).

36. Nash, *Oil Policy*, pp. 250-1.

37. "The Quest for Stability: Domestic Oil Policy, 1919–1935," Ph.D. dissertation (Stanford University, 1970); see also Norman E. Nordhauser, "Origins of Federal Oil Regulation in the 1920s," *Business History Review*, 47, 1 (Spring 1973): pp. 53–71.
38. Gabriel Kolko, *The Triumph of Conservatism* (New York: Glencoe, 1963), pp. 1–10.
39. Nordhauser, "Quest," preface, p. ix.
40. *Ibid.*

CHAPTER II

Pioneering in Natural Gas and Oil: Sun's Early History, 1876–1900

Although the Sun Oil Company's history technically began when J.N. Pew, Sr. and his partner, E.O. Emerson, obtained oil producing leases in 1886, earlier events are relevant to the firm's story. In particular, Joseph Pew's prior excursions into the oil and gas business provide insight into his later business decisions. Pew grew up in the oil regions of northeast Pennsylvania and gained valuable experience in many aspects of the new industry as a young man. Lessons learned in the initial phase of his career influenced the patterns of growth and vertical integration which Sun Oil continued into its modern era after 1900. Moreover, Pew's struggle to survive along with other "independents" in competition with Standard Oil offers a glimpse into the structure of the oil industry during its formative decades in the nineteenth century. One can understand much of Sun's later history in terms of the Pew family perceptions of monopoly, competition, and prudent management that developed out of the rough and tumble days of the early oil industry.

EARLY VENTURES IN OIL

Joseph N. Pew was born on a farm near Mercer, Pennsylvania, on July 25, 1848. Eleven years later, the modern oil industry began with Colonel Edwin Drake's first successful petroleum well in Titusville, Pennsylvania, forty miles away. Pew was the youngest of ten children, and his father's farm was not large enough to support all of them. Joseph attended school in Mercer and as a young man obtained a job teaching school in the nearby town of London. He had held this position for three years when, in 1869, he enrolled in the Normal School at Edinboro, Pennsylvania.[1]

Pew left Edinboro in 1870, but instead of returning to teaching, he opened a real estate office in Mercer. After a short stay, the twenty-two-year-old businessman moved to Titusville, a town still booming ten years after the drilling of the Drake well. After first entering the real estate, insurance, and loan business, he quickly took advantage of the lucrative trade in oil property and leases. By 1874, he had accumulated a personal fortune of $40,000 and had become a leading Titusville figure. That same year, he married Mary Catherine Anderson, a member of a prominent western Pennsylvania family with connections in the oil business.[2]

Soon after his marriage, however, young Pew encountered financial difficulties. Although the record is incomplete, he had apparently been speculating in pipeline certificates, the oilmen's equivalent of warehouse receipts.[3] A system using these receipts developed in the oil regions by the 1870s to facilitate the exchange of oil from wellhead to refinery. An informal "traveling exchange" first developed when the Farmers Railroad opened between Titusville and Oil City in 1866. As the train stopped along the way, refiners' agents would negotiate with the many producers endeavoring to sell their oil. These agents would buy oil for future delivery, but there were usually no contracts signed and the *ad hoc* system proved ineffectual in handling the large volumes of crude needed by the refineries. A better mechanism developed when the pipeline companies began to perform a storage as well as transportation function by issuing certificates for each thousand barrels of crude held for a producer. A storage fee the producer paid enabled him to hold his oil off the market until he found a buyer at a favorable price. The refiners also benefited from the more uniform flow of oil now available. At formal oil exchanges, initially at Oil City, Pennsylvania, in 1869, refiners' agents, brokers, and other middlemen dealt in the purchase and sale of pipeline certificates.[4]

An oil broker could make a large profit by buying certificates at a low price and selling them during a period of scarcity. Reports of a wildcatters's dry hole or of salt water appearing in a pool would precipitate a wave of buying, but if a gusher well came in unexpectedly, it could break the market. This is what happened to J.N. Pew in 1875. He had purchased a large number of pipeline certificates at the Titusville exchange for oil that was still under the ground and not yet held by the pipeline company. To finance this speculation, he used his own savings and also went into debt for another $20,000. When the market went into a nosedive following news of flush discoveries, Pew's receipts greatly depreciated in value and he was hard pressed to pay back his loan. The record does not explain how he extricated himself from this situation, only indicating that he eventually recouped his personal losses and paid back every cent to his creditors.[5]

Pew attempted to continue as a middleman in the nearby Parkers'

Landing and Bullion oil fields along Oil Creek, but did not meet with success and moved northward to Bradford, Pennsylvania, in 1876. Men had drilled in the Bradford pool since 1864, but a real boom did not begin there until 1876–1877. Before long, Standard Oil established itself as the dominant purchaser and shipper of crude in Bradford.[6] No doubt Standard's powerful position influenced Pew's thinking at the time, for it was at this point that natural gas, not oil, captured his attention.

In Titusville, Pew had met Edward Octavius Emerson, a successful banker and oil man fourteen years his senior. Emerson was originally from New England, but had become a banker in Wisconsin before the Civil War. While he was serving in the Union Army, Emerson's Wisconsin operations lost a great deal of money. After the war, he came to Pennsylvania at the peak of the oil boom and made profits dealing in oil leases in the Titusville-Pithole area. In 1876, Emerson formed a partnership with J.N. Pew to exploit the natural gas fields in and around Bradford. The partners entered a business on the fringe of the petroleum industry, one in which Standard had shown little interest, and therefore posed no threat.[7]

EARLY GAS OPERATIONS

Nineteenth century oil men at first ignored the natural gas deposits in the Appalachian field of northwestern Pennsylvania and New York state. Oil drillers wasted millions of cubic feet of gas through inefficient methods of capping wells and controlling output, and when a wildcatter struck gas without any oil, he abandoned the well as a loss. Petroleum was king in the 1860s and 1870s, and experimentation with natural gas remained small. There existed an opportunity for someone with imagination to exploit this untapped resource, but the first efforts to utilize this new energy source occurred in the oil fields themselves.

By the early 1860s, oil field operators had adapted steam power to both percussion drilling of new wells and the pumping of established pools. In 1864, the superintendent of the Ladies Well, near Titusville, first utilized natural gas as fuel in the furnaces of his stationary steam engines. oil men now had a use for the gas that they had wasted for so long. The burning gas produced a high temperature cleanly, but its chief advantage lay in its economy. Most of the early steam engines in the oil regions burned wood and could easily consume a cord in twelve hours at a cost of seven to ten dollars. More expensive coal cost sixty cents a bushel in Oil City and as much as a dollar and twenty-five cents along Oil Creek. By the end of the decade, once operators had developed safety features, they used natural gas widely to fuel the steam-driven drills and pumps that dotted the major producing fields.[8]

Pew and Emerson began piping natural gas to Bradford drilling sites in 1876. Joseph Pew provided technical and managerial skill, while Emerson furnished the necessary cash and credit. The partners were successful and soon expanded into the new but potentially profitable business of supplying natural gas to private homes and industry. On February 1, 1881, they incorporated the Keystone Gas Company to pipe gas for illumination and heating to the town of Bradford. Within a short time, their operations broadened to provide gas to Olean, New York, thirty miles to the north. Edward Emerson was the president and J.N. Pew the treasurer of Keystone, an arrangement that reflected the elder partner's greater financial interest.[9]

Cities in the United States and abroad had illuminated their streets with manufactured coal gas for many years. In 1802, Thomas Murdock had first made gas from coal successfully, and ten years later, London brightened its streets with gas. Baltimore initiated gas street lighting in 1816, and by 1830 New York and Boston had followed suit. By 1837, Brooklyn, Bristol (Rhode Island), Louisville, New Orleans, Philadelphia, and Pittsburgh also boasted of their gas works. By the 1840s and 1850s, improved burners and distribution systems enabled manufactured gas to be used in homes as well as in lighting streets and public buildings.[10]

Gas lighting revolutionized urban living in the nineteenth century, but the majority of Americans lived outside the cities. It was this rural market that absorbed coal oil, and later kerosene, to fuel its lamps. When Pew and Emerson began to pipe natural gas to cities, they participated in a new chapter of energy development. They perceived the great market for illumination and heating, but significantly did not try to compete with Standard for the manufacture of illuminating oil (kerosene). They exploited a new resource, one that required even less capital than oil. Moreover, they could benefit from the distribution technology and the public acceptance of gas created by the manufactured gas industry.

Natural gas seepages had fascinated people for generations, but the first attempt at commercial exploitation of this resource for illuminating purposes had occurred in 1865 with the incorporation of the Fredonia Gas Light and Water Works in Fredonia, New York. Many American and foreign travelers had journeyed to that city to observe the unusual illuminating powers of this gas. The Fredonia Company, however, did not succeed in using natural gas for lighting. People were skeptical about the gas, and the enthusiasm generated by the oil boom soon eclipsed the venture. Another experiment began in New York in 1878 when the Bloomfield Company brought natural gas from Bloomfield, New York, to Rochester via a wooden pipeline. Unfortunately, the wooden pipes split, causing the loss of a large volume of gas. The company also discovered that natural gas did not illuminate properly unless mixed with manufactured coal gas.[11]

A major problem with natural gas resulted from the fact that it was potentially lethal to humans and had no odor. Initially there was some confusion on this point; one writer in the *American Gas Light Journal* in 1878 stated that "men breathe it in their lungs as they do air and there is no danger as there is in manufactured gas."[12] Such notions were soon dispelled.

The greatest danger with the absence of odor was of course the inability to detect leaks of this highly explosive substance. Discussion of this serious situation soon appeared in contemporary journals. In 1885, Irvin Butterworth described the inherent danger in the natural gas then being piped into Pittsburgh. He attributed the frequent explosions in the city to the failure to detect leaks and to the great pressures built up in the lines (550 pounds per square inch). Butterworth recommended the use of automatic valves and governors, improved pipes, and the introduction of odor into the gas.[13] This last practice eventually proved to be one answer to the problem.

PIPING GAS TO PITTSBURGH

After their success in Bradford and Olean with the Keystone Gas Company, Pew and Emerson decided to expand in a significant way. They became the first to supply natural gas to a major city, Pittsburgh, in 1883. Since Pittsburgh's use of natural gas became the object of world-wide acclaim, the partners' contribution was historically very important. They chose Pittsburgh for a combination of reasons. The city had a population of 235,000 people, plus previous experience with the use of manufactured gas, which made it a very attractive market. Even more importantly, an excellent supply of natural gas lay nearby.[14]

In 1878, two oil drillers, Matthew and Obe Haymaker, sunk a deep well in Murrysville, Westmoreland County, Pennsylvania. Although they had originally obtained financing on the basis of the presence of natural gas on the site, it was oil they sought. Drilling to a depth of 1,400 feet, they found not oil but natural gas, the largest well discovered up to that time. The uncapped well spewed forth approximately one million cubic feet of gas an hour, a waste later estimated to be equivalent to over two million tons of coal before it was halted. The escaping gas literally roared under the tremendous well pressure, and it became a popular tourist attraction. Unfortunately, one onlooker got too close with a kerosene latern and ignited a blaze that lasted for a year and a half. There were no serious injuries in the explosion, and the illumination from the well provided almost continuous daylight for miles around. But even after the Haymakers had snuffed out the fire by putting an old forty-five-foot-long smokestack over the well opening, their problems were not over. They now had to find a buyer.[15]

The brothers and their financial backer, a Mr. Brunot, had difficulty consummating a favorable deal. Pew and Emerson were not the first individuals to contemplate piping natural gas to Pittsburgh. The Fuel Gas Company of Allegheny County had earlier obtained a Pennsylvania charter with the Pittsburgh market in mind, and made an offer for the Haymaker well. When discussions broke off, however, the firm purchased an adjoining drilling site.[16]

In 1881, the well owners agreed to sell the Haymaker to a Chicago promoter for $20,000. He put $1,000 down and was to pay an additional $3,000 within thirty days, and the balance in sixty days. A year had passed after the initial transaction with no further word from the Chicagoan, when Brunot and J.N. Pew met on the Buffalo to Bradford train in 1882. Pew took an immediate interest and journeyed to Murrysville to investigate. He made an offer, and the Haymaker interests sold to Pew and Emerson for $20,000.[17]

On February 2, 1882, Pew and Emerson incorporated the Penn Fuel Company "to furnish heat to the public within the city of Pittsburgh by means of natural gas."[18] The wording suggests that the partners looked toward energy uses for gas rather than illumination. Because of its abundant supply, homeowners and factory operators had begun to perceive gas as an economical fuel. Oil field operators had already demonstrated its use in firing boilers, and an 1878 account cited its successful use in manufacturing, heating, and cooking, as well as in illumination. Even critics of the dangerous substance praised its cost benefits and greater caloric power compared to manufactured gas for heating and industrial use.[19]

As early as 1868, some firms in western Pennsylvania had experimented with natural gas for industrial use. The Jarecki Manufacturing Company of Erie (a producer of oil field equipment) burned gas in its steam boilers, and the Rogers and Burchfield Company of Leechburg introduced gas to the manufacture of iron and steel. They used gas as a fuel in their heating and puddling furnaces with much success. Once its use was proven feasible, natural gas contributed significantly to the industrial growth enjoyed by Pittsburgh in the last decades of the century.[20]

After Pew and Emerson obtained a charter, they acquired additional producing acreage and immediately began laying a five and five-eighths-inch pipeline fifteen miles into Pittsburgh. They completed their line into the east end of the city in January 1883. Since their first competitor, the Fuel Gas Company, did not complete its line until November of that year, Pew and Emerson have a rightful claim to being the first to supply natural gas to a major American city. Both firms drilled other wells, and by the summer of 1883 had developed the gas field as far as Lyons Run, two miles south of Murrysville. Initially, the two companies were cautious about their pipelines and gas supplies, and they accepted customers slowly. In a few months, however, both did well, and Penn Fuel started construction of a new eight-inch pipeline.[21]

Although Pew and Emerson were the moving force behind the Penn Fuel Company, they did not have a controlling financial interest in the firm. J.N. Pew acted as the promoter and manager of the enterprise, but a group of Pittsburgh investors were the principal backers. A rift developed between Pew and Emerson on the one hand and the Pittsburgh group on the other. The gas companies had to secure pipeline rights of way and construction permits from the municipal government in order to do business. To facilitate matters, some Pittsburgh politicians requested a financial pay-off for the granting of certain privileges. The cost of the graft was small, but Pew, the managing director, refused to make payment. The other investors then ousted the unacceptably honest Pew as director, and he and Emerson took their money out of Penn Fuel.[22]

The remaining owners sold the Penn Fuel Company to George Westinghouse and Associates in 1884. Westinghouse, already noted for his work in developing train airbrakes and later to become a giant in the electrical industry, was moving toward domination of the Pittsburgh natural gas industry in the early 1880s. The success of Penn Fuel and the Fuel Gas Company had encouraged a rash of new companies in western Pennsylvania, and over five hundred small firms had secured charters.[23]

Westinghouse had himself discovered gas on his property within the Pittsburgh city limits and organized the Philadelphia Company in 1884. Within a short time, he absorbed several of the smaller companies, including Penn Fuel. By 1887, his Philadelphia Company had become the largest natural gas supplier in Pittsburgh. With an authorized capital of $7,500,000, it serviced 16,000 manufacturing establishments and firms, had 520 miles of gas mains, and controlled 68,000 acres of tested producing acreage. The Philadelphia Company's one hundred completed wells provided 350,000 cubic feet of gas a day for its pipelines.[24] Westinghouse had not, however, gained control of all of the natural gas business in Pittsburgh.

THE PEOPLES NATURAL GAS COMPANY

Shortly after leaving Penn Fuel in 1884, Pew and Emerson incorporated a new firm, the Peoples Natural Gas Company, to compete with the Westinghouse interests. The partners had learned a bitter lesson, and this time they assured themselves control by owning most of the firm's stock. Peoples had an authorized capital of $1,000,000, making it a relatively small competitor with the Philadelphia Company and another large firm, the Chartiers Valley Gas Company.[25]

J.N. Pew immediately began pipeline construction in 1884, but encountered opposition from his strong competitors. A period of physical harassment and legal action against Peoples followed. The Philadelphia and Chartiers Valley companies obtained injunctions against Pew and Emerson's construction of gas mains on territory they claimed as theirs. On several occasions, they had Pew's construction gangs arrested when they disobeyed court injunctions. Whenever this happened, Pew directed another gang to begin laying pipe in the next block not covered by the injunction.[26] Despite these constant hindrances, Peoples Natural Gas managed to complete its main pipeline into Pittsburgh, obtain customers, and begin service.

Once Pew and Emerson had reestablished themselves in Pittsburgh, competition took on new forms. Struggles over rights of way continued, and competitive rate-cutting appeared in the mid-1880s. The larger firms, such as the Philadelphia Company and Chartiers Valley, could more easily survive drastic price cuts, while the smallest firms often went out of business. Peoples Natural Gas was just large enough to compete successfully. It had put gas lines into previously unserviced parts of the city and was able to retain most of its customers.

After a period of extensive rate-cutting, the largest competitors came to an understanding to eliminate this unprofitable and potentially suicidal practice. Now that the major interests had carved up the city's territory, they decided to seek an agreement on fixing rates. Pew and Emerson had pioneered in the natural gas industry, but now they had to be content with a minor share of the very lucrative business. Success for them meant that they had survived the intense competition of the 1880s and were included in the oligopoly that controlled the Pittsburgh business.[27]

In 1887, Peoples supplied 4,000 customers, had 100 miles of gas main, and sent 20,000 cubic feet of gas per day from twelve producing wells. The company also claimed to have a delivery system superior to those of all other firms operating in Pittsburgh. This claim rested largely on a pump used to carry gas from the fields after the decline of original rock pressure. The pump, patented by J.N. Pew, helped to maintain controlled pressure and volume to the customer, a safety as well as a convenience feature. While with the Peoples Company, Pew also invented and patented an improved gas meter for measuring deliveries. By maintaining a reputation for safe and efficient service, Peoples Natural Gas retained its relative position in Pittsburgh. The structure of the natural gas industry, however, was rapidly undergoing change.[28]

STANDARD ENTERS THE SCENE

When the Titusville partners first started their new enterprise, there were opportunities for small fringe operators to make a profit. However, after people like Pew and Emerson proved the viability of natural gas service, larger interests decided to enter the market. In Pittsburgh, George Westinghouse had used the large capital available to him to establish a dominant position in that city's industry. As natural gas lighted and heated more and more of Pittsburgh's homes and fired its industry's boilers, the city's fame spread world-wide. Significantly, the Standard Oil Company then turned its attention to the production and distribution of natural gas. By 1886, Rockefeller and his associates had accomplished gains through its subsidiary, the National Transit Company, that led to the formation of the Standard-dominated National Gas Trust.[29]

Ironically, just when Standard Oil was beginning to capture a significant share of the natural gas business, Pew and Emerson made their first joint move into oil in 1886. Although secondary to the partners' main interests in natural gas, this was the beginning of the Sun Oil Company. Pew and Emerson observed Standard's methods and later developed a Sun Oil strategy designed to get around the monopoly's control of the petroleum industry. Yet Standard may also have learned much through observing the independent gas operators like Pew and Emerson. In all their ventures, the Keystone in Bradford and Olean, and the Penn Fuel and Peoples companies in Pittsburgh, the partners integrated their operations fully from wellhead to consumer. They owned and operated their own wells, built and operated their own pipelines, and maintained a complete marketing system.[30]

By the 1890s, however, vertical integration alone was not a guarantee of longevity in the gas business without a continued influx of capital and a willingness to expand. As local gas reserves began to run out, Peoples Natural Gas Company's share of the Pittsburgh market grew less secure. The demand for industrial and home use of gas continued to increase, but the available supplies from the Pennsylvania fields were diminishing. West Virginia fields had recently yielded new supplies, but large investments in exploration, drilling, and pipelines were necessary to develop these resources. Standard established itself as a major gas producer in West Virginia, consolidating its operations there in 1897 with the Hope Natural Gas Company. Standard, with lots of gas but few ready markets, now expressed some interest in buying Peoples Natural Gas.[31]

Meanwhile, Pew and Emerson considered acquiring West Virginia properties in 1898, but the capital demands discouraged them. Emerson proposed in 1899 that Peoples purchase coal leases, since coal would greatly advance in price as gas failed in the Pittsburgh area. He envisioned that Peoples might have to supplement its supplies with manufactured coal gas, and by the turn of the century Peoples had begun to purchase both natural and manufactured gas from others in order to supply its customers.[32]

Amid this uncertainty, conflict developed between the partners in 1899. Pew and Emerson had never been general partners, but had formed specific companies to exploit natural gas and oil. Emerson was involved in many other endeavors, and his joint ventures with Pew were only a part of his wide business interests. A large cause of their dispute stemmed from a disagreement over the desirable depth of their involvement in the oil business, a subject to be explored later. But friction also existed concerning the future strategy of the Peoples Company. Pew argued that prospects appeared bleak without major expansion into production. In order to accomplish this, they would have to plow back a large share of the profits into the business. Emerson, on the other hand, insisted on the payment of full cash dividends "as long as we pay par to the company for stock."[33] The split culminated in April 1899, when Pew bought out all of his partner's interest in their oil operations, but the problem still remained with the Peoples Company. Because he needed cash at the time, Emerson also sold Pew 2,000 shares of his Peoples Natural Gas stock, and began to urge his partner that they should sell out.[34]

The expanding Philadelphia Company had acquired the Equitable Gas Company and the Pittsburgh Street Car Lines in 1899, and Emerson felt that they would now be interested in buying Peoples. In July, he wrote to Pew urging that they sell for $1,800,000, or ninety dollars a share. There was some bitterness in Emerson's tone as he argued that "you ought to get me all there is in it for a good many reasons. One is that you got such a preponderance of the stock yourself at a *very* low price and you also virtually promised to get all you could for me on the shares the same as yourself."[35]

The Philadelphia Company deal did not go through in 1899, but Emerson still advocated the sale of Peoples. He was now sixty-five years old and living in Titusville in semi-retirement. He wrote to Pew in 1902, again urging the sale of the company and citing the inability of Peoples to pay much more than an eight percent dividend. It is not clear whether he was referring to the Philadelphia Company or National Transit, but he wrote that "I think one large company could do the whole business much more economical [*sic*] than we could."[36]

In 1903, Emerson finally succeeded. J.N. Pew, Emerson, and Pew's nephew, Robert C. Pew, sold their interests in Peoples Natural Gas to the National Transit Company, the Standard subsidiary, for $4,483,000. National Transit continued to operate the firm as a marketing affiliate under the name of Peoples Natural Gas Company. Another of J.N. Pew's nephews, John G. Pew, an employee of Peoples since 1886, stayed on as vice-president and general manager of the company. Later he would join Sun Oil.[37] Except for the financial interest that both men retained in the small Keystone Gas Company in Olean, New York, the Peoples sale brought an end to the partnership and J.N. Pew's career in natural gas. Pew was now a fully committed oil man, and Emerson had retired from active business.

THE SUN OIL COMPANY (OHIO)

In the midst of their competitive struggles in the Pittsburgh gas industry, Pew and Emerson had turned their attention for the first time to the production of crude oil. The news of major oil strikes in northwestern Ohio in the fall and winter of 1885 generated great excitement. These new fields, located near Lima and Findlay, represented the first significant new production outside the Pennsylvania-New York region. Although the Ohio crude had a very high sulphur content, the oil industry welcomed the new supplies. The partners sent Pew's nephew, Robert C. Pew, to Ohio to investigate possibilities in 1886. The young man liked what he saw and filed a favorable report. On his recommendation, the Peoples Natural Gas Company acquired two oil drilling leases in Findlay Township, Ohio, for $4,500 on May 5, 1886.[38]

Pew and Emerson thus began their oil operations as small producers in Ohio. They drilled successfully on their leases and shipped crude to refineries in Toledo in wooden barrels on mule wagons. Robert C. Pew managed this small business for the partners. It proved successful and Pew and Emerson moved to strengthen their interests with the incorporation of the Sun Oil Line Company of Ohio in December 1888. With an authorized capital of $300,000, the firm set out to transport, store, and handle crude oil in the forty counties of northwestern Ohio. The name "Sun" appeared here for the first time.[39]

The Sun Oil Line Company purchased leases, storage tanks, and tank cars, and soon became an important supplier of crude in Ohio. J.N. Pew used his knowledge of the gas industry to market some of the firm's crude oil, and Sun shipped Ohio crude to gas manufacturing plants in Pennsylvania, as far west as Des Moines, Iowa, and north to Toronto.[40] A decade before Pew's and Emerson's entry into crude production, the introduction of a new process for manufacturing illuminating gas had opened new petroleum markets.

Beginning in the 1860s, some of the lighter, more volatile fractional petroleum distillates found a use in making illuminating gas. Initially, refiners sold these products, gasoline and naphtha, to the owners of air-gas machines. These devices, placed in individual buildings, produced an illuminating gas from the evaporation of these volatile oils. In the early 1870s, a new use for naphtha resulted from the development of the Gale-Rand process. This procedure made a permanent gas that held its illuminating properties even when transported over great distances by pipeline. The system quickly revolutionized the manufactured gas industry as urban companies used naphtha gas either to replace or enrich their coal gas.[41] Although the market for these petroleum fractions remained much smaller than that for kerosene, these innovations did create a demand that had not existed before.

In 1875, Thaddeus Lowe invented the "carburetion" process to produce water gas, a development that created the expanded market for Pew and Emerson. Lowe could produce a carbon monoxide and hydrogen mixture from heavier petroleum fractions by using his patented "super heater." Gas companies could now use the cheaper fractions that existed in the naphtha-poor Ohio crude. Refineries sold fuel oil and gas oil, a generic name for those fractions boiling at a point immediately above that for kerosene but lower than that for naphtha on the refiner's scale, to the gas companies. The carburetion plants could also use good quality Ohio crude to manufacture water gas, and this is the business Sun had moved into prior to 1890. In the 1890s, Sun made large crude sales to the United Gas Improvement Company of Philadelphia for its carburetion units.[42]

Pew and Emerson expanded their oil venture in 1890 with the formation of the Sun Oil Company (Ohio). Their intention to integrate fully was evident in the stated purpose of the firm to produce, transport, store, refine, pump, manufacture, sell, ship, and market oil. They accomplished much of this intended vertical integration in a few years, but initially the company was primarily a producer and transporter of crude. Sun Oil Line owned 176 thousand-gallon railroad tank cars purchased from the Youngstown Manufacturing Company, but did not yet operate any gathering or trunk pipelines.[43]

Sun's share of Lima crude production in the 1890s remained small compared with that of Standard Oil. Standard had preceded Sun's entry into production by establishing the Buckeye Pipeline Company in 1886. By 1891, Buckeye owned all the gathering lines and part of the trunk lines in Ohio. If Sun or other independent operators wished to pipe their crude, they would have to deal with Rockefeller. Standard had also begun to acquire its own producing acreage, and by 1889 produced approximately two-fifths of the oil from the Ohio-Indiana fields, largely as a result of its purchase of the Ohio Oil Company. By 1891, it raised this percentage to one-half. Moreover, Standard's rights to the newly developed Frasch process for desulphurizing Lima crude allowed the giant company to produce kerosene from this petroleum grade and achieve domination over the "Ohio business."[44]

Sun's experiences with Standard Oil largely shaped its next important move. Under Robert C. Pew's guidance, Sun had acquired leases and had drilled over 2,500 wells in northwestern and southwestern Ohio and

Huntington and Marion Counties, Indiana. Now that they possessed secure crude supplies, Pew and Emerson integrated forward into petroleum refining. In December 1894, Sun formed a partnership with the Meriam and Morgan Paraffine Company of Cleveland to purchase a refinery in Toledo. The plant belonged to the financially troubled Crystal Oil Company, which sold it for $22,200.[45]

The partners formed a new company, the Diamond Oil Company, with an authorized capital of $25,000. Meriam and Morgan and the Pew and Emerson interests each subscribed to 125 shares of $100 par value stock. J.B. Meriam became president and John C. McKisson of Sun Oil the treasurer of Diamond. An item in the *Toledo Evening Bee* of December 13, 1894, suggested the impact that the new refining company made: "The Diamond will manufacture everything from crude to candles just as the Standard Oil Trust has been doing for years. This includes gasoline, naphtha, burning oil, pitch, paraffine, candles, etc. . . . The Sun Oil Company is really Pew and Emerson, two wealthy producers of Pittsburgh . . . [who] own more production in Ohio than any other concern outside the Standard Oil Company."[46] Although this account overstates Sun's competitive position, the firm was one of the largest independent Ohio producers next to Standard.

Meriam and Morgan, a small refiner of specialty products, had seemed the ideal partner with which to launch the venture. Ten years earlier, J.B. Meriam, William Morgan, and Herman Frasch had jointly patented new methods to refine kerosene and to manufacture paraffin. It was following these successes that Standard hired Frasch to develop his famous process to desulphurize Lima crude. For many years, Standard had been a partial owner of Meriam and Morgan, but had severed relations in 1882 when the two partners and Frasch sold exclusive rights to their new processes to the Imperial Oil Company, Ltd., Standard's main Canadian competitor. Thus, although Frasch had later signed on with Standard, Meriam and Morgan possessed impeccable "independent" credentials in 1894. Shortly after the refinery opened in 1895, however, Meriam and Morgan went into bankruptcy. Pew and Emerson purchased their half interest in the Diamond Company for $12,500 and transferred the entire operation to the Sun Oil Company (Ohio) in December 1895. Robert C. Pew became refinery manager on the fourteen-acre site, and the plant resumed production of kerosene, fuel oil, and gas oil, which it sold to gas manufacturing plants.[47]

Vertical integration became a successful business strategy in the late nineteenth century. When Standard Oil became a major crude oil producer in the 1880s and the 1890s, it further extended the monopolistic position it had achieved through its dominance in refining, transportation, and marketing. But vertical integration was a game more than one could play. Sun Oil, even at this early stage of its history, used integration as a weapon in its competition with Standard. It could partially escape the dominance of the Standard group by operating its own refinery, transportation system, and marketing organization. The Pure Oil Company became the most successful non-Standard independent before 1900 as a result of its complete vertical integration, but a few other smaller firms, including Sun and the Pacific Coast Oil Company in California, had matched its degree of integration.[48]

In 1895, Sun improved its transportation facilities now that it had a refinery. First Pew and Emerson formed the Bay Terminal Railroad Company, a spur line to connect the refinery with the tracks of the major railroads coming into Toledo. Good railroad connections already existed between Toledo and the Lima-Indiana production fields. The Bay Terminal purchased the 170 tank cars originally owned and operated by the Sun Oil Line Company. More importantly, Sun constructed a pipeline from Toledo south to the North Lima field, and later it extended this line to Tiffin, Ohio. When completed, the main trunk line and its gathering lines comprised approximately 275 miles of pipe. The company also built small gathering lines and storage tanks in the vicinities of Beaver Dam, Bluffton, and Lima, Ohio.[49] By operating its own pipelines, Sun could circumvent Standard Oil's control of the railroads.

COMPETITION WITH STANDARD

Sun Oil relied primarily on independent distributors to market its products. The firm sold some kerosene by horsedrawn wagon in Toledo and Detroit, but shipped most of its production to distributors in Ohio, Indiana, Illinois, and Wisconsin. Sun also exported small quantities of kerosene in wooden barrels, but this was not a major product. Advertisements in the period 1895–1900 described Sun as "Producers, Refiners, Shippers, Exporters of Petroleum and All its Products." Listed among these products were illuminating oils, gasolines, naphtha, black oils, greases, and all grades of gas-making oils, including prime crude oil, gas oils, and naphthas. This latter group of products, used to enrich manufactured gas, became Sun's most important line. In the 1890s only two firms, Standard Oil and Sun Oil, advertised in the *American Gas Light Journal*, the official organ of the manufactured gas industry.[50]

Sun had many lucrative contracts with the United Gas Improvement Company of Philadelphia (U.G.I.) to supply its gas manufacturing plants. U.G.I. operated plants in thirty cities from Fall River, Massachusetts, to St. Augustine, Florida, and from Omaha to Philadelphia. J.N. Pew knew the U.G.I. people from his gas business in Pittsburgh, where they operated a manufacturing plant on the south side of the city.

Now Sun was in a position to bid on U.G.I. contracts for Lima crude, naphtha, and gas oil. The crude and gas oil went to U.G.I.'s carburetion plants and the more expensive naphtha to other facilities. These contracts were to supply an agreed number of gallons over a one year period at a specified price per gallon, and Sun shipped gallonage by tank car as the various plants required it. Gas oil represented a minor share of Standard's refining, but a small firm like Sun could do very well specializing in this product.[51]

The Standard Oil Company, however, exerted what one might euphemistically call "price leadership." Independents rarely attempted to undercut Standard in direct competition with its products, since retaliation could mean disaster. Standard contented itself with about eighty percent of the market for refined products and let the independent firms compete for the rest, much of it in specialty items. When directly competing with Standard, as in kerosene, the best the small companies could do was to keep a ready eye on the market and hope to sell at top price. Standard, meanwhile, monitored the independents through an elaborate reporting system, the "Domestic Trading Committee." Sun Oil operated in a business world dominated by Standard.[52]

Vice-president J.N. Pew oversaw operations from Sun Oil's main office in Pittsburgh, while president E.O. Emerson stayed at his home in Titusville. Although Emerson had left the day-to-day management of their enterprises to J.N. Pew, he still offered advice, and Pew consulted him on all important matters. His many years in the oil regions had made Emerson wise to the problems faced by the small operator in a monopoly-dominated industry. J.N.'s nephew, Robert C. Pew, had now become second vice-president, with offices in Toledo. His brother, James Edgar Pew, later to become a significant figure in both Sun's and the entire industry's future, joined the Toledo workforce in 1896. J. Edgar, his twin brother John G., and older brother Robert C. Pew, were sons of J.N. Pew's brother, Thomas, who had died in 1891. The twins had worked for Peoples Natural Gas for ten years and J. Edgar now moved to Toledo. As J.N. Pew's interest turned to oil, John G. Pew obtained more autonomy in running the Peoples Gas business. The correspondence between these members of the family tells the story of Sun's business relationships in the period.[53]

One problem was Standard's ceaseless surveillance of the industry, which led to the use of codes by other firms. U.G.I. used ciphers when quoting prices or plant locations to Sun and periodically changed its code books in actions suggestive of international intrigue. Sun also began to use ciphers in all of its telegram messages by 1900. The price that it submitted in a competitive bid to U.G.I., for example, was crucial, and Sun feared, with some cause, that Standard Oil sought to take away its business.[54]

U.G.I. threatened Sun with the Standard Oil bogeyman whenever it became displeased with its service. Most of the complaints related to poor quality gas oil or a failure to ship supplies promptly by tank car. Managers of the gas plants did not have large storage facilities and became nervous when reserves ran low. Sun contracted to supply oil for one year at a specified price, but the market price at some point in that year might move higher. The firm therefore preferred to keep oil in its tanks rather than U.G.I.'s so it could sell oil on the open market for higher profit. This situation created friction. In one typical example, U.G.I.'s Philadelphia purchasing agent wrote to J.N. Pew: "We do not want to go to the Standard Oil Co. for oil for this work, and I do not suppose you want us to do this either, but we will not run any risks as we cannot afford to take the slightest chances."[55] Sun managed to meet its contractual commitments, but not without much gnashing of teeth.

Sun also had interesting relations with Standard Oil in Toledo. J. Edgar Pew had taken greater responsibility in the sales and marketing end of the Ohio business, and his brother Robert concentrated on running the refinery. When Sun's managers began selling some kerosene in Toledo, Standard told them that the larger company would inform them of all changes in tank wagon prices before the fact. The independent was usually powerless to challenge this type of price-fixing. On one occasion in 1899, however, Standard made two price reductions without notifying Sun. When J. Edgar Pew protested, Standard told him to "look after his own business." The next time Standard announced a three-quarter cent price increase per gallon, Pew sent out his wagons at the old price, "filling the Standard's tanks everywhere." He wrote to his uncle that "This has been their game in every city. When a decline in the market was on, they would always endeavor to fill up the oppositions' tanks before any notification was given them. . . . I thought the best way was to settle this at once and the way to do it was to take their own tactics, but we hope this will be effectual."[56]

On another occasion, J. Edgar Pew complained to his uncle about a number of large, "gilt-edged" jobbers who professed to be independents, but actually bought exclusively from Standard or refiners tied to them. These jobbers pressed Sun for price quotations and expressed anti-Standard and anti-monopoly sentiments. Presumably, these people were part of Standard's information-gathering network. Sun Oil covered itself by giving higher quotations to them than it did to customers known to be independents.[57]

The Sun Oil Company (Ohio) came a long way in a short time. Through partial vertical integration and a concentration on special markets, such as supplying U.G.I. with gas oil, the firm survived and appeared to be on the brink of becoming a very profitable venture. But it had not expanded horizontally to any great degree and therefore represented no threat to the Standard Oil Company. Sun survived by dealing in the leftovers bequeathed by Standard. Part of Sun's limited scope stemmed from Edward Emerson's caution. He seemed content with Sun's small operations, did not relish spending capital for major expansion, and favored the same full dividend policy that he had insisted on with Peoples Natural Gas. J.N. Pew, however, was becoming increasingly impatient. Soon, by a combination of bold strokes, he would lead Sun Oil in a new direction.

PEW GAINS CONTROL

J.N. Pew had apparently been thinking about striking out on his own for some time. The natural gas fields around Pittsburgh were diminishing rapidly and Peoples Natural Gas had a very bleak future. Although the Sun Oil Company, on the other hand, looked promising to Pew, he and Emerson differed on the proper strategy for competing in the oil industry.[58] Standard's hold on the transportation system had always been the key to its control of the industry. If one could find a way to get around Standard's control of the railroads and pipelines, an independent could successfully compete. One way of doing this would be to locate a source of crude near deep water and use tank ships to transport it to the large eastern market.

As early as 1898, Pew became interested in the possibilities of Texas crude oil. He sent his nephew, J. Edgar Pew, to Corsicana, the first commercial oil field in that state, to investigate a possible deal with J.S. Cullinan, a business acquaintance. The colorful Cullinan later became the founder of the Texas Company after the Spindletop discovery in 1901. In 1898, however, he was an independent operator looking for financial backing.[59]

Cullinan had been an employee of Standard Oil in Pennsylvania for many years and had reached the position of pipeline superintendent in 1893. He left Standard in 1895 following a dispute with his immediate superior, Calvin Payne, and formed a partnership to manufacture and erect steel oil storage tanks and steam boilers. This company, the Petroleum Iron Works, supplied boilers to Peoples Natural Gas, and the Pews knew him well. When artesian well drillers in Corsicana discovered oil by chance in 1896, the owner sent for Cullinan to investigate.[60]

The Pennsylvanian recognized the potential in Texas and came back east to seek financial backing from independent oil men he knew. Since the Waters-Pierce Oil Company, a Standard subsidiary, was under antitrust investigation in Texas, it looked like a good opportunity for independents to gain a step on the giant.[61] Cullinan saw J.N. Pew in Pittsburgh and arranged for J. Edgar to meet him in Corsicana. When the nephew arrived about midnight of the appointed date, Cullinan informed Pew that he was too late and that he "had fixed it up with the old crowd."[62] Cullinan had met with Standard's C.N. Payne, his old boss, and H.C. Folger, Jr. in St. Louis prior to his return to Texas. Standard agreed to supply the capital to establish the J.S. Cullinan Company, which in turn would control the stock of the Corsicana Petroleum Company.[63]

Because of the hostile climate toward Standard in Texas in 1898, Cullinan kept the names of his backers secret. However, the rumors spread quickly that he was a "Standard agent." Standard's historians and Cullinan's biographer both cite the failure of the Pennsylvania backers (Pew) to come up with the promised capital for the venture. Cullinan's 1913 testimony in the Texas antitrust prosecution against the Magnolia Petroleum Company (another Standard subsidiary in which Cullinan had been involved) was the source of this generally accepted version. At these antitrust hearings, Cullinan argued that he was forced to bring Standard into Texas when his "friends" from western Pennsylvania balked. A full reading of the record indicates, however, that the shrewd Cullinan had played one group off against the other until he obtained the most favorable deal for himself.[64]

J.N. Pew, his appetite whetted by the near success in Corsicana, continued to think beyond the Toledo-based operations of the Sun Oil Company. Although Emerson probably knew about the Corsicana sally, there is no evidence that he was directly involved, and Pew realized that he would need a freer hand with the company before he could make any significant moves to expand in the oil business.

Throughout his association with Pew, Edward Emerson had retained Titusville as his base of operations. He had banking and other business interests there and was active in industry and civic affairs. He served a term as president of the Titusville Oil Exchange, and from 1890 to 1893 was mayor of the city. He was sixty-five years old in 1899, and his health was failing. He remained pessimistic about competition with Standard and limited J.N. Pew's hand by his insistence on a full dividend policy. Prompted by a need for cash in 1899, he agreed to sell his interest in the Sun Oil Company (Ohio) to his partner. The transaction took place in April 1899, and Emerson soon left on a retirement trip to Europe. J.N. Pew became the president and controlling interest in the Sun Oil Company.[65]

LESSONS LEARNED

The long-standing relationship between Pew and Emerson had been fruitful. They had contributed to the development of an important new industry, natural gas, and had successfully carved a small but successful corner out of the oil industry. In the process, both men had accumulated sizable personal fortunes from their endeavors. Yet the most significant result of this early period was the experience J.N. Pew gained. His greatest achievement would be the creation of the modern Sun Oil Company. In the development of this firm, he made use of many of the lessons learned in the nineteenth-century oil and gas industries.

Under Pew's direction, the Sun Company later followed conservative financial policies, rarely going into debt. His bad experience with pipeline certificates influenced this attitude, as did the episode with the Penn Fuel Company. After buying out Emerson, he later insisted on having financial control of his company. Pew would bring in outside interests to obtain financial backing, as he had with Emerson, but he insisted that he maintain majority control. He believed that only total control insured freedom of action, and he passed this lesson along to his sons.

Pew learned the advantages of vertical integration from his competition with Standard. By controlling one's own crude and transportation facilities, a refiner would not be at Standard's mercy. But Pew did develop a healthy respect for large competitors, first dealing with Westinghouse and then Rockefeller in gas and oil. He met initial success entering a specialized area, natural gas, which the large firms at first left alone, and would later concentrate on specialized markets for petroleum products. But Pew had also learned that success breeds competition, and to stay in business, one had to continually develop new strategies and be innovative. Despite the large share of the market commanded by the Westinghouses and Rockefellers, there was business to obtain if one were willing to go after it.

Pew's years of struggle against Standard Oil had also hardened his views against monopoly. He passed these attitudes along to his sons, who in later years admitted only one proper role of the government in the economy, the vigorous enforcement of the antitrust laws. With the other independent oil men, J.N. Pew saw the Sherman Antitrust Act as a tool to weaken Standard and improve his competitive position, and he later cheered the dissolution of Standard Oil by the Supreme Court in 1911. But in 1900, Pew was beginning a new phase of his career. Although he remained in the gas business until the sale of Peoples Natural Gas in 1903, he now considered himself to be an oil man.

NOTES

1. "J.N. Pew: A Biographical Sketch," *Our Sun*, 75th Anniversary Issue, 26, nos. 3 and 4 (Summer/Autumn 1961), pp. 11-2.
2. *Ibid.*, p. 12.
3. *Ibid.*

4. Harold F. Williamson and Arnold R. Daum, *The Age of Illumination, 1859-1899*, Volume I, *The American Petroleum Industry* (Evanston: Northwestern University Press, 1959), pp. 294-5; "Early Oil Exchanges Provided Easy Way to Win or Go Broke," *Oil and Gas Journal*, Diamond Jubilee Issue published jointly with the *Oil City Derrick*, 33 (August 27, 1934), p. 9.

5. "Early Oil Exchanges Provided Easy Way," *Oil and Gas Journal*, p. 13; "J.N. Pew," *Our Sun*, p. 12.

6. "J.N. Pew," *Our Sun*, p. 12; Williamson and Daum, *Illumination*, pp. 381-4.

7. "Two Sun Pioneers," *Our Sun*, 75th Anniversary Issue, 26, nos. 3,4 (Summer/Autumn 1961), p. 15.

8. Williamson and Daum, *Illumination*, pp. 137-41.

9. "Story of Sun," *Our Sun*, 75th Anniversary Issue, 26, Nos. 3,4 (Summer/Autumn 1961), p. 17; "Keystone Gas Company," Acc. 1317, EMHL, File 21-A, Box 315.

10. "Gas Light Companies in the United States," *The American Gas Light Journal* 4 (June 15, 1863), p. 373; Williamson and Daum, *Illumination*, pp. 32-3, 38-41.

11. Alfred M. Leeston, John A. Crichton, and John D. Jacobs, *The Dynamic Natural Gas Industry* (Norman: University of Oklahoma Press, 1963), p. 5; "Wells Drilled for Gas and Product Utilized Many Years Before Drake Drilled his Oil Well," *Oil and Gas Journal*, Diamond Jubilee Issue published jointly with the Oil City Derrick 33 (August 27, 1934), p. 40; "Gas Stock That Didn't Pay," *The American Gas Light Journal* 29, 2 (July 16, 1878), p. 37.

12. "Important Uses of Natural Gas," *The American Gas Light Journal* 30, 7 (October 2, 1878), p. 148.

13. Irvin Butterworth, "Natural Gas: What is It and Can it Become a Competitor of Coal Gas as an Illuminating Agent," *The American Gas Light Journal* 42,8 (April 16, 1885), p. 206; *The Mercantile, Manufacturing and Mining Interests of Pittsburgh, 1884* (Pittsburgh: Chamber of Commerce, 1884), p. 44.

14. "Story of Sun," p. 17; "Pioneer Venture," *Our Sun* 16,1 (Winter 1951), p. 15; Ralph W. Hidy and Muriel E. Hidy, *Pioneering in Big Business, 1882-1911*, Volume II, *History of the Standard Oil Company (New Jersey)* (New York: Harper Brothers, 1955), p. 172.

15. "The Drilling of Haymakers' Well," *Our Sun*, 50th Anniversary Issue, 3,1 (1936), pp. 28-9, 47; "Many Gas Companies Formed and Pipe Lines Extended Over Wide Area During 1874-1889," *Oil and Gas Journal*, Diamond Jubilee Issue published jointly with the *Oil City Derrick*, 33 (August 27, 1934), p. 71.

16. *Pittsburgh: Its Commerce and Industries, and the Natural Gas Interest* (Pittsburgh: George B. Hill and Company, 1887), p. 5.

17. "Drilling of Haymakers' Well," *Our Sun*, p. 47.

18. "Story of Sun," *Our Sun*, p. 17.

19. "Important Uses of Natural Gas," *American Gas Light Journal*, p. 18; Butterworth, "Natural Gas," p. 206.

20. Leeston et al., *Natural Gas*, p. 6; "Wells Drilled for Gas," *Oil and Gas Journal*, p. 41.

21. *Pittsburgh*, pp. 6-7. Penn Fuel's chief source of supply remained the Haymaker well, but a controversy erupted over its ownership. On November 26, 1883, an armed confrontation took place between the Haymaker brothers and employees of the Chicago promoter who had earlier contracted to purchase the well. While Matt and Obe Haymaker were digging a pipeline ditch for J.N. Pew, along with ten others, fifty armed men took control of the well site. A fight ensued and gunshots struck several in the Haymaker party. Obe Haymaker died from gunshot and bayonet wounds he received that day. The local sheriff arrested the invading party, and both its leader and the Chicago entrepreneur went to trial. They received jail terms and issued no further challenges to the Pew and Emerson control of the Haymaker well ("Drilling of Haymakers' Well," *Our Sun*, p. 47).

22. Franklin Waltman to Robert L. Klaus, September 7, 1961, Memo, *Our Sun*, 75th Anniversary Issue, Folder #2, "Sun Oil History," Library File, Sun Oil Company Library,

Marcus Hook, Pennsylvania, p. 2; J.N. Pew to C.D. Gillespie, January 10, 1906, J.N. Pew Papers, Box 300, Sun Oil Company Corporate Headquarters, Radnor, Pennsylvania.

23. "Pioneer Venture," *Our Sun*, p. 16; *Pittsburgh*, p. 8.

24. Leeston et al., *Natural Gas*, pp. 6–7; *Pittsburgh*, p. 33.

25. Waltman to Klaus, Library File, Marcus Hook, p. 2; *Pittsburgh*, pp. 33–5.

26. Porter Howard, "History of the Sun Oil Company," Folder #1, "Sun Oil History," Library File, Sun Oil Company Library, Marcus, Hook, Pennsylvania, p. 7.

27. *Pittsburgh*, Hill and Company, p. 33.

28. *Ibid.*, p. 35; "Pioneer Venture," Our Sun, p. 16.

29. Jno. F. Carll, "The Natural Gas Craze," *The American Gas Light Journal* 44, 6 (March 16, 1886), p. 162; Hidy and Hidy, *Pioneering*, pp. 169–75; Williamson and Daum, *Illumination*, p. 606.

30. "Story of Sun," *Our Sun*, p. 18; see Pittsburgh business correspondence in "Peoples Gas," Acc. #1317, EMHL, File 21-A, Boxes 309–11.

31. Waltman to Klaus, Library File, Marcus Hook, p. 6; "Story of Sun," *Our Sun*, p. 17; Gibb and Knowlton, *Resurgent Years*, pp. 70–2; Hidy and Hidy, *Pioneering*, pp. 383–4.

32. E.O. Emerson to J.N. Pew, October 19, 1898, Acc. #1317, EMHL, File 21-A, Box 315; Arthur E. Pew to J.N. Pew, March 14, 1901, Acc. #1317, EMHL, File 21-A, Box 309; A.E. Pew to J.N. Pew, December 6, 1901, File 21-A, Box 311.

33. E.O. Emerson to J.N. Pew, February 13, 1891, J.N. Pew Papers, Radnor, Pennsylvania, Box 315; E.O. Emerson to J.N. Pew, January 12, 1899, Acc. #1317, EMHL, File 21-A, Box 315.

34. Emerson to Pew, April 17, 1899, April 21, 1899, Acc. #1317, EMHL, File 21-A, Box 315.

35. Emerson to Pew, July 11, 1899, EMHL, File 21-A, Box 315.

36. Emerson to Pew, October 4, 1902, EMHL, File 21-A, Box 315.

37. Hidy and Hidy, *Pioneering*, p. 390; "Story of Sun," *Our Sun*, p. 21; "Twin Sun Leaders," *Our Sun* 11, 4 (November 1945), pp. 14–15.

38. Williamson and Daum, *Illumination*, pp. 589–94; "Story of Sun," *Our Sun*, p. 18.

39. *Forty Years Growth in the Oil Industry* (Philadelphia: Sun Oil Company, 1926), p. 2; H.O. Camerson to Frank Cross, February 12, 1936, "Sun Oil History," Library File, Sun Oil Company Library, Marcus Hook, Pennsylvania. In a letter written in 1920, Robert C. Pew explained that "The names of several of the heavenly bodies were considered; Mr. J.N. Pew (Sr.) liked the word 'Sun' on account of the sun's size; hence the name 'Sun' was adopted for the new company" ("Story of Sun," *Our Sun*, p. 18).

40. Cameron to Cross, February 12, 1936, Library File, Marcus Hook; Waltman to Klaus, Library File, Marcus Hook, p. 3.

41. Williamson and Daum, *Illumination*, pp. 234–38.

42. *Ibid.*, pp. 680–1, 37; Sun Oil to U.G.I., September 1, 1897, Acc. #1317, EMHL, File 21-A, Box 312.

43. Cameron to Cross, February 12, 1936, Library File, Marcus Hook; "Story of Sun," p. 18.

44. Williamson and Daum, *Illumination*, pp. 598–600, 602–7; Hidy and Hidy, *Pioneering*, pp. 165–6.

45. Cameron to Cross, February 12, 1936, Library File, Marcus Hook; "Story of Sun," *Our Sun*, p. 18.

46. "Story of Sun," *Our Sun*, p. 18; Waltman to Klaus, Library File, Marcus Hook, pp. 1–2.

47. Hidy and Hidy, *Pioneering*, pp. 43, 105, 160; "The Toledo Refinery Looks Back Over 50 years," *Our Sun* 2, 3 (July/August 1945), pp. 1–2; J.N. Pew to J.B. Meriam, E.B. Meriam, and Fred Berg, June 5, 1895, J.N. Pew Papers, Radnor, Pennsylvania, Box 284.

Sun's present trademark, a diamond pierced by an arrow, is the only remnant of the old Diamond Oil Company.

48. John G. McLean, and Robert W. Haigh, *The Growth of Integrated Oil Companies* (Boston: Harvard University Press, 1954), p. 11; Williamson and Daum, *Illumination*, pp. 576-81.

49. Waltman to Klaus, Library File, Marcus Hook, p. 4; Cameron to Cross, February 12, 1936, Library File, Marcus Hook; "Toledo Looks Back," *Our Sun*, p. 2.

50. "Story of Sun," *Our Sun*, pp. 18-9; "Toledo Looks Back," *Our Sun*, pp. 2-4; see the *American Gas Light Journal*, Volumes 69-78 (1890-1899).

51. See "U.G.I. Contracts," Acc. #1317, EMHL, File 21-A, Box 312.

52. Williamson and Daum, *Illumination*, p. 695.

53. "Notes Made by J. Edgar Pew, 1942," "J. Edgar Pew," Folder #2, Library File, Sun Oil Company Library, Marcus Hook, Pennsylvania; "Twin Sun Leaders," *Our Sun*, p. 14; see general business correspondence, Acc. #1317, EMHL, File 21-A, Boxes 1-4.

54. Lewis Lillie, U.G.I. Comptroller, to J.N. Pew, February 19, 1900, Acc. #1317, EMHL, File 21-A, Box 312; Walton Clark, U.G.I. General Superintendent, to J.N. Pew, December 13, 1902, File 21-A, Box 313; ciphered telegrams, R.C. Pew to J.N. Pew, File 21-A, Box 281; E.O. Emerson to J.N. Pew, March 8, 1899, Acc. #1317, EMHL, File 21-A, Box 315. In one letter to Pew, Emerson commented on the difficulty of striking the top market price and speculated: "I doubt if even the Stand'd know themselves just what they will do a week hence." He then informed Pew of the ciphers they would use in telegraphing instructions to each other. In this instance, he suggested the color gray for "sell" and red for "buy" since the number of letters in each word was the same. (Emerson to Pew, October 30, 1898, Acc. #1317, EMHL, File 21-A, Box 315.

55. J.A. Pearson to J.N. Pew, October 30, 1899, Acc. #1317, EMHL, File 21-A, Box 312; see also Folders: "U.G.I., 1898, 1899," Box 312.

56. J. Edgar Pew to J.N. Pew, September 6, 1899, Acc. #1317, EMHL, File 21-A, Box 4.

57. J. Edgar Pew to J.N. Pew, November 25, 1899, File 21-A, Box 4.

58. E.O. Emerson to J.N. Pew, October 19, 1898, February 20, 1902, Acc. #1317, EMHL, File 21-A, Box 315.

59. "J. Edgar Pew Notes, 1942," Library File, Marcus Hook, p. 11.

60. *Ibid.;* John O. King, *Joseph Stephen Cullinan* (Nashville: Vanderbilt University Press, 1970), p. 6.

61. "J. Edgar Pew Notes, 1942," p. 12; King, *Cullinan*, pp. 29-46.

62. "J. Edgar Pew Notes, 1942," p. 12.

63. Hidy and Hidy, *Pioneering*, p. 276; King, *Cullinan*, pp. 29-46.

64. King, *Cullinan*, pp. 29-46; "J. Edgar Pew Notes, 1942," pp. 11-2.

65. "Two Sun Pioneers," *Our Sun*, p. 15; E.O. Emerson to J.N. Pew, April 19, 1899, April 21, 1899, Acc. #1317, EMHL, File 21-A, Box 315.

CHAPTER III

Integration and Independence: The Evolution of the Modern Firm, 1900–1920

In 1900, the petroleum industry was on the eve of a new era in its history. The discovery of huge crude deposits on the Texas Gulf coast in 1901 was the first of several major strikes that would soon occur in the western half of the United States. New companies like Gulf and Texaco emerged to compete with Standard, Pure, Sun, and the older firms whose roots lay in the tumultuous past. The incandescent lamp was already spelling doom for kerosene and gas oil sales, but that new upstart, the automobile, was beginning to show signs that it just might be the savior of the industry. Many of the independent firms soon realized the potential for gasoline and were quicker to adapt to a changing market than was the Standard monolith.

The Standard Oil trust was also burdened with a growing number of antitrust suits that culminated in the break-up of the huge corporation by the Supreme Court of the United States in 1911. Yet, historians have argued convincingly that the monopolistic position Standard enjoyed for so many years had already become weakened due to the shifts in the industry accompanying new crude supplies and increased sales of refined products by independents. Those companies that enjoyed the most success were the ones that had fully integrated forward and backward, and competition in the post-1911 period became more oligopolistic than atomistic.[1] By 1920, therefore, an "independent" was no longer a Standard competitor, but a non-integrated crude producer, refiner, or marketer. Many independents now viewed some of the older integrated firms like Sun as representing some of the same evils they had identified with the old Standard Oil trust: bigness, control of transportation facilities, and aggressive marketing. Sun still remained relatively small, but events occurring in the first two decades of the twentieth century crucially shaped the modern corporation.

THE SUN COMPANY (NEW JERSEY)

The turn of the century found Joseph N. Pew fully committed to oil. With a slight trace of bitterness, Edward Emerson wrote him from London in 1899 that "perhaps I was foolish to sell at the price, just as we were on the eve of a paying business, yet I hope you will make the business pay as you deserve to after taking the risk of buying your partner out."[2] Pew had indeed taken a risk, but he had not acted blindly. Consultations with the United Gas Improvement Company (U.G.I.) had assured him of continued business and promised financial assistance. U.G.I. had become the single largest purchaser of crude and gas oils from Toledo, but transportation difficulties still caused Sun's tank cars to arrive frequently behind schedule. U.G.I. continually complained of poor service.[3] Pew thought that if he could build a refinery near Philadelphia with adequate storage facilities, he could give U.G.I.'s large gas plants improved service at cheaper cost.

A full year before the Spindletop discovery, Pew had started to put his plan into operation. U.G.I.'s Samuel T. Bodine, a personal friend of Pew's, assisted in the preliminary negotiations for financial help in building the refinery. As early as December 1899, Bodine had begun investigating sites on the Delaware River near Philadelphia, and sent Pew information on the Atlantic Refining Company's property in Chester, Pennsylvania, and on three other locations. Although Pew's initial strategy was to transport Mexican crude by tank ship to Philadelphia, his earlier interest in Corsicana shows that he had considered Texas as a source of crude. However, he could hardly have envisioned the events of January 10, 1901, and their consequences. On this day, Captain Anthony Lucas brought in the biggest oil gusher found up to that time. Spindletop, located about four miles north of Beaumont, Texas, near the Gulf Coast, changed the course of petroleum history. Sun's modern era really began here, along with that of the Texas Company, Gulf Oil, and others.[4]

Harold F. Williamson and Ralph Andreano cite four main reasons to explain the important changes wrought by Spindletop. In the first place, the Gulf region was virgin territory, especially attractive to new firms entering the business. Second, the poor quality of the crude actually aided new entries into the market. A heavy, asphaltic-based, and high sulphur oil, it made adequate fuel oil but poor kerosene, lubricants, and gasoline. Since there was an established market for fuel oil and it was inexpensive to manufacture, the new firms could easily integrate forward into transportation, refining, and marketing. The geographic location of the oil field represented a third important factor. Because it was near deep water, small investments in pipelines enabled operators to move crude to ocean vessels. Lastly, the small Standard presence in the Gulf region allowed the new independents to thrive. A hostile legal climate for Standard Oil in Texas partially explains its limited involvement, but poor management decisions also inhibited the firm from gaining a larger share of the Texas business.[5] All four of these factors were relevant to Sun's experience.

Soon after hearing of the Spindletop strike, J.N. Pew wired his nephew, Robert C. Pew, to go to Texas and investigate, much as he had done in the Lima fields in 1885. When Robert arrived in Beaumont on January 16, 1901, the sight of Spindletop overwhelmed him. He reported the enormous size of the well and speculated that there had to be great quantities of oil under the entire county. Preliminary examination of the crude indicated that it would make good fuel oil and, more importantly for Sun, gas oil. Robert quickly grasped the significance of transportation as the key to Spindletop operations. In a letter written the day after his arrival, he recommended the acquisition of a pipeline to the coast and the immediate purchase or charter of steam vessels. He enthusiastically exclaimed that "We could put oil into *all* of the Atlantic coast cities for less than one fraction of the cost of transportation from Ohio."[6]

Although enthusiastic about the Texas business, Robert Pew did not think as well of Texas itself. He complained about the weather, the people, the town of Beaumont, a series of personal maladies, and requested to return to Toledo. He left Beaumont on March 6 after negotiating purchase of five large leases on and in the vicinity of Spindletop. His younger brother, J. Edgar, succeeded him in Beaumont later that month, but before he left Toledo, J. Edgar had supervised tests of the Spindletop crude and had become convinced that Sun could manufacture gas oil from it.[7]

A Beaumont newspaper was close to the mark when it prophesied in 1901 that "The Beaumont oil will eclipse in significance, extent, and human interest the Indian stocks, the Mississippi Bubble, the gold fields of '49, and even that pathetic melodrama of the Klondike."[8] J. Edgar Pew arrived in Beaumont carrying a .41 caliber revolver, which his uncle and brother insisted he bring for protection. He encountered ham and eggs at a dollar a plate, a cot for five dollars, and a four-mile ride to the well site for eighteen to twenty dollars. J. Edgar also found wild speculation in drilling leases and a number of slick stock promoters waiting to hustle the "suckers" arriving daily by train.[9] The traditional pattern of oil field development begun in Pennsylvania continued here. In the first four months after the Spindletop discovery, 132 companies filed for charters to do business in Texas. Drillers sank wells so close together that they built plank runways between them as an escape from fire. As dictated by the "rule of capture," everyone wished to put in a well before his neighbor drained the oil under the lease. Often the derrick legs of one well overlapped those of another, and operators dug earthen storage tanks to hold the overwhelming flow of crude. Supply so far exceeded demand that oil reportedly sold for three cents a barrel, while a cup of drinking water went for five cents in Beaumont. J. Edgar Pew later commented that "You could buy millions of barrels of oil at five cents a barrel, but the point was to know what to do with it."[10] All operators quicly realized the necessity of obtaining tank sites, storage tanks, pipeline connections, shipping facilities, and new markets.

Events moved simultaneously for Sun in Beaumont and back east. On May 2, 1901, J.N. Pew incorporated a new enterprise, the Sun Company of New Jersey, with an authorized capital stock of $1,000,000. Pew personally subscribed to forty-five percent of the company's common stock and U.G.I. took another forty-five percent. The remaining ten percent went to the other members of the Board of Directors, all family or close associates. When Sun increased the authorized stock of the company on different occasions over the next few years, U.G.I. continued to retain forty-five percent interest in the New Jersey Corporation.[11] Pew had exchanged one partner, Emerson, for another, but U.G.I. capital now enabled him to expand in new directions and to lay the foundations of the modern, fully integrated Sun Oil Company.

At the incorporation meeting in Jersey City on May 3, 1901, J.N. Pew, his eldest son Arthur E. Pew, legal counsel W.S. Miller, and Pittsburgh associates Frank Cross and Joseph V. Clark became directors of the new corporation. The board met in Pittsburgh on May 8 and elected J.N. Pew president, Arthur E. Pew vice-president, Miller treasurer, and Cross secretary. Meanwhile, J.N. Pew remained president of the Sun Oil Company (Ohio). At its June meeting, the Sun Company Board voted to purchase drilling leases held in the name of Robert C. Pew and J. Edgar Pew for $500,000. Robert received 4,000 shares of Sun $100 par value common stock, and J. Edgar 1,000 shares. The board appointed J. Edgar Pew general agent of the company in Texas, authorizing him "to buy and sell oil, oil leases, and lands and transact such other business for the company as he may be directed to by the President of said company."[12]

Also in May, J. Edgar Pew obtained forty-two acres of land near Smith's Bluff on the Neches River, near Beaumont. He constructed earthen reserviors and storage tanks on the site, later named Sun Station. A four-inch pipeline, operated under the corporate name of Texas Transit, connected Sun Station with the Spindletop field. Located near tidewater, Sun could now ship by water, but its first consignments of crude went to the Toledo refinery by tank car. However, these first crude shipments proved so high in sulphur that U.G.I. could not use it in its carburetion plants. This meant that Sun would have to refine the crude into gas oil, and the rapid construction of an eastern refinery became an essential priority in J.N. Pew's plans.[13]

In October 1901, Pew purchased an eighty-two acre site for his refinery from a group of Philadelphians for $45,000. Located on the Delaware River in Marcus Hook, Pennsylvania, the property had formerly been a popular bathing resort and picnic ground for Chester area residents. He chose this location for essentially three reasons. There was deep water for tankers, Pennsylvania and Reading Railroad lines were nearby, and since other refiners operated in the area, a pipeline from the Pennsylvania fields provided the insurance of a secondary crude supply.[14]

Construction began at Marcus Hook in November, and Pew purchased a one-year-old Great Lakes ore carrier which he had converted for carrying oil. This ship, the S.S. *Paraguay*, had an 18,000 barrel capacity and represented the first in a long line of Sun tankers. The first Texas crude oil shipment arrived in Marcus Hook on March 31, 1902, but since Sun did not complete the refinery until June, the crude went into temporary earthen storage. The original refinery consisted of eight 800-barrel crude stills to refine gas oil and fuel oil, three receiving tanks, and two crude tanks.[15]

By the end of 1902, the *Paraguay* had carried some 400,000 barrels of Texas crude to Marcus Hook, about ten percent of the total shipments from Gulf ports that year. Pew also organized a small subsidiary, the Delaware and Union Railroad, to provide rail transportation within the refinery and connection with the main trunk lines in Marcus Hook.[16]

Meanwhile, the Sun Oil Company (Ohio) continued to supply U.G.I. and its other customers with Ohio crude and refined products. However, U.G.I.'s purchasing department complained often about the Ohio Company's mediocre performance in fulfilling its contracts, and further implied threats to give its business to Standard Oil indicated that the Sun-U.G.I. relationship was not entirely harmonious.[17]

The gas company also expressed concern with the Marcus Hook situation. In March 1902, U.G.I. had conducted experiments with Texas crude at its Des Moines, Iowa carburetion works. As a result of the information learned from these tests, U.G.I. management claimed that Sun had a jump on all of its competitors in refining Texas crude. Yet, U.G.I. was not pleased with the first batches of gas oil received from Marcus Hook in June and July. It had a high sulphur content and was of inferior quality. U.G.I. indicated that it would not pay Ohio oil prices for

Texas oil unless the quality was up to par. The Pews had been shipping approximately equal amounts of Texas gas oil and Ohio gas oil to U.G.I., but both at the price originally bid for Ohio oil. Furthermore, in September, U.G.I. expressed concern that the Marcus Hook refinery would not be able to supply its 25th Ward Works in Philadelphia adequately, as Pew had promised. J.N. Pew, still based in Pittsburgh, came to Philadelphia in the fall of 1902 and successfully smoothed out these matters.[18]

In a relatively short time, J.N. Pew had accomplished much after gaining control of Sun. By the end of 1902, he had obtained a crude source near deep water and built a refinery near a population center also on deep water. His plan had become a reality. Now that the basic structure of the firm was in place, Pew moved to consolidate and expand his operations. Sun stood on the brink of a major expansion phase.

REFINING AND MARKETING

Once the Marcus Hook refinery became established, Sun refined almost all of the incoming Texas crude into gas oil and sold it to U.G.I. However, after running the gas oil from its stills, a significant amount of heavy, black, asphaltic "residuum" remained. Most of the other refiners of Spindletop crude made various grades of fuel oil, but Sun had to fulfill its U.G.I. contracts to make gas oil, a lighter fraction, so was left with more residuum from its refining process. Pew had originally hoped to obtain lubricating oils from the Texas crude, and he now turned his attention to this project.

The Sun Oil Company (Ohio) had sold some lubricating oil both at home and abroad, and J.N. Pew sought to expand this market. Lubricants had long been attractive to independents in competition with Standard Oil, since small refiners could specialize in these premium quality products and cultivate their business with good customer service. The high prices charged for specialty items helped defray the cost of transporting relatively small quantities. Although Standard had not totally ignored these specialty markets, it had concentrated on the large volume market for kerosene. J.N. Pew now hoped to capture part of the lubricant market with Texas crude. If he could somehow upgrade the residuum into lubricating sock, he could develop an efficient refining system. At first, however, the Spindletop crude defied the efforts of the Toledo refinery. J. Howard Pew, the founder's second eldest son, began his long Sun career by working on this project.[19]

J. Howard had graduated from tiny Grove City College in northwestern Pennsylvania at the precocious age of sixteen in 1900. J.N. Pew had helped start the school, and it is still heavily supported by the Pew family. J. Howard spent the academic year 1900–1901 at "Boston Tech" (M.I.T.), taking engineering training. When he reported to Sun Oil in Pittsburgh in September 1901, his father sent him to work for a month in a Franklin, Pennsylvania refinery, in the heart of the old oil regions. From there he went to the Toledo plant to work with a group trying to develop lubricants from Spindletop crude.[20]

After months of hard work, they devised a vacuum process for distilling a lubricating stock that Sun named "summer black." By heating the oil in a vacuum under lower temperatures, the risk of breaking down the lubricating properties of the product in the distillation process was reduced. When treated in an agitator, this stock produced a high viscosity lubricating oil that retained its fluid properties at lower temperatures than ordinary oils. This product marked the beginning of the firm's highly successful "Sun Red Oils."[21]

J. Howard Pew arrived in Marcus Hook early in 1902, faced with duplicating the Toledo pilot plant and developing a commerical facility for refining lubricating stock from the heavy black residuum that Sun produced in great abundance there. After some initial difficulty, the Marcus Hook team succeeded in developing a successful commercial process in 1903. Because of its high viscosity, Sun Red immediately became popular for use in industrial machinery. Sun advertised only in trade journals, but this product was largely responsible for making the firm's reputation for marketing high quality, specialized products.[22]

Continued research at Marcus Hook and Toledo broadened the range of Sun products. In 1903, a team headed by J. Howard Pew developed a process for manufacturing high quality asphalt from the Texas crude residuum. Named "Hydrolene," this proved commercially successful and was Sun's first trademarked product. The process consisted of blowing air through hot oil to oxidize and thicken it. Unfortunately for Sun, the procedure infringed on a patent granted to a similar process earlier.[23]

In 1894, Francis X. Byerley had patented an oxidation process for the manufacture of asphalt and other products from petroleum. In 1904, he brought suit against Sun, and litigation dragged on for many years. In the course of the long court battle, Byerley died and his son, Francis A., continued the fight on behalf of the estate. Sun maintained that its process differed from Byerley's, that Byerley himself had originally conceded this fact, and that the air oxidation process had been general knowledge among early Pennsylvania refiners and was used before the granting of Byerley's patent. Sun lost the case, appealed, and finally agreed to pay damages and royalties to the Byerley estate in 1912. In the interim, Sun had continued to market Hydrolene, a product that found an increasing use in the rapidly expanding highway building program of the early twentieth century.[24]

The Hydrolene suit was not the only difficulty that grew out of the early Marcus Hook operations. Most residents of Marcus Hook and its environs welcomed the jobs and tax revenues the new refinery provided.[25] However, some Chester County citizens did not appreciate the replacement of a once popular resort with a polluting oil refinery. J. Howard Pew wrote to his father in 1902 that "the townspeople are beginning to talk quite badly of us. . . . It will be, I think, bad policy to run any more oil until the sewer is finished and we get fixed to burn the gas."[26] Howard also reported the deaths of birds and fish in Marcus Hook as a result of Sun's pollution. In August 1902, a group of Marcus Hook property owners sought an injunction against Sun as a public nuisance. This action resulted in Sun employing more efficient methods to dispose of sulphur and other wastes with U.G.I.'s help.[27]

In the fall of 1904, another court case arose from complaints about the noxious fumes emanating from the refinery. Several Sun employees testified in the firm's defense, one boiler maker stating that he liked to come to "the Hook" because it benefited his health. Although these complaints influenced Sun's utilization of cleaner processing, they were not sufficient to stop the rapid expansion of the facility. Sun also realized the importance of good community relations and began programs in this area. In 1908, for example, Sun financed most of the construction of the Marcus Hook Fire House, a project that helped both the firm and the community. After the dedication of the building, Sun held a community picnic and transported the celebrants by steamer from the Sun dock to nearby Woodland Beach.[28]

Although the Marcus Hook refinery remained the key to Sun's expansion with Texas crude oil, the Toledo plant was still an important cog in J.N. Pew's overall operation. In one instance, the influx of Gulf Coast oil accounted for a specific expansion of the Toledo plant. Robert C. Pew realized that the rush by refiners to use Texas crudes might result in less production of paraffin-base oils and therefore increase the demand for waxes. The high asphaltic Texas crude contained no paraffin at all. While not as good as Pennsylvania crude for this purpose, Ohio-Indiana oil did possess some paraffin. To dispose of its accumulations of wax oils in Toledo and exploit this market, Sun built a new wax plant in 1902. The Toledo refinery now sent waxes to manufacturers of gunpowder and matches both home and abroad. Toledo also expanded its lubricant production, and by 1904, the original fourteen-acre site had grown to fifty-six acres.[29]

In addition to gas oil, asphalt, and Sun Red Oils, Sun's lubricating stills, completed in Marcus Hook in 1904 and 1910, refined other successful products. Among these were industrial oils: Sun Pale, Sun Spindle, Newport Pale, and Newport Red Oils. In a related product development, the Toledo refinery, under Robert Pew, developed a successful petroleum turpentine substitute in 1905. Sun originally marketed this from Toledo and Marcus Hook as "Number 17 and Number 18 Special," later changing the name to "Sunoco Spirits." This was the first use of the "Sunoco" trademark which later appeared on all Sun Oil products. In 1911, Sun developed a new machine cutting oil which it first trademarked as "Emulso" and later as "Sunoco Emulsifying Oil." Although the Sun organization continued to sell gas oil, fuel oil, and other refined products, the firm increasingly became known as a lubricant specialist.[30]

FOREIGN MARKETING

From 1861, when the first shipment of American petroleum crossed the Atlantic, foreign markets accounted for an increasing share of total United States petroleum sales. By far, the largest part of this total was in kerosene, the product that accounted for fully seventy percent of all refined product sales in the 1870s and 1880s. Although efforts to sell kerosene to Asia, Africa, Latin America, and the Middle East met with growing success, the European market constituted the overwhelming percentage of foreign purchases.[31]

American firms, including Standard Oil, at first relied on foreign import houses to distribute and market their goods. Usually located in a major marketing center, these firms arranged to handle purchases and obtain shipping space and marine insurance through resident representatives or commissioned agents in the United States. These firms either marketed the oil themselves in Britain or the continent, or sold it to European distributors. The system worked well for a long time. It benefited the oil companies by enabling downturns in the market to be absorbed by various middlemen as well as providing a way to distribute oil to the many small consumers in Europe.[32]

Since Standard was the largest U.S. refiner of kerosene and other products, it naturally had the dominant share of the foreign market. However, it too refrained from opening its own marketing affiliates and worked through foreign firms. The first exception to this occurred in trademarked lubricating oils and greases, specialty products that competed on the world market. In the late 1870s, Standard bought into mercantile firms with the aim of copying the successful marketing operations of the Chesebrough Manufacturing Company. This firm, makers of "Vaseline," had sales offices in Montreal, London, Paris, Barcelona, Hamburg, Rio de Janeiro, and Buenos Aires. So impressed had Standard been with Chesebrough's success that the giant firm absorbed the company in 1880. In 1888, Standard organized the Anglo-American Oil

Company, Ltd., the first of its major foreign affiliates. This company was largely a response to the successful marketing efforts of the Kerosene Company, the major marketing arm of the Rothschild's in Europe. By 1900, Standard had already become a giant multinational enterprise.[33]

Sun had exported some gas oils, lubricants, and other refined products prior to 1900, but its entry into the European market did not blossom until it began to concentrate more on lubricant production. The firm followed the general industry pattern and sold its first trademarked products to European jobbers, who in turn distributed them abroad. Sun found very favorable markets for its lubricating oils, and both J.N. Pew, Sr. and Arthur E. Pew made frequent trips to Europe in pursuit of this business.[34]

In 1903, J. Edgar Pew had met William Smellie, a representative of the British distributors Meade, King, and Robinson in Texas. The firm purchased a large order of Hydrolene from Sun which it marketed abroad, and for the next several years, Sun sold many of its products, particularly lubricating oils, through this Liverpool firm. An increased European demand for Sun Red and other lubricants prompted Sun to establish its first foreign marketing subsidiary, the British Sun Company, in 1909. The new company opened offices in Liverpool with William Smellie as manager.[35]

Although Sun acted later than Standard to establish overseas marketing subsidiaries, its pattern resembled that of the other firms that had prospered since Spindletop. The Texas Company, for example, opened its first foreign sales agency in Europe in 1905, and by 1913, it marketed in Europe, Latin America, Australia, Africa, and several Asian countries. After the 1911 Standard Oil dissolution, many of the existing "independents," like Sun, saw even more possibilities abroad. Gulf Oil increased its foreign sales, but, unlike Sun, did not develop its own distribution system until the 1920s. The exception to the rule was Pure Oil. It had established foreign sales agencies in the 1890s in competition with Standard, but later gave them up. In 1905, Pure sold some of its marketing affiliates to Shell, and in 1911, Jersey Standard bought the rest.[36]

In the years prior to World War I, Sun greatly increased its European business. In addition to British Sun, J.N. Pew, Sr. established two other marketing subsidiaries, Sun Oil (Belgium) and Netherlands Sun. Sun also marketed in Germany through an affiliate, Mineraloelwerke Albrecht & Company of Hamburg. In 1910, J.N. Pew wrote to his son Arthur from London that Sun could have all of the foreign orders that it wanted. He reported especially high demand for Sun Red and American Cylinder Oil, commenting that there was no competition for these grades. Pew further wrote that "They all speak in the highest terms of the grades we are making, and they tell me that the Sun Company are delivering more lubricating oil on the continent than any other American company."[37]

Companies still exported oils in wooden barrels, and Sun moved to integrate its marketing operations by acquiring extensive timber lands in West Virginia and organizing another subsidiary, the Hardwood Package Company, at Marcus Hook in 1912. This plant manufactured barrels until 1927, when the general use of steel drums made them obsolete.[38]

WARTIME EXPANSION

A growing demand for gasoline and fuel oil had already begun to fill the gap left by decreasing kerosene sales in the industry due to the popularity of first gas and then electricity, but the outbreak of World War I further stimulated all aspects of the oil business. Key to the expanded demand had been the conversion of most naval vessels prior to the war to the burning of fuel oil rather than coal to heat their boilers, and the increasing use of internal combustion engines in military vehicles. However, lubricating oils and greases also played an important role in keeping both naval and land vehicles moving during the war. In addition, petroleum products in the form of kerosene, paraffin, mineral spirits, and other distillates also performed useful tasks.[39]

Although American firms made profitable sales to the Central Powers during the period of United States neutrality, 1914–17, the largest share of petroleum shipments went to the Allies. Sun, for example, continued to market lubricants through its German affiliate, Mineraolelwerke Albrecht & Company, up to the entry of the United States into the conflict. One study estimates that direct industry shipments to the Allies, including the United States military, totaled 133 million barrels; 60 million barrels between June 30, 1914, and December 31, 1916, and 70 million in 1917 and 1918.[40]

These large-volume sales brought about significant growth in the individual firms throughout the oil industry, and Sun benefited from wartime expansion. The firm supplied the European belligerents with many essential petroleum products during the neutrality period. Lubricating oils headed the list of Sun exports, and Sun later claimed that France was obtaining fully one-third of her lubricant needs from the Pew organization at one point of the war. In addition to these lubricant sales, Sun exploited new markets. The company contracted to sell low-grade fuel oil to the British Admiralty prior to 1917, and later sold this product to the United States Navy. To keep pace with increased demand for its products, Sun constructed a third, larger refining plant on its site in Marcus Hook.[41]

Sun had not yet begun to refine and market gasoline, even though its importance was growing daily. To a large extent, this resulted from contractual agreements with its financial partner, the United Gas Improvement Company (U.G.I.), to supply gas oil. It was relatively easy and highly profitable to distill the petroleum fractions in high-grade gas oil into gasoline, but Sun was committed to gas oil manufacture. For this reason, Sun's entry into the gasoline market was late compared with other Spindletop-era firms like Gulf and Texas, which had marketed gasoline extensively before the war. Friction with U.G.I. over this and Sun's financial policies eventually led to a parting of the ways during the war. Lubricant and fuel oil sales were excellent, however, and Sun emerged from the war much stronger than it had been in 1914. Total sales volume had increased more than three-fold to 31 million dollars from 1914 to 1918.[42]

THE SUN SHIPBUILDING COMPANY

In the midst of World War I, the Sun Company expanded into a unique phase of vertical integration. The Pews began to construct oil tankers, a business related to oil transportation, but not one of the primary forms of oil industry integration. This tertiary integration not only represented a singular development in the industry, but further illustrated Sun's desire for independence and self-sufficiency. Many of the major firms, including Jersey Standard, Atlantic, Gulf, Texas, Vacuum, and Sun had become alarmed at the loss of tankers to German U-Boats and had chartered the construction of several new vessels by 1916. Following the entry of the United States into war in 1917, some of them even turned their machine shops over to the manufacture of parts for ships. However, the formation of the Sun Shipbuilding Company, later the Sun Shipbuilding and Drydock Company, enabled Sun to build its own tankers, and it has remained a valuable subsidiary up to the present day.[43]

On a business trip to Europe in 1915, J. Howard Pew learned of German plans for U-Boats first hand. While visiting Sun's Hamburg affiliate, Mineraloelwerke Albrecht, Pew observed the submarines under construction in the shipyards. The Germans also boasted of their air power, and they began the first Zeppelin raids on London while Pew was in Hamburg. Impressed with German naval and air strength, and concerned for his wife's safety in London, Pew left for the United Kingdom at once. After Howard had returned to the United States, he and J.N. Pew, Jr. concluded that the German navy would probably sink a large percentage of Allied shipping, including oil tankers. Accordingly, they commissioned the Cramp Shipyard in Philadelphia to construct two tankers for Sun.[44]

Shortly thereafter, the German Navy commenced its intensive U-Boat campaign. The sharp increase in the number of sinkings accelerated ship construction, and soon every yard in the country was working at full capacity. Sun sold its rights to the two tankers on the Cramp Shipyard's ways at a handsome profit and used this money to launch the Sun Shipbuilding Company in 1916.[45]

The Sun Company incorporated the subsidiary in May 1916, and construction began on a site in Chester, Pennsylvania, on April 24. Chester, located next to Marcus Hook, had good deep water facilities on the Delaware River. Sun built five shipbuilding ways capable of handling vessels up to 500 feet long and weighing 15,000 deadweight tons. In July 1916, the yard acquired the Chester factory of Robert Wetherill and Company, a manufacturer of stationary steam engines since 1872, and Sun converted the facility into the production of marine engines.[46]

Sun Ship completed its first vessel, the S.S. *Sun*, on October 30, 1917. At the time of the armistice, the firm had completed six tankers, another was three-quarters complete, and five others were approximately half finished. In addition, the yard built three mine sweepers for the United States Navy. At the height of its operations in the fall of 1918, the shipyard employed 10,000 men—9,000 on the shipbuilding payroll, and another 1,000 on housing construction for the ship workers.[47]

Federal government participation in the wartime shipbuilding effort proved a mixed blessing for Sun Ship. Soon after the United States entered the war, the Wilson administration chartered the Emergency Fleet Corporation, a subsidiary of the United States Shipping Board. With a stated purpose to build ships faster than German submarines could sink them, the agency became intimately involved in ship construction on the eastern seaboard. In August 1917, J.N. Pew, Jr. expressed his willingness to cooperate with the war effort, but protested that the government's shipping representatives "have adopted, seemingly a policy of confiscation."[48]

The head of the Shipping Board, former Federal Trade Commission Chairman E.N. Hurley, and his hand-picked head of the Emergency Fleet Corporation, Charles F. Priez, had indeed ruffled many of the feathers of private enterprise. Production figures lagged far behind projections, and the Shipping Board placed the blame on the shoulders of the private shipyards. In April 1918, Hurley sent a report to President Woodrow Wilson on the disappointingly low production. He cited careless workmanship, material shortages, and the failure of ship owners to fight for increased production. Hurley recognized that someone with the confidence of the industry was needed to bail out the shipbuilding program. After Henry Ford declined the job, President Wilson persuaded a former political opponent, Bethlehem Steel Chairman Charles Schwab, to take the position.[49]

Within two months after taking the dollar-a-year job, Schwab had indeed improved the situation. By June 1918, there was a marked increase in the number of keels laid, ships launched, and ships delivered to final destination. Schwab openly disapproved of the government's previous wartime threats to nationalize industries that did not "voluntarily" agree to stabilize prices, but at the same time moved to cut costs and speed delivery by going from cost-plus to lump sum ship contracts with the yards. Thus, although J. N. Pew, Jr. found Schwab's philosophy more appealing, the new contract policy continued to place pressure on the individual shipyard. It received the same payment whether the work was completed on schedule or not, and it behooved management to eleminate waste and inefficiency.[50]

Sun did encounter some difficulties with the Emergency Fleet Corporation in obtaining complete payment for contracted vessels. However, these contracts brought revenue to Sun Ship, and therefore to the consolidated profit statement of Sun Oil. As for the total national program, Schwab had arrived too late on the scene, and at the end of the war, the Shipping Board had fallen far short of its goal to build 15,000,000 tons of new shipping.[51]

In the beginning, the Sun Shipbuilding project was largely in the hands of J.N. Pew, Jr., the president and board chairman of the subsidiary. In 1919, however, John G. Pew became president and chief operating officer, with J.N., Jr. remaining as chairman. The total output of the Chester concern was not very significant compared to the total United States wartime shipbuilding effort, but Sun had established the foundations of an important business. After some lean years in the 1920s and 1930s, Sun Ship emerged in World War II as the largest private shipyard and single biggest producer of oil tankers in the United States.[52]

SUN ENTERS THE AUTOMOBILE AGE

In the post-war period, Sun began to market automotive products for the first time. The sudden increase in automobile ownership after 1900 had created a new demand for gasoline, which early refiners had discarded as waste from their kerosene production. Because Standard Oil had moved slowly into this market, post-Spindletop independents like the Texas and Gulf Oil companies established strong footholds in this business. Until the end of the war, however, Sun concentrated on producing lubricating and industrial oils and sold its lighter distillates to U.G.I. and other customers in the form of gas oil. The Toledo refinery had always refined small quantities of gasoline which it sold locally, but it had no incentive to increase the output of this fraction from its normal distillation procedures. Now in the post-war period, a combination of circumstances found Sun with large quantities of gas oil which it could use successfully to manufacture gasoline.[53]

As early as 1908, a number of individuals had worked on the development of a new refining process for obtaining a higher yield of gasoline from the charging stock, either crude or heavy distillate that went into the front end of the still. Crude oil production had failed to keep up with the tremendously increased demand for gasoline, and the cracking processes then used yielded small percentages of gasoline. In 1913, Dr. William Burton of Standard (Indiana) patented a new thermal cracking process capable of doubling the output of gasoline from a barrel of crude. Burton's breakthrough provided the technological revolution necessary to meet the demand created by the automobile. The charging stock Burton used in his high pressure and high temperature stills had great relevance for Sun's development. Rather than introducing residuum into the stills, as in earlier cracking processes, Burton used gas oil.[54]

Because of financial and business conflicts, Sun had purchased U.G.I.'s interest in the firm during the war. Now the Pews had large quantities of gas oil on hand and a plant capacity for refining more of it. With this gas oil as a base, Sun was in a good position to move into gasoline refining and marketing. The firm licensed thermal cracking processes and later developed its own process in the twenties to expand this business. First, however, Sun decided to enter the automotive field with motor oil, hoping to bank on its reputation for lubricants.[55]

"Sunoco Motor Oil," the first Sun product intended for a mass consumption market, appeared in 1919. A full advertising campaign, including full page messages in the *Saturday Evening Post*, heralded its introduction. In January 1919, the Sun Company board of directors adopted two resolutions reflecting this trend toward product identification. The first resolution was one to market all lubricating and industrial oils as "Sunoils" and all motor oils as "Sunoco Motor Oils." The second resolved to change the name of the corporation from the Sun Company (New Jersey) to the Sun Oil Company (New Jersey). The board did not amend Sun's corporate charter to accomplish the latter end until 1922.[56]

In December 1919, the board approved an expenditure of $20,000 to construct Sun's first gasoline filling station at Ardmore, Pennsylvania. The station opened in 1920, inaugurating the firm's gasoline marketing program. Meanwhile, Sun directed research into gasoline refining and initiated an eastern marketing network. General Sales Manager Samuel Eckert lined up distributors to handle Sunoco gasoline and oil, since many stations at that time retailed the products of more than one refiner. In those cases, Sun installed its own pumps, while in other instances it moved to establish exclusive dealerships.[57]

Although Sun had now entered the competitive market for automotive products, it did not abandon its traditional base of operations. When the war ended, Sun continued to pursue its lucrative business in industrial lubricants. A Sun tanker loaded with lubricating oil was one of the first commercial ships into Hamburg harbor following repeal of the Trading With the Enemy Act. The Pews reestablished their connections with Mineraloelwerke Albrecht in Germany and were well organized through their subsidiaries in Britain, Belgium, and the Netherlands. Because of the high volume business done with the French government during the war, Sun now wished to organize a more permanent marketing organization in that country.[58]

In 1919, J. Howard Pew traveled to Europe and entered into an agreement to purchase a one-third interest in the Societé Anonyme d'Armement d'Industrie (SAIC), a French-based marketing organization. Sun bought 8,996 shares in the firm for $3,500,000. The Swiss Nobel organization owned the largest block, 9,283 shares, and a French group backed by the Rothschilds the remaining 8,711 shares. Both the Nobels and Rothschilds owned large international petroleum operations and were rapidly expanding in the post-war era. SAIC marketed Sun oils in Europe until Sun sold its interests to the Gulf Oil Company in 1925. Gulf was late in establishing its own marketing affiliates and had been distributing its products in Europe through the Jersey Standard sales organization. By 1925, Sun had found its wholly-owned subsidiaries sufficient to handle its European sales and was willing to sell to Gulf for a substantial profit.[59]

As of January 1, 1920, Sun's refining capacity ranked twenty-first in the United States. Standard (New Jersey), with nine and one-half percent of total capacity, led seven other former firms from the old trust ranking ahead of Sun. Sun had only nine-tenths of one percent of total capacity, but by 1920, the firm had fully begun to gear its refining and marketing for a successful entry into the automobile age.[60] (See Appendix, Table 2).

CRUDE OIL PRODUCTION AND TRANSPORTATION

After establishing deepwater and storage facilities at Sun Station on the Neches River in 1901, J. Edgar Pew had moved to strengthen Sun's relative position in Texas on several fronts. By the beginning of 1903, it became obvious that Spindletop production was declining, and he acquired drilling leases at the new Sour Lake and Batsun fields located west of Beaumont. Sun had to send its oil to Port Arthur from these fields through pipelines operated by others. One of the firms Sun dealt with was the Lone Star and Crescent Oil Company, a supplier as well as pipeline company from whom U.G.I. had purchased oil. U.G.I. received Lone Star deliveries at Marcus Hook, where it contracted with Sun to unload the crude from ships by pipeline.[61]

Lone Star and Crescent was one of the first successful organizations in the Spindletop field. Organized by a New Orleans group, Lone Star owned seven and one-half producing tracts on the "top," producing acreage and two 25,000-barrel tanks southwest of the field, and a six-inch pipeline to Sabine Pass, on the coast near Port Arthur. At Sabine Pass, the company possessed two 50,000-barrel tanks, a wharf, and a ten-inch loading pipeline for vessels. In addition, it owned a pumping station at Spindletop, south of Gladys City—in total, an investment of over one million dollars.[62]

In early 1903, however, the company ran into difficulty. Lone Star had signed one-year contracts to supply crude to U.G.I., Standard Oil, and several other firms. The company entered into these agreements in the spring of 1902, but by January 1903 found that it did not have sufficient crude to fulfill them. A victim of the dwindling Spindletop pool and its own overzealousness, the company went into receivership. When the receiver tried to settle the contracts by having Lone Star supply its customers on a percentage basis of its production rather than the stipulated quotas, all of the firms except U.G.I. agreed to the compromise plan.[63]

Given ensuing events, there is reason to believe the U.G.I.'s hard line may have indicated collusion with J.N. Pew against Lone Star and Crescent. However, on several occasions, J. Edgar Pew, the man in the field, recommended to his uncle that his "friends in Philadelphia" settle, since they would soon lose all claims when the Spindletop pool dried up completely. With Lone Star in the receiver's hands, several companies including Sun expressed interest in acquiring the bankrupt firm's pipeline to Sabine. Among those showing an interest were Gulf, the Texas Company, and Security Oil (Standard). When the receiver was unable to arrange a reorganization of the company, the court ordered the sale of all Lone Star and Crescent property in April 1904. On May 30, 1904, J. Edgar Pew, acting for his uncle, purchased the Texas properties of the company at a sheriff's sale for $100,000.[64]

To coordinate its pipeline operations, Sun formed the Sun Pipe Company (Texas) on July 1, 1904. Capitalized at $400,000, Sun Pipe absorbed all feeder lines running to Sun Station and Sabine Pass. Included was the former Lone Star line, a six-inch pipeline extending twenty-seven miles from Gladys City to the Sabine Pass docks. By 1904, four firms—Gulf, Texas, Security, and Sun—had come to dominate the Gulf Coast crude oil market. The pipelines of these companies ran about three-quarters of the entire output of the field. Sun extended its line to Sour Lake in 1905 with the purchase of an eight-inch pipeline from the United Oil and Gas Company, and when the huge Humble field began producing in 1905, it laid a six-inch line sixty miles from Humble to Sour Lake. In addition to running its own crude that it produced or purchased, Sun acted as a common carrier and piped oil for other producers. Since their pipelines operated intrastate only, Sun's managers did not feel it necessary to file tariffs with the Interstate Commerce Commission after passage of the Hepburn Act in 1906.[65]

Now that the Marcus Hook refinery had begun full operations and Sun had excellent pipeline and loading facilities in Texas, the Sun Company increased its tanker fleet. In 1902, Sun purchased its second tanker, the SS. *Toledo*, a vessel with a carrying capacity of 28,500 barrels. Soon after, it added a carrying barge with a 2,380-barrel capacity and chartered several other vessels. The *Thomas W. Lawson*, a seven-masted schooner-rigged sailing vessel, was the most interesting of the chartered ships. Sun converted it into a tanker and towed it with a seagoing tug. It also bought a new 55,000 barrel tanker, the S.S. *Sun*, in 1907, and had the S.S. *Paraguay* cut in half and lengthened in 1919.[66]

Although Sun remained a significant producer in the southwest, its relative position slipped as other firms, such as the Texas Company and Gulf Oil, acquired larger shares of the Humble and new Oklahoma fields that opened after 1905. By 1908 and for the next decade after, Oklahoma production exceeded that of Texas. But Sun had acted slowly and did not extend its pipelines into the Mid-Continent (Oklahoma) fields nor obtain much producing acreage there. Unlike Gulf and Texaco, Sun had not built a refinery in the southwest and it was able to purchase any additional crude it needed for the Marcus Hook refinery. Moreover, since it was not yet refining gasoline, the firm had not felt the need for additional crude supplies. In 1910, Sun finally did form a fully-owned subsidiary, the Twin State Oil Company, to expand operations in Oklahoma. Also in 1910, Sun drilled a successful well in Louisiana's Vivian field, the first of several wells developed in that state. The firm continued, however, to supplement its own production with crude purchases, particularly from the Humble field.[67]

Twin State acquired valuable Oklahoma leases in the Cushing and Healdton fields after 1912, but did not have its own pipeline to carry oil out. Within a few years, the company had seventy-three wells in operation in Healdton alone. Production increased so rapidly that Sun did open a small 3,000-barrel-a-day refinery in 1915 on a sixty-five acre tract at Yale, Oklahoma. This refinery was only a skimming plant to heat the crude and distill the lighter fractions of kerosene and spirits, "water-white products," from the top. Although this process was wasteful, the tremendous oversupply of crude made it economical in the short run. Sun organized a subsidiary, Sun Oil Company of Delaware, to run this refinery.[68]

The traditional pattern of glut and scarcity had repeated itself with the opening of each new producing tract. Soon after a new field opened, the price structure for crude began a downward trend. The Oklahoma boom, beginning with the Glenn Pool in 1905 and later in Cushing and Healdton, highlighted the waste and inefficiency of rampant production as never before. At one point, Sun literally had more oil than it knew what to do with. The situation was complicated by the fact that the valuable leases in these pools were located on Indian lands granted to the "Five Civilized Tribes" (Creek, Choctaw, Seminole, Chickasaw, and Cherokee) and the Osage Nation. The Curtis Act of 1898 had designated the United States Department of the Interior to supervise and administer the leasing of all mineral rights to this land. But rather than inaugurating a period of rational planned conservation, the government administration lost the opportunity.[69]

Despite intense opposition from the Cherokee and Delaware tribes, the Secretary of the Interior granted twelve separate leases totaling 11,500 acres to the Cherokee Oil and Gas Company in October 1901. This was only the first of many grants given to oil companies which overrode the sovereign rights claimed by Indian tribal councils. When Oklahoma became a state in 1907, all tribal claims based on political sovereignty came to an end and further oil exploitation rapidly increased in the abundant Mid-Continent field. Although many federal officials tried to protect the rights of the tribes in the period from 1900–1919, rampant dishonesty and government inefficiency characterized the exploitation of the Indian lands. As one scholar has recently argued, the moral and ethical principles on the side of the American Indian were overwhelmed by the prevailing zeal for industrial growth and economic progress.[70]

Because of its successful Twin State operations, Sun expanded production very little during this period. It did build a dock and storage facility in Baytown, Texas, supplied by a pipeline from the Goose Creek Field in 1916. The Toledo refinery still relied primarily on crude from Ohio, Indiana, and Illinois wells, but did receive some Texas and Oklahoma oil by railroad tank car. Although Sun did not supply all of its crude needs and periodically purchased oil, it had established a strong production organization and was in an excellent position to exploit new fields discovered in the next decade.[71]

MANAGEMENT

In 1904, J.N. Pew, Sr. moved the Sun Company's corporate headquarters from Pittsburgh to Philadelphia. Now fully committed to the oil business, Pew built in Bryn Mawr, Pennsylvania, a family home, which he named Glenmeade. Robert C. Pew, vice-president of the Sun Oil

Company (Ohio) remained in Toledo to head those operations, while J.N.'s oldest son, Arthur E. Pew, vice-president of the Sun Company (New Jersey), concentrated on marketing. Arthur, along with his father, had established the network of customers for Sun lubricating oils before World War I. The second son, J. Howard Pew, continued to work in the refining end of the business. In 1906, Howard, then assistant refinery manager in Marcus Hook, became a director and second vice-president of the Sun Company. He became Marcus Hook general manager in 1910, succeeding O.C. Pudan, who went to Oklahoma with Twin State and later built the Yale refinery. The founder's youngest son, Joseph Newton Pew, Jr., joined the firm in 1908 after earning an engineering degree from Cornell. At his father's urging, J.N., Jr. went to work at a variety of jobs within the organization to prepare himself for a management position. J. Edgar Pew remained general agent for Sun in Texas until 1913.[72] J.N. Pew, Sr. had established a pattern of family management which was to continue for many years, but Sun still faced a problem common to all family firms. Who would succeed the founder at the head of Sun's operations upon his death?

The elder Pew had experienced declining health in 1911 and 1912. He suffered from angina pectoris, and on October 10, 1912, had a heart attack while working at his office desk in downtown Philadelphia. That afternoon he experienced another attack, which proved fatal.[73] By the time of his death, Sun had established itself as a sound business organization. Moreover, the United States Supreme Court had broken up the Standard Oil Trust the previous year and the relative position of the old "independents" appeared better than ever. On the brink of an important stage of its development, the Sun Company faced a crisis of leadership.

Arthur E. Pew, J.N.'s eldest son by seven years, would have seemed the logical successor to his father. But earlier in the year, on February 28, 1912, the board of directors had accepted his resignation as vice-president and director of the company. Arthur was a very sick man himself, and within four years died of a liver ailment aggravated by heavy drinking. Arthur's resignation indicated that his father had already decided that J. Howard would succeed to Sun's presidency.[74]

J. Howard Pew, thirty years old, became president on October 23, 1912. His younger brother, J.N. Pew, Jr., assumed the position of second vice-president and remained at his brother's right hand for the next fifty years. Both men continued actively in Sun management until their deaths, J.N. first in 1963 and Howard in 1971. Although Howard assumed the superior position in 1912, both men received a salary of $7,500, one indication of the dual leadership that they gave to Sun. In later years, J.N. Pew, Jr. cultivated other interests, particularly in Republican politics, while Howard remained primarily involved with managing

Sun. J.N. Pew, Jr. continued, nevertheless, to be intimately involved in all major business decisions.[75]

J.N. Pew, Sr.'s will was significant for Sun's future development. His estate went to his wife, three sons, and two daughters in trust for a period of twenty years. Since the estate owned more than fifty percent of the Sun Company, the trust arrangement helped to insure the family control and management that the founder had developed. It would also be in the sons' best interests to improve and expand the business that their father had left them.[76]

J. Edgar Pew had left the Sun Company soon after J.N. Pew, Sr.'s death. He resigned his position as Texas general agent in March 1913, but accounts differ on his reasons for leaving. Company publications refer to his period away as a "special leave of absence," and the historians of Standard Oil cite the lure of economic opportunity in the rich Oklahoma fields. The timing of his resignation, the absence of his branch of the family from J.N. Pew, Sr.'s will, and the elevation of his two younger cousins to top managment suggest another reason. J. Edgar was only forty-three years old, but had already amassed a wealth of experience in all phases of the oil business. In particular, he had become an expert on the competitive western crude production fields. It appeared that his future with Sun offered little room for immediate advancement and he struck out on his own. His older brother, Robert C. Pew, remained with Sun as a vice-president in charge of Toledo operations.[77]

After leaving Sun in 1913, J. Edgar formed a partnershp with two others in a Tulsa production company. He remained an independent producer for almost two years before beginning a new career with the Carter Oil Company, a wholly-owned subsidiary of the Standard Oil Company (New Jersey). Jersey Standard was anxious to build up its crude production in the years immediately following the 1911 dissolution. Impressed with Pew's reputation, Carter hired him to head its western division in 1915.[78]

Armed with $34,000,000 of Standard money, Pew established Carter Oil firmly in the Oklahoma fields. Despite what proved to be a short tenure with Carter, Pew earned an excellent reputation in the Standard organization before returning to Sun in 1918. George Gibb and Evelyn Knowlton assert that Pew was "too dynamic" for the conservative Jersey Standard directors. The strong-minded Pew had also alienated his immediate superior, Arthur F. Corwin, by occasionally going over his head to Walter C. Teagle and A.C. Bedford, Standard executives in New York whom he knew personally.[79]

In June 1918, J. Edgar Pew and his twin brother, John G., joined Sun as vice-presidents and members of the board of directors. John G. had carved out a successful career with Jersey Standard's natural gas subsidiaries. J. Howard and J. N., Jr. patched up any existing family differences and induced their experienced cousins to assume the important positions in the Sun organization that they would hold for many years. J. Edgar now headed the production department from Dallas, and John G. stayed in Philadelphia, first with the parent firm and later with the Sun Ship Company. J. Howard and J. N. Pew, Jr. brought their cousins into the firm primarily because of a need for their experience during Sun's wartime expansion. However, there is evidence to suggest an added reason. Both Howard and J. N. were young men eligible for military service. If the government drafted them, the presence of their older cousins in the firm would insure continuity of family management.[80]

Sun allotted shares of stock in the closed corporation to key employees, but provision existed for the company to purchase these shares when the employee left the firm. J. Edgar had divested himself of his Sun stock in 1913, and now he and John G. Pew each purchased 1,000 of the shares obtained from U.G.I.[81] Arthur E. Pew had been in bad health for some time and died in 1916, but the presence of J. Edgar and John G. Pew, along with their younger cousins, insured that family members remained at the top management level in the post-war era.

FINANCE

The Sun Company remained a closed corporation, with U.G.I. continuing to own forty-five percent of Sun's capital stock, and the Pew family retaining majority control. In 1906, the board voted to increase the authorized stock from $2,000,000 to $3,000,000, or 30,000 shares of par one hundred dollar common. It raised this authorization to $4,000,000 in 1909 to acquire the stock and property of the Sun Oil Company (Ohio) and place the Ohio operations and J.N. Pew, Sr.'s new ventures under the same corporate roof.[82]

J.N. Pew, Sr. had fostered a pattern of conservative financial policies for his firm. Sun generally paid a modest quarterly cash dividend of one and one-half percent and plowed much of its profits back into the business by issuing stock dividends to its shareholders. To finance special expansion programs, Sun raised funds through voting increased stock authorizations, on which the members of the closed corporation had first options based on a percentage of their holdings. Pew resisted going into long-term debt financing and maintained majority control of the corporation in his family. Although these cautious policies insured his personal control of the corporation and reduced risk, they also served to inhibit Sun's expansion. Sun did not, for example, enter the gasoline marketing field until after it had acquired the necessary capital from successful sales of lubricants and fuel oils in World War I.[83]

In 1910, the year the Sun Company acquired the Sun Oil Company (Ohio), the six members of the family active in the firm owned slightly more than 54.5 percent of the firm's outstanding common stock. U.G.I. held its 45 percent, and the remaining small amount was in the hands of Pew's close associates. J.N. Pew, Sr. himself held the single biggest block (20,111 shares), and U.G.I. owned 18,000 shares.[84] Sun's dividend policy suited the family, which had a stake in long-term expansion of the firm, but it soon caused friction with U.G.I., a company whose affiliation with Sun Oil had always been limited to its need for gas oil.

A large organization with investments in many parts of the country, U.G.I. desired a greater cash yield from its Sun holdings in order to use the funds in other areas of its business. In 1910 and again in 1913, Sun moved to accommodate U.G.I. by voting its partner an additional cash dividend rather than the normal stock dividend. The authorized stock of the firm was now $6,000,000, but U.G.I.'s holdings had decreased to only thirty-five percent of the total. In 1915, the Sun board voted to increase the capital stock to $9,600,000 and pay a sixty percent stock dividend on all shares except the 21,000 that U.G.I. then owned. On this they voted an equivalent cash dividend. When U.G.I. continued to make known its need for funds, Sun purchased 2,500 shares from them after first paying a cash dividend on the stock. Sun made similar arrangements with U.G.I. in 1916, but it was becoming increasingly clear that it needed a more permanent solution.[85]

U.G.I. informed Sun that it wished to liquidate its holdings, but a serious difficulty existed in determining the actual value of U.G.I.'s shares. After a period of negotiations, Sun agreed to purchase U.G.I.'s remaining shares for two and one-half times their par value or $250 a share, a total of $4,200,000. The Pews had an excellent opportunity to gain complete control of Sun, but they found themselves short of funds. Sun paid half of the principal, $2,100,000, in cash and the other half with a personal note signed by J. Howard Pew to U.G.I. at six percent interest. Under the agreement with U.G.I., the 16,800 shares of Sun stock remained in J. Howard Pew's name, not Sun's, until payment of the note. Howard made the final payment in January 1918, and turned the shares over to the Sun Company.[86]

The sons had achieved the total independence through financial control that their father had urged. In this instance, conflict with U.G.I. brought the situation to a head and precipitated the sale sooner than it otherwise would have occurred. However, it seems likely that Sun eventually would have moved to acquire the U.G.I. interests, just as J.N. Pew, Sr. had bought out Edward O. Emerson in 1899.

J.N. Pew, Sr. had never forgotten his early difficulties with the Penn Fuel Company and his later disputes with Emerson. He had inculcated his sons with the idea that only complete financial control insured independent action, and after their father's death, Howard and J.N., Jr. continued his cautious financial policies.[87] J. Howard Pew later stated his view of this policy in a 1939 statement to the Temporary National Economic Committee (T.N.E.C.):

> The return of earnings into the business and retention of control in relatively few hands has made it possible for the managment of Sun Oil Company to follow a progressive policy in the constant improvement of methods and facilities. Such a procedure is not always possible when a managment represents widely scattered and uninformed stockholders.[88]

Of course, by this time, the reinvestment strategy also held strong tax advantages for the Pew family, but significantly, its adoption long predated the era of high personal income tax.

The demands of post-war expansion did, however, force Sun to execute some long-term debt financing for the first time. To help finance its refining and marketing of motor oils and gasoline, the company issued $6,000,000 of ten-year six percent debenture bonds on May 1, 1919.[89] Although this action marked a break with the previous exclusive policy of reinvestment of earnings, J.N. Pew's sons would use this tool selectively, only in response to particular investment needs which promised quick profit returns.

Sun issued these debentures through the Philadelphia investment bank of Brown Brothers, but the bankers advised that certain corporate moves be made to strengthen Sun's financial position prior to actual sale of the bonds. Specifically, Brown Brothers suggested that the Sun Company purchase controlling interest in a number of small subsidiaries which J.N. Pew, Sr. had financed or acquired over the years with his own capital and which his estate now owned. Among these companies were Sun Oil Line (Ohio), Sun Pipeline (Texas), Hardwood Package (Pennsylvania), and Sun Oil Company (Delaware). In April 1919, the Sun Company board of directors approved the acquisition of a controlling interest in these subsidiaries for a total of $731,000 to clear the way for the sale of the bond issue.[90] Although Sun now carried a debt, the Pews kept it well within the limits of what they viewed as prudent finance.

CONCLUSION

Sun was no longer a minor petroleum company in 1920. The firm produced oil in several states, owned and operated pipelines and tankers, constructed tankers in its own shipyard, and processed petroleum products in three refineries. In addition to its rapidly growing marketing network for specialized products, Sun was now moving into the refining and marketing of gasoline and motor oils. In shaping his business at the turn of the century, J.N. Pew Sr. had planned his strategy around the Standard Oil company. The monopoly's presence influenced the decisions of all the smaller independent firms. However, the successes of Pew and other independents had combined with other legal and market factors to bring about a fundamental change in the industry. An oligipoly of vertically-integrated firms now competed vigorously for the old Standard Oil business.

Sun Oil also joined the ranks of the large companies at a time when new issues were arising. The independents had earlier united under the banner of antitrust. Now the issues were more complex. The Standard Oil monopoly developed as a response to economic instability. The government had dissolved the trust, but the causes of instability remained. Moreover, the tremendous demand for petroleum created by the war highlighted a growing awareness of long-term scarcity. The next two decades would see both the government and the industry formulate policies to face these problems.

NOTES

1. Harold F. Williamson, Ralph L. Andreano, Arnold R. Daum, and Gilbert C. Klose, *The Age of Energy, 1899-1959*, Volume II, *The American Petroleum Industry* (Evanston: Northwestern University Press, 1963), pp. 4-7; Gerald D. Nash, *United States Oil Policy, 1890-1964* (Pittsburgh: University of Pittsburgh Press), p. 8. See also Williamson and Andreano, "Competitive Structure of the American Petroleum Industry, 1880-1911: A Reappraisal," in Editors of Business History Review, *Oil's First Century* (Cambridge: Harvard Business School, 1960), pp. 71-84.
2. E.O. Emerson to J.N. Pew, May 29, 1899, Acc. #1317, EMHL, File 21-A, Box 315.
3. See U.G.I. folders, Acc. #1317, EMHL, File 21*8*A, Boxes 312-3.
4. S.T. Bodine to J.N. Pew, December 28, 1899, Acc. #1317, EMHL, File 21-A, Box 312; Walton Clark, U.G.I. General Superintendent, to J.N. Pew, January 15, 1900, File 21-A, Box 313; Williamson et al., *Energy*, pp. 79-81.
5. Williamson et al., *Energy*, pp. 74-6.
6. R.C. Pew to J.N. Pew, January 17, 18, 1901, Acc. #1317, EMHL, File 21-A, Box 281.
7. R.C. Pew to J.N. Pew, January 28, 1901, February 12, 27, 18, 1901, March 6, 13, 1901, File 21-A, Box 281; "J. Edgar Pew Notes, 1942," p. 11.
8. *J. Edgar Pew, His Life and Times, 1870-1946* (Philadelphia: Sun Oil Company, 1947) (reprinted from *American Petroleum Institute Quarterly* [April 1946], p. 1).
9. *Ibid.*
10. "Sun Rise in the Southwest," *Our Sun* 16, 1 (Winter 1951), p. 19.
11. Board Minutes, The Sun Company (New Jersey), May 8, 1901, Minute Book #1, pp. 1-22; May 36, 1902, Minute Book #1, pp. 80-1; March 10, 1903, Minute Book #1, pp. 87-8; October 3, 1906, Minute Book #1, pp. 159-63; May 10, 1909, Minute Book #1, pp. 222-5, Office of the Assistant Secretary, Sun Oil Company, Radnor, Pennsylvania; "Story of Sun," *Our Sun*, p. 20; S.T. Bodine to J.N. Pew, August 24, 1901, Acc. #1317, EMHL, File 21-A, Box 312.
12. Board Minutes, May 3, 8, 1901, Minute Book #1, pp. 1-22.
13. "Sun Rise in Southwest," *Our Sun*, pp. 19-20; "Story of Sun," *Our Sun*, p. 20; S.T. Bodine to J.N. Pew, August 30, 1901, Acc. #1317, EMHL, File 21-A, Box 312.

14. Board Minutes, February 25, 1902, Book #1, p. 70; "Story of Sun," *Our Sun*, p. 20; "The First Fifty Years," *Our Sun* 16, 4 (Autumn 1951), p. 2.

15. "First Fifty Years," *Our Sun*, p. 3; "Turning Crude into Sun Products," *Our Sun*, 50th Anniversary Issue, 3, 1 (1936), p. 8; "Tankers Under the Sun Flag," *Our Sun*, 50th Anniversary Issue, 3, 1 (1936), pp. 20–1; Waltman to Klaus, *Our Sun*, Library File, Marcus Hook, p. 7; A. MacMurtrie, Ed Hughes, John Marshall, Robert Cross, George Lampugh, and Murry Herman, "Refinery Notes" (June 1, 1936), J. Howard Pew Papers, Box 58, p. 5.

16. MacMurtrie et al., "Reginery Notes," p. 5; John L. Kelsey, "A Financial History of the Sun Oil Company" (M.B.A. Thesis, Wharton School, University of Pennsylvania, 1950), p. 10; "First Fifty Years," *Our Sun*, p. 3.

17. J.A. Pearson, U.G.I. Purchasing Agent, to J.N. Pew, November 7, 1901, Acc. #1317, EMHL, File 21-A, Box 312.

18. Walton Clark to J.N. Pew, March 11, 1902, July 3, 1902, E.H. Earnshaw to J.N. Pew, September 6, 1902; telegram, Samuel T. Bodine to J.N. Pew, September 27, 1902, J.A. Pearson to J.N. Pew, June 4, 1902, Acc. #1317, EMHL, File 21-A, Box 313.

19. "Toledo Looks Back," *Our Sun*, p. 4; Williamson et al., *Energy*, pp. 465–6; "Refiner, Administrator, Statesman, Leader," *Our Sun* 16, 4 (Autumn 1951), p. 16.

20. "Refiner, Administrator," Our Sun, p. 16; Isaac C. Ketler, Grove City College, to J.N. Pew, March 1, 1900, Acc. #1317, EMHL, File 21-A, Box 236; Jno. G. "Jack" Pew, retired vice-president and member of the board, interview held at Dallas, Texas on November 17, 1975.

21. "Refiner, Administrator," *Our Sun*, p. 16; J. Howard Pew to J.N. Pew, Sr., February 8, 12, 1902, Acc. #1317, EMHL, File 21-A, Box 4; "First Fifty Years," *Our Sun*, p. 3.

22. "Story of Sun," *Our Sun*, pp. 20–, pp. 20–1; "First Fifty Years," *Our Sun*, p. 3.

23. "Story of Sun," p. 21; Waltman to Klaus, *Our Sun*, p. 10; *Hydrolene*, Volume 8 of a series of pamphlets issued monthly (Philadelphia: Sun Company, 1917), pp. 229–36.

24. J.N. Pew to A.E. Pew, July 14, 1910, Acc. #1317, EMHL, File 21-A, Box 206; "Reply Brief for Complainant, District Court of the United States," 21-A, Box 58; see also "F.X. Byerley vs. Sun Oil, 1903-1912," 21-A, Boxes 296-301; Williamson et al., *Energy*, p. 202; *Hydrolene*, pp. 236–55.

25. *Chester Times* (March 28, 1902), p. 1.

26. J. Howard Pew to J.N. Pew, Sr., 1902 (no other date), Acc. #1317, EMHL, File 21-A, Box 11.

27. S.T. Bodine to J.N. Pew, Sr., August 4, 6, 1902, Acc. #1317, EMHL, File 21-A, Box 313.

28. "Sun Oil Case is Still on Trial," *The Chester Morning Republican* (November 1, 1904), p. 1; "First Fifty Years," *Our Sun*, p. 4.

29. R.C. Pew to J.N. Pew, Sr., May 28, 1901, Acc. #1317, EMHL, File 21-A, Box 281; "Toledo Looks Back," *Our Sun*, pp. 2, 5.

30. "Story of Sun," p. 21; "First Fifty Years," *Our Sun*, p. 4; "Looking Back Over Fifty Years," *Our Sun*, 50th Anniversary Issue, 3, 1 (1936), p. 5.

31. Harold F. Williamson and Arnold R. Daum, *The Age of Illumination, 1859-1899*, Volume I, *The American Petroleum Industry* (Evanston: Northwestern University Press, 1959), pp. 322-36, 488-96.

32. *Ibid.*, pp. 498-502; Mira Wilkins, *The Emergence of Multinational Enterprise: American Business Abroad From the Colonial Era to 1914* (Cambridge: Harvard University Press, 1970), pp. 62-3.

33. Ralph W. Hidy and Muriel E. Hidy, *Pioneering in Big Business, 1882-1911*, Volume I, *History of Standard Oil Company (New Jersey)* (New York: Harper and Brothers, 1955), pp. 146-7; Williamson and Daum, *Illumination*, pp. 498-9; Wilkins, *Emergence of Multinational Enterprise*, pp. 64, 82.

34. "Toledo Looks Back Over 50 Years," pp. 4-5; Waltman to Klaus, *Our Sun*, Library File, Marcus Hook, p. 8.

35. J. Edgar Pew to J.N. Pew, Sr., March 28, 1903, Acc. #1317, EMHL, File 21-A, Box 211; "Golden Anniversary in London," *Our Sun* 24, 2 (Spring 1959), pp. 23-4; "First Fifty Years," *Our Sun*, p. 5.

36. Marquis James, *The Texaco Story: The First Fifty Years, 1902-1952* (New York: The Texas Company, 1953), p. 31; Wilkins, *Emergence of Multinational Enterprise*, pp. 86-7.

37. "Golden Anniversary," *Our Sun*, p. 23: "Mineraloelwerke Albrecht & Companie, A Progressive Sun Oil German Affiliate," *Our Sun* 3, 4 (January 30, 1926), pp. 3-7; J.N. Pew, Sr. to A.E. Pew, July 5, 1910, Acc. #1317, EMHL, File 21-A, Box 206.

38. "First Fifty Years," *Our Sun*, p. 5.

39. Williamson et al., *Energy*, pp. 261-2; 291-2.

40. *Ibid.*, pp. 291-2.

41. "Looking Back Over Fifty Years," *Our Sun*, p. 5; Board Minutes, July 8, 1915, Book #2, p. 121; MacMurtrie et al., "Refinery Notes," J. Howard Pew papers, Box 58, p. 5.

42. *Fuel Oil*, Volume 11 of a series of pamphlets issued monthly (Philadelphia: Sun Company, 1918), pp. 325-50; "Story of Sun," *Our Sun*, p. 22.

43. John G. McLean and Robert W. Haigh, *The Growth of Integrated Oil Companies*, (Boston: Harvard University Press, 1954), pp. 9-11; "Sun Ship's 30 Years," *Our Sun* 12, 2 (July/August 1946), p. 4; Williamson et al., *Energy*, pp. 2647; George Sweet Gibb and Evelyn H. Knowlton, *The Resurgent Years, 1911-1927*, Volume 11, *History of Standard Oil (New Jersey)* (New York: Harper and Brothers, 1956), pp. 221-6: Paul H. Giddens, *Standard Oil Company (Indiana) Oil Pioneer of the Middle West* (New York: Appleton-Century-Crofts, 1955), p. 196.

44. "Story of Sun," *Our Sun*, p. 20; Robert G. Dunlop, interview held at St. Davids, Pennsylvania, October 31, 1975; Clarence Thayer, interview held at Media, Pennsylvania, November 25, 1975; Waltman to Klaus, *Our Sun*, pp. 14-5.

45. J.N. Pew, Jr. to Eugene Hollon, Standard Oil (New Jersey), July 3, 1941, Acc. #1317, EMHL, File 21-B, Box 29.

46. "Sun Ship's 30 Years," *Our Sun*, pp. 4-5.

47. J.N. Pew, Jr. to Hollon, July 3, 1941; Board Minutes, June 6, 1916, Book #2, p. 165; J.N. Pew, Jr. to Lt. Page A. Watson, September 14, 1918, Acc. #1317, EMHL, File 21-B, Box 1.

48. Arthur S. Link, *The American Epoch* (New York: Alfred Knopf, 1963), p. 210; J.N. Pew, Jr. to Lt. Samuel B. Eckert, August 7, 1917, Acc. #1317, EMHL, File 21-B, Box 1.

49. Robert Hessen, *Steel Titan: The Life of Charles M. Schwab* (New York: Oxford University Press, 1975), pp. 235-9.

50. *Ibid.*, pp. 241-3.

51. Thayer interview; Waltman to Klaus, *Our Sun*, Library File, Marcus Hook, p. 16; Link, *Epoch*, p. 210.

52. "Story of Sun," *Our Sun*, p. 22; "Sun Ship's 30 Years," *Our Sun*, p. 4. 53. Williamson, et al., *Energy*, pp. 17, 112, 214, 217-30; R.C. Pew to J.N. Pew, Sr., February 16, 1900, Acc. #1317, EMHL, File 21-A, Box 280; Thayer interview.

54. Williamson et al., *Energy*, pp. 135, 136; see also John Lawrence Enos, *Petroleum Progress and Profits* (Cambridge: M.I.T. Press, 1962), Ch. 1, "The Burton Process."

55. Thayer interview; "Story of Sun," *Our Sun*, p. 22; Waltman to Klaus, *Our Sun*, p. 17.

56. Board Minutes, January 29, 1919, Book #2, p. 181; "Story of Sun," *Our Sun*, p. 22.

57. Board Minutes, December 15, 1919, Book #3, p. 30; Thayer interview; "Advertising Reflects the Progress of Sunoco," *Our Sun*, 50 Anniversary Issue, 3, 1 (1936), p. 36.

58. Waltman to Klaus, *Our Sun*, p. 17; "Mineraloelwerke Albrecht & Company," *Our Sun*, p. 3.

59. Board Minutes, December 5, 1919, Book #3, pp. 13-4; October 16, 1925, Book #3, p. 325; Mira Wilkins, *The Maturing of Multinational Enterprise: American Business Abroad from 1914-1970* (Cambridge: Harvard University Press, 1974), p. 85.

60. McLean and Haigh, *Integrated Oil Companies*, p. 528.

61. "Sun Rise in Southwest," *Our Sun*, p. 21; S.T. Bodine to J.N. Pew, Sr., October 31, 1902, Acc. #1317, EMHL, File 21-A, Box 313.
62. "Sun Rise in Southwest," *Our Sun*, p. 20.
63. J. Edgar Pew to J.N. Pew, Sr., January 7, 9, 29, 1903, March 28, 1903, May 13, 1903, Acc. #1317, EMHL, File 21-A, Box 211.
64. J. Edgar Pew to J.N. Pew, Sr., September 30, 1903, October 28, 1903, December 20, 1903, January 8, 1904, April 16, 25, 1904, May 18, 30, 1904, Acc. #1317, EMHL, File 21-A, Boxes 211-2.
65. J. Edgar Pew to J.N. Pew, Sr., June 25, 1904, Acc. #1317, EMHL, File 21-A, Box 213; Williamson et al., *Energy*, pp. 78, 87; Arthur M. Johnson, *Petroleum Pipelines and Public Policy, 1906-1959* (Cambridge: Harvard Univeristy Press, 1967), p. 38.
66. "Tankers Under the Sun Flag," *Our Sun*, pp. 20-1.
67. Williamson et al., *Energy*, p. 87; "Sun Rise in the Southwest," *Our Sun*, p. 21; Board Minutes, January 12, 1909, Book #1, pp. 210-1, February 9, 1909, Book #1, p. 214.
68. Porter Howard, "History of Sun," p. 13; Daniel Bergen, "Oklahoma," *Our Sun* 13, 1 (July 1947), p. 3.
69. Williamson et al., *Energy*, pp. 51-4; "J. Edgar Pew Notes, 1942," Library File, Marcus Hook, pp. A2-A5.
70. Craig H. Miner, "The Cherokee Oil and Gas Co., 1889-1902: Indian Sovereignty and Economic Change," *Business History Review* 46, 1 (Spring 1972), pp. 45-66.
71. Board Minutes, April 1, 1913, Book #2, p. 52; "Sun Rise in the Southwest," *Our Sun*, p. 21.
72. "J.N. Pew: A Biographical Sketch," *Our Sun*, 75th Anniversary Issue, 26, 3, 4 (Summer/Autumn 1961), p. 13; Waltman to Klaus *Our Sun*, Library File, Marcus Hook, p. 11; "The First Fifty Years," *Our Sun*, pp. 4-5; "Joseph N. Pew, Jr. Marks Half a Century of Service with Sun," *Our Sun* 23, 3 (Summer 1958), p. 12; "J. Edgar Pew, 1870-1946," *Our Sun* 12, 4 (January 1947), p. 14.
73. J.N. Pew Biographical Sketch," *Our Sun*, p. 13.
74. Board Minutes, October 23, 1912, Book #2, pp. 42-4. Arthur's tragic illness largely explains the life-long temperance of J. Howard Pew. Howard not only refrained from personal use, but he forbade alcoholic beverages to be served at Sun functions or on Sun property (see Dan Rottenberg, "The Sun Gods," *Philadelphia Magazine*, [September 1975], p. 115).
75. Board Minutes, October 23, 1912, Book #2, pp. 42-4; "Joseph N. Pew, Jr., Marks Half a Century of Service," *Our Sun*, p. 12.
76. Kelsey, "Financial Study," p. 11.
77. Jno. G. "Jack" Pew, interview held at Dallas, Texas, November 18, 1975; J. Edgar Pew, 1870-1946," *Our Sun* 12, 4 (January 1947), p. 14; George Sweet Gibb and Evelyn H. Knowlton, *The Resurgent Years, 1911-1927*, Volume II, *History of the Standard Oil Company (New Jersey)* (New York: Harper & Brothers, 1956), p. 61.
78. "J. Edgar Pew, Sun Oil Company, Dies," *National Petroleum News* 38, 48 (November 27, 1946), p. 16; "H.H. Boyer Notes," Folder #2, "J. Edgar Pew," Library File, Sun Oil Company Library, Marcus Hook, Pennsylvania; Howard, "History of Sun," p. 13; Gibb and Knowlton, *Resurgent Years*, pp. 61-2.
79. Gibb and Knowlton, *Resurgent Years*, pp. 61-2, 65, 67, 426.
80. "Twin Sun Leaders," *Our Sun*, p. 15; Board Minutes, July 15, 1918, Book #2, pp. 250-1; J.N. Pew, Jr. to Samuel Eckert, June 3, 1918, Acc. #1317, EMHL, File 21-B, Box 1; "Resolution to Draft Board," Board Minutes, October 1, 1918, Book #2, p. 261; Dunlop interview.
81. Board Minutes, April 1, 1913, Book #2, p. 52; June 25, 1918, Book #2, pp. 236-48.
82. Board Minutes, April 10, 1906, Book #1, p. 145; October 3, 1906, Book #1, pp. 159-63; December 10, 1906, Book #, p. 165; May 10, 1909, Book #1, pp. 222-5; June 16, 1910, Book #1, p. 252.

83. On dividend policy followed in this period, see Minute Books #1 and 2, The Sun Company (New Jersey).

84. Board Minutes, April 12, 1910, Book #1, p. 243.

85. Board Minutes, October 5, 1910, Book #1, pp. 314–9; November 26, 1913, Book #2, p. 81; December 15, 1915, Book #2, p. 135; December 28, 1915, Book #2, p. 147; "Story of Sun," p. 22.

86. Board Minutes, March 8, 1917, Book #2, p. 199–209; June 25, 1918, Book #2, pp. 236–48.

87. Dunlop interview; Thayer interview.

88. TNEC Statement, Acc. #1317, EMHL, J. Howard Pew Papers, Box 45.

89. Board Minutes, March 18, 1919, Book #2, p. 283; April 23, 1919, Book #2, p. 287.

90. Board Minutes, April 23, 1919.

CHAPTER IV

Expansion in the Automobile Age, 1920–1930

Sun's entry into the automotive products market coincided with three interrelated industry-wide developments in the post-war period: a general concern about oil scarcity, the spread of thermal cracking technology, and the emergence of engine knock as a serious challenge to refiners and automotive engineers. These factors greatly influenced the company's history in the 1920s.

Through a combination of circumstances, Sun Oil developed its marketing program in ways that affected later company policy and technological development. The Pews had initially decided to market gasoline because of the large gas oil capacity they had following the break with United Gas Improvement (U.G.I.) in 1918. Sun's gas oil was an excellent charging stock for thermal stills and it yielded high octane gasoline when cracked at high temperature and pressure. But Sun now had to decide which of several competing cracking processes it was going to adopt.

CRACKING TECHNOLOGY

Dr. William Burton's initial development of thermal cracking with Indiana Standard for the first time had raised petroleum refining from a process of separating crude oil into its constituent parts to one of transforming and rearranging hydrocarbon molecules in order to upgrade the end product. His 1913 plant had doubled the amount of gasoline obtained from a barrel of crude, but as the demand for gasoline expanded during the war years, competing research groups made improvements on the basic process. Indiana Standard itself patented several advances which enhanced its leading licensing position in thermal cracking, and between 1914 and 1919, fourteen companies licensed the improved Burton process at a royalty agreement of four-tenths of a cent per gallon.[1]

Two new similar processes, the Dubbs and the Texas Company's Holmes-Manley system, appeared during World War I to challenge the Burton monopoly. Although these systems did not produce gasoline of a quality equal to Burton's, they had one major difference that promised greater efficiencies. They were both early tube processes in which cracking occurred on a continuous circulating basis within narrow cracking chambers. A true continuous process offered a larger capacity and a shorter time cycle for distillation than the semi-continuous system employed by the most advanced Burton equipment. The Burton-Standard of Indiana patents still held a virtual monopoly on thermal cracking, but the new processes were on the verge of commercial success in 1919–1920.[2]

Competition among these cracking systems laid the groundwork for what was later termed the "patent club." As early in 1916, the Universal Products Company, then holder of all Dubbs patents, brought suit against Indiana Standard for infringement. Soon after the Texas Company first began to operate the commercial Holmes-Manley process in 1920, it sued Indiana for violations of certain patents it had obtained in 1915. After countersuits and a period of negotiation, Texaco and Indiana Standard reached an agreement in August 1921, whereby each would license the other's process without royalty payment, and exchange immunities under their patents, "past and future."[3]

In 1922, two newcomers, Jersey Standard's Tube and Tank process and the Cross system, appeared to challenge the status quo. The advantage they offered was the use of thinner walled tubes rather than heavy shells for the cracking chamber. Walter and Roy Cross patented a reaction chamber that allowed tremendous gains in pressure resulting in increased cracking efficiency. Where the Burton process operated at pressures of 75 to 100 pounds per square inch, the Cross units ranged from 600 to 750 pounds. In 1921, the Cross brothers assigned their patents to the Gasoline Products Company to license the system to other refiners.[4]

Jersey Standard had originally licensed the Burton process from its former sister company, Indiana. In 1917, however, Jersey opted to develop its own technology and established a research and development organization to coordinate these efforts. By 1920, this new subsidiary had grown into the Standard Oil Development Company (S.O.D.). This company became one of the first large-scale research and development organizations in American industry and the Tube and Tank process was one of its earliest achievements.[5]

The Tube and Tank process also operated at much higher pressures than the Burton process and Jersey Stanford feared that the Cross organization would sue them for infringement. Therefore, Jersey took the initiative in 1922 by suing Pure Oil, a licensee of the Cross Process from Gasoline Products Company. In a complicated series of legal negotiations, Gasoline Products, the Texas Company, and Indiana Standard entered into an agreement in March 1923 to extend the patent pool initially begun by Indiana and Texas. Soon after, Jersey Standard dropped its suit and became the fourth member of the club.[6]

This did not end all problems for the companies, however, for in 1924, the government challenged the "patent club" in the courts. Finally, in 1929, the United States District Court in Chicago found Standard of Indiana et al. guilty of restraint of trade and ordered dissolution of the pool. The companies appealed and the Supreme Court reversed the decision in 1931. That same year, Universal Products Company dropped its long-standing suit against Indiana and entered the club.[7]

When Sun Oil embarked on its program of gasoline refining in the early twenties, it encountered this competitive, uncertain situation in thermal cracking patents. Because it began late, however, Sun benefited from the latest technological developments, and its first crackers were continuous stills. Sun considered three processes seriously, the Dubbs, the Cross, and a similar high-pressure tube process, the Fleming, licensed by the independent M.W. Kellogg Company. However, the first gasoline sold from the Marcus Hook refinery in the period 1920–1923 was directly distilled from Spindletop and other Texas, Mexican, and California crudes. The high price of gasoline made it profitable, but the small percentage obtained per barrel, five to seven percent, was not sufficient for larger scale gasoline marketing.[8]

In 1922, therefore, Sun conducted cost studies of the three processes, and decided on licensing Cross units from the Gasoline Products Company. In all categories considered, cost of installation, net profit per barrel, return on investment, and space required for construction, the Cross seemed the best investment. On August 14, 1922, Sun signed an agreement with Gasoline Products to install the Cross Process at Marcus Hook, and agreed to pay a royalty on each barrel of cracked gasoline. Sun began construction of its cracking plant at Marcus Hook in 1922 and put it into operation in 1923. This plant consisted of six Cross units and represented Sun's first gasoline cracking facility.[9]

Although Sun's contract with the Gasoline Products Company predated the "patent club" agreement of 1923, the firm became a secondary defendent in the 1924 antitrust prosecution since it was a licensee. When the courts ordered the patent club dissolved in 1929, Sun faced the prospect of contempt citations if it maintained relations with Gasoline Products. Since the firm had entered into an agreement prior to the pooling arrangement and did not avail itself of any patents other than those originally obtained, Sun felt that it should never have become a defendant in the 1924 suit. J. Howard Pew commented later that Sun should have taken this stand from the first, but did not do so.[10]

Thus, in early 1930, Sun was in a difficult position even though the courts had suspended direct action against the patent club pending appeal to the Supreme Court. Although in legal difficulty due to the circumstances of its licensing agreement, Sun's philosophical position remained consistent with its earlier stands against monopoly in restraint of trade. J. Howard Pew expressed these views in a letter sent to Sun attorney R.E. Lamberton in January 1930, soon after the initial court decision. He wrote: "It is my conviction that the 'patent club' is in fact a vicious conspiracy in restraint of trade and as such should be broken up into its various component parts." Sun's legal if not ethical dilemma was resolved when the Supreme Court ruled in favor of the oil companies in 1931.[11]

The Cross units were the heart of Sun's expanded gasoline business. The firm sold a blend of cracked gasoline and straight-run gasoline to a small, mid-Atlantic marketing region. This gasoline had a high antiknock rating by the standards of the time. Moreover, as early as 1924, Sun began to purchase substantial quantities of high octane cracked gasoline from Standard of California. The California gasoline was particularly high in aromatics, which made it a good antiknock fuel.[12]

ANTIKNOCK

All oil companies faced the challenge of upgrading the quality of their gasolines. The knock problem had first appeared a few years prior to World War I, when the increased demand for gasoline prompted refiners to stretch their production of straight-run gasoline. There is only a fixed amount of good quality gasoline in each barrel of oil, but it was possible to increase total yield by blending some heavier fractions with the gasoline, a practice that also reduced its volatility and antiknock properties.

The main problem with engine knock was that no one really knew what it was. It sounded like a metal-on-metal ping that occurred when the piston engine was straining at peak efficiency. It seemed to be related to the fuel used, but also to the engine, since fuels knocked in some motors and ran smoothly in others. The problem of knock increased as people demanded more powerful engines for their automobiles. The way to achieve this improved performance was to increase the compression ratio of the piston strokes in the engine. High compression meant increased power, but it also invited knock.[13]

Early research into knock accelerated during World War I, since knocking in aircraft engines was a particularly urgent problem. Piston-driven aircraft engines had to operate at high compression to develop power for take-off, and these engines ran at close to peak performance for the duration of the flight. Research proceeded on both sides of the Atlantic. In England, a team led by Harry Ricardo began the investigation of knock from three points: engine, fuel, and the scientific basis of knock. In the United States, most research centered on the fuel side.[14]

American researchers gathered in the laboratories of major corporations like General Motors, DuPont, and Standard (New Jersey). Charles F. Kettering, already known for his invention of the self-starter for automobile engines, had begun searching for antiknock remedies at his Dayton Engineering Laboratories Company (DELCO) in 1916. Kettering produced electric lighting systems for farms that employed generators using internal combustion engines. Because of safety regulations, his equipment burned kerosene instead of gasoline, but it produced tremendous knocking problems. Kettering hired Thomas Midgely, Jr., a Cornell mechanical engineer, to work on this project with him.[15]

Kettering and Midgely determined that knock did not occur before ignition, but was a violent disturbance following ignition by the spark plug. Concluding that insufficient volatility was the problem, they dyed the kerosene red with iodine, thinking that the color would absorb heat and make the fuel vaporize better. It did reduce the knock, but they quickly realized that the iodine rather than the color red had done the trick. During the war in 1917–1918, Kettering and Midgely diverted their attention to a joint project with the U.S. Bureau of Mines to develop an aviation fuel for Liberty engines. DELCO succeeded in hydrogenating benzene for such a fuel, but the war ended and it scrapped the project.[16]

Post-war predictions of an oil shortage by the United States Geological Survey stimulated the industry to develop effective antiknock agents, since higher compression promised greater fuel economy as well as better performance. In 1919, General Motors purchased the DELCO lab and Kettering's other interests in order to concentrate efforts on the antiknock problem. Proceeding from iodine, Kettering's group tried a number of other antiknock ingredients, including various aniline compounds, selenium, tellurium, and finally tetraethyl lead (TEL), all elements near the bottom of the periodic table. Lead proved by far the most effective, exhibiting fifty percent more antiknock action than aniline. The next problem was the establishment of a commercial process for producing TEL and a way to combat the accumulation of lead oxide in engines. They partially solved the latter difficulty by including ethylene dibromide in the additive. This combined with the lead, and the resulting combinations exited in the exhaust to pollute the environment rather than the engine. However, the use of TEL continued to cause problems of lead buildup in cylinders, valves, and spark plugs.[17]

The DuPont Company and Jersey Standard developed processes for producing quantities of tetraethyl lead. DuPont, at the time the controlling interest in General Motors, began producing TEL in 1922, and General Motors Chemical Company began marketing Ethyl (leaded) gasoline in 1923. One year later, in August, 1924, General Motors formed the Ethyl Corporation with Jersey Standard. The two giants each owned fifty percent of the new company which manufactured, licensed, and sold tetraethyl lead to petroleum refiners and distributors. Indiana Standard was the first large-scale Ethyl marketer, securing an exclusive five-year contract for sales in its midwest marketing area. The Ethyl Corporation from the first exhibited the pattern of price influence that was to receive great criticism over the years. Not only did it have a monopoly on the sale of Ethyl fluid, but the company maintained an industry price differential of three to five cents between "regular" and "Ethyl" brands.[18]

Soon after the sales of Ethyl fluid had begun, a serious setback occurred. In October 1924, forty-five cases of lead poisoning and four fatalities at Esso's (Jersey Standard's) Bayway production plant threw the public into a panic. Other fatalities had also occurred in the DuPont plant and at the Dayton Laboratories. The poisonous TEL fluid is readily absorbed through the skin, and proper safeguards had not been taken. Ethyl Corporation stopped all sales in May 1925 and suspended operations for a year while the U.S. Surgeon General completed an investigation. Sales resumed in 1926 after Ethyl had implemented certain safety measures. In the future, all Ethyl fluid would be blended at distribution centers rather than at service stations. Up to that time, Standard of Indiana used fluid cannisters that were attached to the gasoline pump, and the station attendant blended it on the spot. All Ethyl gasoline would also now be dyed red to distinguish it from non-leaded gasoline, and the government imposed an allowable maximum of 3 cc. of TEL to be added per gallon.[19]

TEL had proven to be an effective antiknock agent, but controversy swarmed around its use for years. Many feared the poisonous effects of the lead, and there was strong evidence that it damaged automobile engines. Moreover, the Ethyl Corporation, controlled by Jersey Standard and General Motors, symbolized monopoly control both to oil companies and consumer groups.

The successful introduction of Ethyl gasoline, however, influenced refiners without Ethyl contracts to re-examine their marketing strategies. These refiners increasingly turned their attention to the production of premium antiknock gasoline without the addition of tetraethyl lead. This could be achieved through increased thermal cracking, blending stocks of natural gasoline (gasoline absorbed from natural gas through various processes), or adding benzol. In July 1925, Gulf announced a new fuel, "NoNox," offered at a three cent premium over its regular grade of gasoline. The Texas Company introduced a "new and better gas" in May 1926, and Sinclair brought out its premium "H.C." gasoline that same year. The Roxana Petroleum Company introduced "Super-Shell," and by the summer of 1926, almost every refiner in the country offered a second grade of antiknock gasoline of some kind, with or without TEL.[20]

BLUE SUNOCO

The Sun Oil Company decided to avoid using lead and attacked the problem of knock through the upgrading of motor fuel through thermal cracking. The gasoline obtained from its six Cross units charged with gas oil was a successful antiknock fuel. Tests conducted at Sun's Norwood, Pennsylvania, laboratory in 1925 with the Midgely "bouncing pin" apparatus for determining antiknock quality proved encouraging. Sunoco gasoline consistently demonstrated a higher antiknock performance than competitors' unleaded gas.[21] The fact that Sun obtained this quality without lead largely determined the Pews' decision to concentrate on improving their cracking. Rather than purchase Ethyl fluid from the Standard-General Motors combine, Sun began to develop a marketing strategy of its own.

As early as 1925, Sun's managers explored the idea of using dye to identify their product as a special antiknock fuel. Although one rarely notices the color of gasoline today, older pumps had a large clear glass reservoir on the top, and one could readily see the fuel being gravity-fed into his gasoline tank. Sun decided to adopt a blue color to match a piece of tile seen by Mr. and Mrs. J.N. Pew, Jr. on their honeymoon to China. In the spring of 1925, Dr. Gerhard Alleman of Sun's Norwood laboratory conducted experiments with different dyes, and Sun decided to adopt astrol blue, an anthraquinone oil-soluble blue dye, for its gasoline. This was the color used two years later with the birth of "Blue Sunoco."[22] First, however, Sun needed a larger and better cracking capacity before it could expand its marketing network.

Sun's original Cross units consisted of two 600-barrel-a-day and four 1,000-barrel-a-day stills, and in 1926, the firm drew up plans for a new 4,000-a-day unit. William D. Mason, who became refinery manager that year, hired Clarence Thayer from Standard of California, an industry leader in thermal cracking, to assist at Marcus Hook. The regular Cross units were running at approximately 650 lbs. pressure, but Sun increased the pressures in its new unit to 1,100 lbs., and later to 1,200 lbs. Thayer also increased operating temperatures from the previous high of 885 degrees Fahrenheit to 905 degrees. Higher temperatures and pressure increased the yield of gasoline from the charging stock and also upgraded its octane. But because of the high pressures used in the new still, maintenance and safety precautions became paramount, and Thayer introduced regular programs of tube inspection and replacement as well as design modifications on existing equipment.[23]

Sun's attack on antiknock through cracking technology received continued help from its research laboratory at Norwood, Pennsylvania. In particular, a new star arose in the form of another Pew, Arthur E., Jr. (son of the founder's eldest son, Arthur E., who had died in 1916). Young Pew was a graduate engineer from the United States Naval Academy who joined Sun in 1921 and commenced work in refinery engineering at Marcus Hook in 1922. Pew and Gerhard Alleman carried out research in improving cracking in the laboratory, while Clarence Thayer provided the practical know-how to incorporate improvements into commercial operation.[24]

The success of Sun's own refinery innovations assured it of a sound base from which to launch a major gasoline marketing drive. The present system of octane ratings did not develop until the federal government undertook a program of measurement standards in 1928. The government then developed an octane scale using two reference points: heptane, a hydrocarbon with low antiknock, and iso-octane, one having a very high antiknock rating. The octane number of the gasoline reflected the percentage of iso-octane in a mixture of iso-octane and normal heptane. Although modified slightly over the years, this is still the basic octane scale. Sun's cracked gasoline in 1926 had an octane rating of 73 and its straight-run gasoline a rating around 70. Both of these were very high for their time. Most other firms could only match these figures by the addition of TEL.[25]

In January 1927, A. Ludlow Clayden of Sun's Norwood Laboratory sent six samples of gasoline to Graham Edgar, Director of Research for the Ethyl Corporation. Sun requested tests run to determine how much Ethyl fluid had to be added to bring each sample up to Ethyl antiknock standards. The samples were labeled "A" through "F" and Edgar did not know which sample was which. The results were extremely encouraging. The Sunoco gasoline sample, consisting of thermally cracked Spindletop crude, required no Ethyl fluid, and Edgar conceded that it was "perhaps very slightly better than Ethyl gasoline." A mixture of fifty percent Sunoco and fifty percent cracked gasoline from Sun's Toledo refinery required .75 cc. of lead per gallon, while straight-run California gasoline used as Sun's standard measure required .5 cc. of lead. In contrast, regular Sinclair gasoline required 4 cc. of lead and cracked Mid-Continent antiknock gasoline .65 cc.[26] Armed with this confirmation of their own findings, Sun proceeded with a bold new marketing concept.

In April 1927, Sun announced the sale of its new antiknock gasoline "Blue Sunoco," "The High Powered Knockless Fuel at No Extra Price."[27] Dyed blue for product identification, this fuel sold at the price of regular gasoline but possessed as good an antiknock quality as Ethyl, or better. Since the introduction of Ethyl, the industry practice was to market two grades of gasoline, one "regular" and one "premium" or "Ethyl." Those companies that did not use lead still for the most part sold their antiknock gasoline at a premium price. The price differential was always between three and five cents per gallon between regular and premium. Sun sold only one grade of gasoline, "Blue Sunoco," and at regular rather than premium price.[28]

This product met immediate success in the marketplace. Not only did the low price attract customers, but Sun capitalized on the persistent public worry about TEL. The firm's introductory advertisements, for example, said that "Blue Sunoco is absolutely the same knockless, nonpoisonous, pure petroleum motor fuel as the colorless Sunoco which carried you over high hills, shot you out in front at the traffic signal, and gave you the most mile for your dollar.[29]

To bolster its sales in Canada, Sun published an advertisement in July 1927 reporting results of tests conducted by the Canadian Automobile Association. These concluded that Blue Sunoco was superior to all other regular gasolines and to three antiknock gasolines selling for three cents more a gallon. Between 1927 and 1929, the amount of gasoline Sun sold increased 135 percent. Sun marketed Blue Sunoco in Pennsylvania, New York, New Jersey, Delaware, Maryland, Ohio, Illinois, Michigan, Florida, New England, and Canada. By 1930, the company owned or controlled 500 filling stations and distributed its gasoline through an additional number of independent stations.[30]

All of Sun's competitors continued to sell two grades of gasoline, and in the thirties, many began to sell three. Sun, on the other hand, sold only one at a regular price, until the fifties.[31] The Pews boasted of their poison-free gasoline and the fact that it produced no harmful lead oxides to gum up spark plugs and valves. They also could take pride in the fact that Sun was the only major oil company that did not "pay tribute" to the Ethyl Corporation by purchasing tetraethyl lead to upgrade its gasoline.[32]

Such considerations had played a role in Sun's original decision to refrain from using lead, and aided thereafter in expanding its market.

Sun banked on a reputation for quality products and could back up its claims for antiknock effectiveness. But the main reason that the company began to expand with Blue Sunoco was the simple fact that it did not have to use lead. Sun's straight-run Spindletop gasoline in the early 1920s was of adequate quality, but to its great pleasure it found that when gas oil refined from Spindletop crude was put into the firm's Cross cracker, a superior grade of antiknock fuel resulted. Other refiners had stuck with selling most of their Spindletop production as fuel oil. Because of its connections with the United Gas Improvement Company, Sun had developed an antiknock gasoline equivalent to adding fifty percent benzol to straight-run Pennsylvania crude.[33]

A major challenge to Sun and other refiners who marketed unleaded "premium" gasolines was the introduction of higher compression engines by the automobile industry. In 1924, Walter P. Chrysler, at that time head of the Maxwell Motor Company, brought out the Chrysler Six, the first medium-priced car with a high compression engine. Consumer acceptance of this efficient high performance automobile and the profits made by Maxwell stimulated the industry to respond with more advances in compression in subsequent years. More and more refiners turned to tetraethyl lead as the cheapest and most efficient way to meet the fuel demands of these new automobiles. By 1929, fifty refiners were adding Ethyl Fluid to their premium motor fuels.[34]

The initial cost of Ethyl Fluid in 1924 had been one cent per cc. and it took, on the average, two to three cc. of fluid to bring "regular" gasoline up to "premium" octane grade. Although the cost had dropped to only one-half cent per cc. by 1929, there still remained blending and distribution costs that cut into profit margins. If a refiner like Sun could continue to achieve sufficient quality without lead, these savings were money in the bank. The success that Sun and other companies marketing grades of unleaded gasoline were having in the marketplace provided serious competition for the Ethyl Corporation.[35]

Sun's battle with Ethyl was just taking shape by the end of the decade, and more skirmishes would emerge in the thirties. But for Sun, the success of Blue Sunoco had started the firm on a course it was reluctant to leave. As demand increased for Blue Sunoco, Sun found that it could not supply its needs from its own cracking plants. The firm had been buying a large amount of cracked gasoline from Standard of California, and in 1929, it negotiated to buy 5,000,000 barrels (forty-two gallons each) of gasoline cracked from California crude. The combination of high grade crude and cracking gave Standard a product that was equal to Blue Sunoco standards. This was, of course, unleaded gasoline and Sun simply dyed it blue before distribution. Sun also purchased 18,000,000 barrels of gasoline from the Atlantic Refining Company and

1,000,000 barrels from Humble Oil in 1929. This was all thermally cracked gasoline, usually of 70 plus octane. However, Sun could blend purchased gasoline with its own product to achieve desire octane levels.[36]

The handwriting, however, was on the wall. Sun would either have to turn to Ethyl or develop some other process to achieve necessary antiknock quality. In the next decade, the firm's response was to develop a significant new process, catalytic cracking, to continue its marketing of one grade of unleaded antiknock gasoline. Thus the marketing strategy of the twenties placed an indelible stamp on the future decisions of the firm.

In 1929, Sun sold more gasoline than Union Oil, Standard of Ohio, and Cities Service, and was rapidly becoming a major gasoline marketer in the east. This success prompted the company to rebuild its largely neglected Toledo refinery in 1928. In addition to constructing two 5,000-barrel-per-day cracking units, Sun built a compression absorption gas plant to reclaim gasoline from the refinery gas system. The "natural gasoline" refined in this and similar processes was a very volatile and high octane substance valuable in blending motor fuel.[37]

The charcoal or absorption process for extracting gasoline from natural gas containing pentane and heavier paraffins had been in existence since 1918. The Toledo plant was able to use refinery gases to manufacture gasoline, and later improvements enabled Sun to market commercial propane gas. Toledo's crackers largely used fuel oil charging stock refined fromMid-Continent crude oil piped from Oklahoma. Sun was now able to supply Blue Sunoco to its midwest marketing area in adequate quantities.[38]

SUN OILS

The expansion of gasoline marketing did not cause Sun to neglect the facet of its business that had built its reputation, lubricating oils. On the contrary, Sun undertook several projects to improve the quality of its oils, both industrial lubricants and its new Sunoco motor oil. The company began an intensive search for processes to replace the conventional refining methods of batch distillation, batch reducing, acid treating, and neutralizing. Pennsylvania oils with high paraffinic and low ashpaltic residue still dominated the market. Sun had earlier developed a batch distillation process using high temperatures in a vacuum for making their Sun Red oils out of Texas crude, but on the whole the industry had not widely adopted the vacuum process because of the expense involved.[39]

In 1920, however, the Vacuum Oil Company developed a continuous process using a high vacuum distillation for making lubricants out of non-Pennsylvania crudes. Two years later, the continuous Schulze process appeared, which operated at an oil temperature of 635 degrees under a vacuum of 29.9 inches of mercury. The Red River Refining Company of Chicago acquired all rights to the Schulze process, and Sun obtained a non-exclusive license for the process in February 1924. It paid $200,000 in advance royalties to Red River and agreed to pay a license fee of twenty-one cents per barrel on all stock run through Schulze process stills.[40] Sun achieved only limited success with the Schulze process, and the royalty situation provided a stimulus for developing its own process.

A research group headed by Arthur E. Pew, Jr. began work on a revolutionary new system using a mercury boiler, and they developed a new mercury vapor process in the experimental laboratory in 1924. The prevailing method of extracting lubricating oils was to heat the oil to a high temperature, at which the lubricating oil boils, and condense the lubricants from the resulting vapors. But excessive or prolonged heating decomposed the oil, making it unsuitable for lubricating purposes. By heating the oils in a vacuum, earlier processes had reduced the decomposition danger from excessive heat.[41]

Sun's new process employed mercury vapor as an indirect heating agent to distill the oil. Since mercury is not adversely affected by high temperatures, Sun applied direct fire heat to the mercury in a boiler, and brought the mercury into indirect contact with the oil at high velocity over an inclined flat plate. The hot mercury vapor was admitted in compartments under the plate at controlled rates, where it transferred its heat to the oil to accomplish distillation. Distillation thus occurred at reduced pressure as well as indirect heat, but more importantly, the process allowed temperatures to be controlled precisely. The oil was exposed to heat for only forty-six seconds. The oil vapors were then piped to a fractionating tower and then to a deoderizing plant.[42]

The first commercial mercury vapor plant began operation at Marcus Hook in 1926. The results were so impressive that Sun constructed a second, larger plant in the following year. Indirect heating with mercury vapor permitted the high-temperature distillation of lubricating oils from crude without cracking taking place. Cracking would rearrange the hydrocarbon molecules and weaken the lubricating properties of the oil. The oils produced were of very high quality, particularly in terms of their consistency and uniform viscosity. They were also very light in color, so light in fact that Sun found it had to discolor it slightly for marketing purposes. The consumer was so used to darker oils that he assumed that this oil had little "body" to it. The process made it possible to produce a Sunoco Mercury-Made Motor Oil and such other high grade specialty items as transformer oils, refrigeration oils, switch oils, and turbine oils.[43]

Sun's advertising campaign for Sunoco Motor Oils had from the beginning stressed the wholly-distilled characteristic of its oil, which was produced from a vacuum process in comparison with competitors' compounded oils. Prior to 1920, most oils came from Pennsylvania crude as the fractions below kerosene and above waxes and cylinder stock. These oils were compounds of the lighter oil stocks and some cylinder stock, a heavier, green residue added to improve the "body" of the oil. Sun claimed in its advertising from 1921 through 1926 that these "compounded oils" produced more carbon deposits in the pistons of the internal combustion engine and dirtied spark plugs. Sun advertised that Sunoco oil "eliminates hard carbon deposits, induces maximum engine power, prevents excess friction, stops waste of gasoline, and flows in the coldest weather."[44] Although it began using the mercury vapor process in 1926, Sun did not market a trademarked "Mercury-Made Motor Oil" until 1931. This delay was due to a patent litigation begun in 1929.[45]

In May of that year, the Red River Refining Company, holder of the Schulze patents, sued Sun for an accounting of royalties owed. Red River claimed that all oil run through the mercury vapor plant was subject to the 1924 royalty agreement and Sun owed them twenty-one cents per barrel. The company based its claim on the fact that the Mercury vapor plants ran on the same reduced pressures, below twenty-five mm., as the Schulze process. The processes were different, however, in that Schulze did not use mercury vapor as a heat conducting medium. Red River claimed that the heart of the mercury vapor process was the high vacuum and low pressure system pioneered by Schulze.[46]

Sun won the case in 1931 and began marketing its trademarked "Mercury-Made Oil" in June of that year. Red River appealed, however, and the case was not settled in Sun's favor until 1940. At that time, Sun's patent attorney wrote to J. Howard Pew that "Therefore, as you say, this completes our long continued effort to make lubricating oil without paying tribute to others."[47]

In retrospect, Sun's claims for the "Mercury-Made" oils appear overstated since the vacuum really was the crucial element in their process. But the ability to control precise temperatures was valuable and the quality of Mercury-Made Oil compared favorably with that of competitors'. Sun continued to use the Mercury Vapor plants until the 1950s, but after World War II, the company had already begun using other vacuum processes that produced as good quality oils without the use of mercury.[48]

Refinery production increased dramatically during the decade to meet the growing demands of the automotive market, and motor oil and gasoline became Sun's main products in the United States. From 1925 to 1930, total sales of refined products increased one hundred percent, and gasoline sales increased approximately two hundred percent over the same period. By 1930, monthly sales of gasoline and naphtha exceeded 30,000,000 gallons a month and sales of lubricating oils approximated 7,500,000 gallons per month. Although an increasingly larger share of these lubricants was made up of motor oil, Sun still maintained highly lucrative foreign and domestic markets for industrial lubricants. Sun advertised these items extensively in trade papers and magazines while taking ads in national magazines for gasolines and motor oils.[49]

DOMESTIC PRODUCTION OF CRUDE

Sun fed its expanding refinery stills with oil produced from its own wells and purchased from others. In 1926, the firm owned 23,000 acres of producing oil lands in the states of Oklahoma, Kansas, Texas, Ohio, Louisiana, and Arkansas, which produced on the average 20,000 barrels of crude a day. Sun also controlled leases for an additional 800,000 acres of undeveloped land in Texas, Oklahoma, Arkansas, Kansas, Ohio, Louisiana, Colorado, New Mexico, and California. The center of Sun's production activities shifted with the new discoveries. Originally centered in Beaumont, Texas, then in Oklahoma from 1910 through the war, its western headquarters moved to Dallas in 1918 to manage the Ranger Field in that area. Between 1920 and 1924, Sun brought in wells in the Mexia, Powell, Richland, Currie, and other fields located in central Texas.[50]

The Beaumont office and the Tulsa-based Twin State Oil Company continued to operate successfully during the decade. In Oklahoma, following development of the Cushing strike, Sun brought in wells in the Burbank, Vristow, Manion, Chickashaw, Graham, Thomas, Davenport, Winfield, and Oxford fields. In 1926, Twin State also moved into the important Seminole field, the discovery of which had profound effects for the industry. The exploitation of the rich Seminole pool followed on the heels of a major concern for shortages in the mid-1920s. Through its Twin State subsidiary, Sun expanded into other Oklahoma fields following Seminole.[51]

With the decline of Spindletop and the other Gulf Coast fields, Sun's production there became less significant, but the firm still produced and purchased through the Beaumont office. During the month of July 1925, for example, Sun produced over 77,000 barrels of crude in the Beaumont area and purchased an additional 169,000 barrels. This pattern of purchases exceeding production is also reflected in gross figures for the firm. In 1929, Sun Oil produced only 3,000,000 barrels of crude but purchased 11,000,000 barrels. With the exception of 1930, when uncertain market conditions prompted Sun to cut back on purchases, Sun bought more crude than it produced.[52] In the uncertain crude markets of the 1920s and 1930s, this was not a disadvantaged position. During periods of peak production, Sun could purchase crude cheaply and cut back according to changing conditions for refined products. A chief reason for increasing exploration and development was to insure future supplies as well as to reduce costs.

TRANSPORTATION

Although by 1920 Sun was the twenty-first largest refiner in the United States, the firm maintained only intrastate pipelines and thus did not have to file reports with the Interstate Commerce Commission as required by the Hepburn Act. This did not mean that Sun's lines were all short. In the middle of the decade, Sun operated 260 miles of pipeline in Ohio, Texas, and Oklahoma. The Ohio lines fed the Toledo refinery, the Sun Pipeline Company (Texas) carried crude to Sun Station and Sabine Pass, and the new J. Howard Pew Pipeline extended from the Yale refinery to producing fields in nearby Oklahoma counties. By 1930, excluding all gathering lines and pipelines associated with the refineries, Sun operated 300 miles of trunk pipeline and maintained a working storage capacity of about 5,000,000 barrels.[53]

Sun's investment in pipelines was small compared to those of the largest firms, but its lines fit well into an integrated transportation network. From 1920 to 1929, five more ships joined the Sun fleet of tankers, and by the end of the decade, the company owned and operated seven large seagoing tankers totaling more than 100,000 deadweight tons as well as thirteen seagoing and river barges. Sun also operated two other large seagoing tankers under lease-hold agreement. In 1922, the firm owned 850 railroad tank cars used in conjunction with its pipelines. In 1923 and 1929, Sun added more cars, making a total of 1,643. In addition to these, the firm operated 208 tank cars leased from outside companies.[54]

FOREIGN OPERATIONS

In the twenties, the multinational operations of American oil companies took on added dimensions. Jersey Standard obtained market-oriented refineries in Poland, Italy, Norway, and Belgium to complement its more established facilities in England and Germany, and became by far the largest United States oil enterprise in Europe. The Vacuum Oil Company operated lubricating oil refineries as well as a marketing system in England and the continent, and Texaco, Cities Service, Tidewater, and Sun owned smaller European distribution organizations. Standard Oil of New York opened offices in Bulgaria, Greece, and Yugoslavia.[55]

Sun sold lubricants abroad through distributors either owned by or affiliated with the parent firm. In most cases, the firm did not market these products under the Sun trade name, but sold them in bulk form to be blended with other oils by European marketers. The subsidiaries, British Sun Oil Ltd., Netherlands Sun, and Sun Oil (Belgium), handled large shares of this business, as did the German affiliate Mineraloelwerke Albrecht & Company. The large French market had been particularly attractive to American firms, and Sun had purchased a one-third interest in the Societé Anonyme d'Armement (SAIC), a marketing concern, in 1919. The French market, however, had rapidly become highly competitive.[56]

When Jersey Standard moved to enlarge its distribution network in France, the French government countered by purchasing twenty-five percent interest in the private firm, Compagnie Francaise des Pétroles. Gulf Oil had been selling in Europe through the Jersey Standard organization, but in 1925, it moved to obtain its own distributorship. It approached Sun, and the Pews agreed to sell their one-third interest in SAIC to Gulf for a profit of $1,000,000. By the end of the decade, Gulf expanded its financial interest in SAIC to seventy-five percent. By its agreement with Gulf, Sun continued to market some lubricants through SAIC, and was free from a somewhat difficult business relationship.[57] Its partners in SAIC had been the Nobel and Rothschild interests and Sun had been unable to act in its accustomed independent fashion. Moreover, Sun had no foreign refineries, its European sales were limited to specialized lubricants, and it had no immediate need for further growth. Gulf was now ready to launch a major expansion of its European operations and was therefore willing to commit the type of capital that the Pews were not. All in all, it appears to have been a beneficial arrangement for both parties.

Sun did organize one other new marketing subsidiary in 1923, Sun Oil Company Ltd. in Canada. Originally established to market lubricants, this wholly-owned organization proved to be very valuable by the end of the twenties in the sale of Blue Sunoco north of the border.[58]

The expansion of distribution facilities and a growing concern for the depletion of domestic crude supplies also focused on the need for additional crude reserves, and the 1920s saw the beginning of a thrust for foreign production by American firms. Frequently, American companies asked Washington for diplomatic assistance in foreign ventures and the government usually was cooperative. Here also, Jersey Standard took the lead and became dominant in Middle Eastern and Latin American exploration and production.[59]

Concerns about shortages propelled Sun into a limited program of foreign exploration during the decade. Immediately following the war, warnings of world shortage encouraged many American firms, supported by the national government, into undertaking expansion abroad. Sun briefly became involved with plans for joint exploration in Mesopotamia (Iraq) and in Latin America in 1919–1920, but they never got off the ground. Although Sun expressed a passing interest in other Latin American areas, including Cuba, Brazil, and Mexico, the only country where the firm actually began operations was Venezuela.[60]

Oil companies had known about Venezuelan oil since 1880, but it was not until 1912 that Royal Dutch Shell obtained the first lease concession. American firms became interested in Venezuela after the war for three reasons. Most important was the world-wide shortage scare which made foreign investments more attractive. In Venezuela, moreover, the passage of a relatively liberal land law in 1918 and the presence of apparent political stability in the form of a dictator, Juan Vincente Gomez, made that country particularly attractive. Venezuela's political situation was appealing when compared, for example, to the chaotic situation in Mexico at the time.[61]

Standard Oil (New Jersey) had been negotiating with Gomez in 1919 for a blanket lease over a large area south of Lake Maricaibo. On the very day it filed for these five concessions in January 1920, Julio Mendez, Gomez's son-in-law, filed for the same property. Mendez obtained the concessions and sold them to Addison McKay, one of several middlemen who operated in Venezuela. These men in effect became agents of the companies. In this case, McKay sold his concession to the Sun Oil Company. Although Standard's historians brand McKay "the secret agent" of the Sun Oil Company, he was only one of many successful operators who offered their services to all interested companies. In addition to Sun and Jersey, Maracaibo Oil Exploration, New England Oil, Texas, Gulf, and Sinclair were active in the politics of obtaining drilling concessions.[62]

The 1918 Venezuelan land law limited each foreign firm to a maximum of 80,000 hectares (over 196,000 acres) per lease. The American companies finessed this restriction by creating many subsidiaries to operate in Venezuela. By January 1925, Sun had acquired leases amounting to 800,000 hectares (approximately 1,963,000 acres) at a total cost of $750,000. The company registered eleven separate U.S. corporations in Venezuela to operate its leases. Most of these companies were in turn owned by a holding company, the Andean Sun Company, but other leases were tied up with another firm, the Beacon Sun Company. Beacon Sun's corporate organization was typical of the confusing tangle of American firms that operated in Venezuela at this time. The company was owned fifty percent by Sun Oil and fifty percent by the Beacon Oil Company, a firm controlled by Standard Oil (New Jersey), its majority stockholder. To further complicate matters, however, Beacon Sun's holdings were operated by the Richmond Petroleum Company, a subsidiary of the Standard Oil Company of California.[63]

Although Sun had edged Jersey Standard out of some attractive leases and appeared to be in a good position, the firm's experiences during the decade were disappointing. From 1920 to 1929, for example, Beacon Sun drilled ten wildcat wells and found not one drop of Venezuelan oil. In 1929, there were 107 foreign companies, mostly American, registered in Venezuela, but only five of them exported crude. Shell, Gulf, Standard of Indiana, Standard of New Jersey and Standard of California produced more than ninety-eight percent of all Venezuelan oil.[64]

Development costs were enormous. In addition to bringing all supplies either from the United States or Europe, the American companies found that they had many other problems. They had to construct roads, railroads, houses, warehouses, hospitals, machine shops, power plants, pipelines, tank farms, and loading piers in addition to their investments in exploration and drilling. Standard of New Jersey alone spent millions of dollars drilling dry holes before it brought in successful strikes. In the long run, the large firms were the only ones able to compete.[65] Sun's initial failure in Venezuela was buffered by the success of domestic strikes, particularly Seminole in 1926–27, but the costly Venezuelan venture served to discourage Sun from any active foreign operations for many years.

SUBSIDIARIES

As of March 1930, the Sun Oil Company owned total or controlling interest in twenty-one subsidiary companies involved in the production, transportation, refining, or marketing of petroleum and its products. The total capital stock owned in these subsidiaries amounted to over $5,000,000. By far the largest of these was the Sun Shipbuilding and Drydock Company, which had a total book value over $3,000,000. Sun Oil had founded the shipyard in 1916 and owned over eighty percent of Sun Ship stock, with the Pew family owning the remaining twenty percent. In 1928, the Sun Oil Company acquired one hundred percent stock ownership in the company and operated it as a wholly-owned subsidiary.[66]

The shipyard constructed seven vessels for the Sun Oil Company from 1920 to 1930. These were the primary orders that kept the yard in business in the shipbuilding industry's lean post-war years. As the demand for new ships declined during the decade, Sun added two drydocks, and ship repair work became an important part of the firm's business. The Sun Shipbuilding Company became the Sun Shipbuilding and DryDock Company, the corporate name of the Chester concern today. The yard also undertook the building of special machinery, steel plate work, and some refinery equipment construction to stay busy. The fabrication of heavy refinery equipment became an important part of the yard's work in the 1930s.[67]

Sun Ship also made innovations in tanker construction. The firm introduced a device for controlling pressures and vacuums in oil tankers in 1920 called the U-gauge, a new venting system that was used for twelve years. Sun also obtained the exclusive American rights to the Doxford opposed piston Diesel engine. Invented by Professor Hugo Junkers of Germany, and developed by William Doxford & Sons, Ltd. of England, Sun Ship employed the new engine in its vessels. The company marketed the engine under the licensed trade name of the Sun-Doxford Diesel engine.[68] The shipyard's various activities barely kept the plant alive during the twenties decade, however, because the demand for ships had taken a decided downward turn following the war.

Sun's next largest subsidiary was the Sun Pipe Line Company (Texas), which operated the company's crude oil pipelines in that state. The designated book value of this concern in 1930 was approximately $400,000. The British Sun Company, Ltd., the firm's marketing subsidiary, had an equal net worth. After Sun sold its one-third interest in SAIC to Gulf Oil in 1925, British Sun became the center of the firm's European marketing operations. Most of the remaining subsidiaries were small companies that had been in the organization prior to 1920. Among these were the Sun Oil Line Company (Ohio), Twin State Oil Company, Delaware River & Union Railroad Company, and the Sun Company of Delaware. In addition to these, Sun had acquired some small marketing affiliates like the Hearn Oil Company and Greenslade Oil Company during the decade.[69]

Two smaller subsidiaries, originated during the 1920s, however, were important for Sun's later growth. One of these, the Peninsular State Oil Company (organized in 1925) distributed the company's products in the state of Florida, and thus represented a major expansion of Sun's marketing territory. The other new company was the Sperry-Sun Well Surveying Company, organized as a wholly-owned subsidiary in 1929.[70]

Sperry-Sun arose directly out of Twin State's experiences drilling in the Seminole field in 1926. Although operators knew the depth of the oil sands, the depth of drilled wells varied considerably. Tests verified that many wells were crooked, a fact that added expensive feet to the length of drilling. Moreover, if a well were too out-of-line, it might miss the oil pool completely and result in a dry hole. A new service, well-surveying, came into existence as a result of this problem. The Sun-Sperry gyroscope, developed jointly with the Sperry Corporation, was an instrument used to aid in drilling straight wells, and became the basis of the new Sperry-Sun Company. Later instrument modifications aided the drilling of off-shore pools, selecting drilling sites, and avoiding undesirable strata. In 1930, the firm was small, but it came to occupy a significant place in the well-surveying field and is today still a profitable arm of the Sun organization.[71]

FINANCE

To finance its expansion in the 1920s, Sun could raise funds from three potential sources. It could follow its traditional policy of relying on the reinvestment of earnings, issue bonds, or offer stock for sale to the public. Traditionally, the integrated oil companies had refrained from selling securities on the open market and had remained closed corporations. This was particularly true of the Standard Oil Trust prior to 1911. After the dissolution of Standard in 1911, however, some independent oil men, including Harry Sinclair, E.L. Doheny, and Harry Doherty made use of outside investors in building organizations that successfully exploited new fields in Oklahoma, Mexico, and California respectively. In 1920, the Atlantic Refining Company was the first of the old Standard firms to go public and be listed on the New York Stock Exchange. Between 1920 and 1930, Standard (New Jersey), Standard (California), Standard (New York), Prairie Oil and Gas, Standard Oil (Kansas), and the Ohio Oil Company followed suit and also listed their stock on the big board.[72]

This alternative was attractive to Sun for two reasons. Not only would "going public" open the door for outside investment, it could also solve another problem that had plagued Sun management for some time: how to place an accurate value on Sun's capital stock and provide for an orderly sale of shares held by the Pew family or outside investors. Sun's negotiations with the United Gas Improvement Company in 1917 had highlighted this problem. The Pews had ultimately had to pay U.G.I. a price two and one half times the par value of its shares of Sun common stock.

Since the end of the war, Sun had continued to wrestle with the problem that its authorized par value stock did not accurately reflect the company's worth. During the boom war years, the firm's net income had risen from $1,885,000 in 1915 to a healthy $10,106,000 in 1920. "To more accurately reflect the real value of the business," Sun's directors acted in December 1919 to increase the corporation's authorized capital stock from $7,920,000 to $60,000,000 or 600,000 shares of par $100 common stock. This plan, however, did not come to fruition because of events in 1921.[73]

In that year, Sun experienced the only deficit in its history, a loss of $136,000. This unique but relatively small loss resulted from the general decline in the American economy in this post-war depression year. Sun was also affected by a sharp decline in crude oil prices, particularly in Oklahoma crude, which brought about a corresponding decrease in the prices of its refined products. The company had purchased large stocks of crude at higher prices and was stuck with this inventory during the price decline of 1921. As a consequence of the unfavorable economic conditions of 1920–1921, the Sun board voted in January 1921 not to implement the stock increases authorized in 1919.[74]

In the economic upturn of 1922, Sun earned a net income of $1,658,000 and the board then acted again to increase the firm's authorized capital stock. As of November 30, 1922, Sun's books showed a surplus of $23,600,000 carried over from the boom war years. The board voted to rescind the 1919 resolution to increase the firm's capital stock to $60,000,000 and instead authorized an increase from $7,920,000 to $32,000,000. This authorization allowed the board to vote an astounding three hundred percent stock dividend on its $100 par common stock.[75] This unusually high dividend resulted in a more readily accurate measure of the intrinsic worth of the company's stock, but Sun still lacked an orderly procedure for the buying and selling of its securities.

In 1923, the firm attempted to attack the problem in a compromise fashion. The board voted to increase Sun's capital stock from $32,000,000 to $44,000,000, authorizing for the first time 120,000 shares of $100 par-value preferred stock. Sun would list this preferred stock on the New York Stock Exchange and, therefore would provide something of a yardstick to measure Sun's unlisted common shares. The firm would offer $4,000,000 of this new preferred stock to the public through the investment bank, Brown Brothers of Philadelphia, and sell the remainder to Sun's common shareholders (largely the Pews). A general stockholders meeting approved the action in May 1923, but the board deferred acting in June, citing that "the market for Preferred Stock has slumped." In lieu of the preferred stock, the Board authorized the issuing of $4,000,000 in gold debentures placed with Brown Brothers.[76]

During this period, management had continued its conservative policy of reinvesting earnings. For the ten-year period from 1914 to 1924, Sun showed a net profit figure of $34,000,000 after depreciation, depletion, interest, federal taxes, and all other charges. Of this amount, only

$5,500,000 was paid out in cash dividends, while reinvestment from earnings amounted to over $28,000,000. Sun took no further action on the postponed issuing of preferred stock in 1924, and the board instead approved an issue of $10,000,000 of fifteen-year, 5.5 percent debentures to retire the ten-year, 6 percent bonds issued in 1919 and an issue of ten-year, 7 percent bonds sold in 1921. This left an outstanding debt of $14,000,000. On June 30, 1925, Sun retired the $4,000,000 issue of two-year, 6 percent notes issued in 1923, leaving a debt of $10,000,000.[77] Although Sun now carried a bonded debt as a result of specific expansion activities, the Pews sought to keep it minimal.

Sun scrapped the plan for issuing preferred stock in 1925 when J. Howard Pew reported to the board in May on negotiations to list Sun common stock on the New York Exchange and sell it publicly at no par value. Sun's stockholders voted to increase the firm's total capital stock from 320,000 shares of $100 par common to 1,250,000 shares of stock with no nominal or par value. This increase allowed a three-for-one-exchange with all holders of the old par stock and the sale of 158,000 shares to the public. Sun offered this stock to the public at $36.50 a share and listed it on the New York Stock Exchange. The press reported the "Wall Street view" that Sun could have used a higher figure for Sun stock, but the conservative J. Howard Pew insisted that the new shares be sold at "book value." During the first half year of public participation, Sun's common stock sold at prices ranging from a low of $34.50 to a high of $46.25 a share. Sun began issuing annual reports in 1925, at the time it made available its first public offering of stock.[78]

In 1927, Sun issued 50,000 shares of par $100 cumulative preferred stock, the first preferred it had ever sold. The firm used the proceeds from this stock to finance the expansion of the Toledo refinery in 1928. Although Sun sold only 50,000 shares of this preferred, the board had authorized the sale of 100,000 shares. At the same time, the company increased the authorization of its non-par common stock to 1,500,000 shares. In 1930, this authorization increased to 1,800,000 shares of common stock which combined the 100,000 shares of preferred for a total capitalization of 1,900,000 shares.[79]

The action of "going public" opened the door for moderate influx of outside capital, but the Pews had no thoughts of eventually relinquishing their control of the Sun Oil Company. By listing on the New York Stock Exchange, they now had a vehicle for selling Sun shares rapidly at a market-determined price. This would aid all those stockholders, family included, who wished to sell shares. The company had also started a profit-sharing plan to reward company executives with shares of Sun Common, and planned to launch an employee stock purchase plan. For both of these plans to become successful, it was necessary to have a procedure for uniform disposal of Sun securities. Although Sun now carried a small bonded indebtedness and had access to outside investors through the New York Stock Exchange, its primary source of capital in the twenties remained the reinvestment of earnings.

LABOR RELATIONS AND WELFARE CAPITALISM

In the 1920s, many businessmen adopted management techniques that had been pioneered in the late nineteenth century by firms like the Boston department store of William Filene and Sons and the National Cash Register Company of Dayton, Ohio. The common theme in all of these approaches was an improvement in employee welfare initiated by management and aimed at an ultimate increase in productivity. In practice, this "welfare capitalism" became a viable alternative to the adversary relationship that had arisen between worker and manager as a result of increased unionization of the work force in the Progressive era. Indeed, most historians of the labor movement in America have argued that the main purpose of welfare capitalism was specifically to undermine union growth. In the 1920s, the widespread popularity of various welfare schemes offered unilaterally by management does suggest partial explanation for the decline in union membership during the decade. However, one thing is certain. Even if the programs volunteered by business fell short of the goals desired by progressive union leaders, workers in America benefited directly from the variety of plans made available to them.[80]

When the Sun board of directors voted to go public in 1925, it also approved a new stock-sharing plan for all Sun employees to go into operation on July 1, 1926. Employees could purchase Sun common stock with payroll deductions up to ten percent of their earnings. For every dollar they contributed toward this purchase, the company added fifty cents, and a board of three trustees used the total amount to buy the stock either directly from the company (from both treasury and unissued stock) or on the open market, whichever was lower. After a five-year period, employees became owners of their stock. If they retired or left the firm before that, they would receive only the amount they had invested plus six percent interest.[81]

Although the stock purchase plan did provide some financial advantages to the company, the main reason for its adoption appears to be the sound business reason that employees who owned a stake in the business would become better workers, take a greater interest in their jobs, and stay longer. This plan fit into the general pattern of welfare capitalism heralded by businessmen in the twenties. J. Howard Pew suggested this rationale in a memo sent to all management personnel in 1926:

> While the Company is giving something to the employees, and while in doing so the Company had the welfare of its employees in mind, the fundamental reason for the plan is a business reason. The Company believes that the plan will be worth while to it, because of the added interest which the participating employees will have in the welfare of the Company when they feel that they are partners in the business.[82]

The stock purchase plan also had potential as a method of finance for the company, though this was rarely used. Despite broadened stock participation, the Pew family still maintained dominant control of the firm. After almost three years of functioning as an "open" corporation, the family still held eighty percent of Sun's common stock, the employees and investment plan trustees six percent, and outside interests fourteen percent.[83]

Although Sun did not adopt a pension plan for its employees, it did make available an optional group life insurance plan in 1927. Without a physical examination and with no regard for age, an employee could purchase insurance at the rate of sixty cents per month for each $1,000 of coverage. The company paid any excess costs over that amount. The plan of course cost Sun little, and the insurance company required that at least seventy-five percent of all eligible employees subscribe to the plan.[84]

Although they did not go as far as some other employers in the twenties, the Pews' adoption of a stock purchase plan and sponsorship of a group life insurance plan showed that they were concerned with new management concepts and understood the good that the firm could derive from a policy of employee benefits. Their labor policies may be correctly labeled paternalistic, but they were not harshly exploitive. There were many very dirty jobs in an oil refinery, and by today's standards the work was difficult, but Sun paid competitive wages. Sun's "welfare capitalism" also extended into a wide program of athletic, social, and family activities for employees, all predicated on the belief that a satisfied work force would prove most valuable to the firm in the long run.[85]

MANAGEMENT

Changes at the top management level occurred during the decade, but members of the Pew family continued to dominate the firm. One major link with the past ended on June 26, 1925, when Robert C. Pew, vice-president of Sun and director of the Toledo operations, died at the age of sixty-three.[86] The nephew of J.N. Pew, Sr. and brother of J. Edgar and John G. Pew, he had played an important role in Sun's history from his early days with Pew and Emerson in the natural gas business and the first efforts at oil procurement in the Lima, Ohio, fields. Robert had also been responsible for directing Sun's first refinery efforts in Toledo in 1894–1905, and it was he who first went to Beaumont in 1901 to lay the foundation for Sun's Spindletop activities. Only J. Edgar Pew and his twin brother John G. Pew now remained from the early days of Pew and Emerson in the gas and oil fields.

A new generation of family members joined Sun during the twenties who would have an important influence in the future development of the company. Arthur E. Pew, Sr., the eldest of J.N.'s sons, who had died in 1916, had two sons, both of whom now joined the firm. The elder, Arthur E. Pew, Jr., began work in 1921 and had quickly established a reputation for innovation in the refining end of the business. It was he who had directed research into the mercury vapor process and later into thermal cracking. In the next decade, he was to make substantial contributions in the field of catalytic cracking. On March 11, 1930, Arthur, Jr. became Sun vice-president in charge of manufacturing and assumed general supervision over all three of Sun's refineries.[87]

In March of 1927, Samuel T. Eckert, a close college friend of J.N. Pew, Jr., who had joined the firm after the war, became vice-president in charge of marketing. Eckert had been largely responsible for launching Sun's successful gasoline and motor oil sales. That same year, Walter C. Pew, Arthur E. Pew's younger brother, became general sales manager. Walter worked for Sun in the sales and marketing division until he left for military service in World War II.[88]

In the fall of 1924, Jno. G. ("Jack") Pew, J. Edgar Pew's son, began work in the production department in Richland, Texas, part of Sun's Dallas division. He performed in various capacities until December 1925, when he was transferred to the Beaumont Division to supervise all Louisiana operations. In 1926, he moved to Dallas as an assistant to the vice-president in charge of production, his father, J. Edgar Pew, and in 1930, he became one of a "committee of three" which managed the Dallas Division. He assumed more and more responsibility and eventually succeeded his father as vice-president in charge of production.[89]

By the end of the decade, family management was assured. J. Howard and J.N. Pew, Jr. were firmly in control as president and vice-president. Their cousins, J. Edgar and John G., ran the production department and the Sun Shipbuilding company, respectively. Although both J. Howard and J.N. Pew, Jr. were comparatively young men, there were now representatives of a new generation of Pews in the three main arms of the business: Arthur E. Pew, Jr. in refining, Walter C. Pew in sales and marketing, and Jack Pew in production.

It becomes increasingly difficult to label the Pew brand of family management by the 1920s. Upon first examination, Sun's managers, particularly in the persons of J. Howard and J.N. Pew, Jr., appear to be very conservative. Although Sun technically went public in 1925, the Pew family still controlled most of Sun's stock, and the company in many ways resembled a nineteenth-century partnership rather than a modern twentieth-century corporation. The brothers' conservative financial policies, characterized by reinvestment of earnings and a small bonded indebtedness, reinforced the firm's reputation in financial circles as a conservatively managed company. Moreover, during the policy debates on how to cope with the problems of uneven production and scarcity during the decade which are discussed in the next chapter, the Pews became champions of the *laissez faire* segment of the industry. Rather than joining the movement for regulatory activity, they insisted that the "natural laws" of the marketplace be allowed to determine the price and quantity of crude oil produced in the United States.

Yet, in other ways, the Pews seemed not so old-fashioned. Their innovations in both petroleum cracking and lubricant development demonstrated leadership in refining technology, a practice which they followed in later decades. The Blue Sunoco marketing campaign and their stubborn reliance on lead-free gasoline exhibited both conservative and innovative tendencies, but it clearly demonstrated a willingness to take independent and risky business stances. Their experimentation with welfare capitalism also indicates that they were willing to adapt to modern managerial technique.

One other area of management philosophy demonstrates the difficulty in labeling J. Howard and J.N. Pew, Jr.—public relations. Public relations as a management device was only coming of age in the 1920s with the work of Ivy Lee, Edward Bernays, and a handful of others. Certain men in the oil industry, like A.C. Bedford and Walter Teagle of Jersey Standard, realized the value of creating a good public image, while others like Robert "Bob" Welch, the executive secretary of the American Petroleum Institute, the largest oil trade association, resisted public relations strongly. Despite Welch's objections, Bedford foisted Ivy Lee, with whom he had worked on the wartime National Petroleum War Service Committee, on the A.P.I. Welch largely ignored him, however, and Lee left after only one year of consulting for the trade association.[90]

In 1925, other more public relations-minded oil men created an A.P.I. Public Relations Committee to help improve oil's tarnished image following such difficulties as the Teapot Dome scandal and a growing public uncertainty concerning the true size of oil reserves. Each "major" A.P.I. member paid a subscription of two thousand dollars, and smaller members paid either one thousand or five hundred dollars to support the committee's activities. J. Howard Pew was a member of the new Public Relations Committee, and Judson C. Welliver, a newspaperman and former White House spokesman for Presidents Harding and Coolidge, became its staff director. He was assisted by Ellen Talbott, an extremely competent public relations expert.[91]

After two years of trying to work with the stubborn Welch, Welliver left the A.P.I. in 1927. Soon after, J. Howard Pew hired him as public relations director of the Sun Oil Company, and he brought Ellen Talbott along with him. Welliver played an important role with Sun for many years and handled public relations during Sun's many critical struggles with the government and other firms in the oil industry during the New Deal.[92]

By the end of the 1920s, both J. Howard and J.N. Pew, Jr. had evolved a business philosophy composed of many complex elements. On some issues, they were and would remain for many years extremely cautious and reluctant to change. Sun still remained a family-dominated firm, followed conservative financial policies, refrained from becoming too large, and only hesitantly entered the competitive realm of foreign operations. In other matters, however, they were quite willing to adapt new managerial techniques, develop new technologies, and experiment with new marketing strategies.

SUMMARY

Sun Oil had entered the decade a neophyte in the burgeoning automotive products market. By employing shrewd marketing techniques and reinvesting large amounts on refinery plant improvement, it captured a significant share of the eastern marketing territory. Sun built on its established reputation as a lubricant specialist by marketing high quality motor oil, and it maintained its reputation by developing improved vacuum processes, particularly the mercury vapor system. Not only did Sun make use of the latest developments in thermal cracking to produce high quality antiknock gasoline, it defied the Ethyl Corporation and made use of a unique marketing strategy of selling one grade of unleaded gas at regular prices. During the next decade, Sun continued the same pattern of growth, making advances in refining technology that assured it of sales leadership within its marketing territory.

NOTES

1. John T. Enos, *Petroleum Progress and Profits* (Cambridge: M.I.T. Press, 1962), Chapter 1, "The Burton Process," pp. 1-59.
2. Harold F. Williamson, Ralph L. Andreano, Arnold R. Daum and Gilbert C. Klose, *The American Petroleum Industry,* Volume II, *The Age of Energy, 1899-1959* (Evanston: Northwestern University Press, 1963), pp. 150-63.
3. *Ibid.*, pp. 375-6.
4. *Ibid.*, pp. 379-82.

5. *Ibid.*, pp. 383-5; see Enos, *Petroleum Progress*, Chapter 3, "The Tube and Tank Process."

6. Williamson et al., *Energy*, pp. 385-9.

7. *Ibid.*

8. Henry Thomas to J. Howard Pew, December 13, 1922, Accession #1317, Eleutherian Mills Historical Library, J. Howard Pew Papers, Box 39; Clarence Thayer, retired chief engineer and vice-president, Sun Oil Company, interview held at Media, Pennsylvania, November 25, 1975; Board Minutes, The Sun Company (New Jersey), Office of the Assistant Secretary, Sun Oil Company, St. Davids, Pennsylvania, April 17, 1924, Minute Book #3, p. 237.

9. Board Minutes, April 17, 1924; Thomas to Pew, December 13, Acc. #317, EMHL 1933, J. Howard Pew papers, Box 39.

10. Robert E. Lamberton, Sun Attorney, to J. Howard Pew, January 22, 1930; J. Howard Pew to Lamberton, January 23, 1930, Acc. #1317, EMHL, J. Howard Pew papers, Box 39.

11. Pew to Lamberton, January 23, 1930.

12. Clarence Thayer, interview held at Media, Pennsylvania, November 25, 1975; Enos, *Petroleum Progress*, pp. 144-5; "The First Fifty Years," *Our Sun* 16, 4 (Autumn 1951), p. 6; the octane number of a gasoline is the percentage of high antiknock iso-octane in a mixture of iso-octane and normal heptane. The government established this standard in 1928 and it remains today the basic antiknock measurement.

13. T.A. Boyd, "Pathfinding in Fuels and Engines," *Society of Automotive Engineering Transactions (SAE) Quarterly Transactions* 4, 2 (April 1950), pp. 182-3; Lynwood Bryant, "The Problem of Knock in Gasoline Engines" (unpublished manuscript, 1972), in author's personal file, p. 1.

14. Bryant, "Problem of Knock," pp. 4-7; Boyd, "Pathfinding," pp. 183-4; Williamson et al.; *Energy*, pp. 410-1.

15. Boyd, "Pathfinding," pp. 182-3; Bryant, "Problem of Knock," pp. 7-8; Williamson et al., *Energy*, pp. 409-10.

16. Boyd, "Pathfinding," pp. 183-4.

17. *Ibid.*, pp. 184-9; Williamson et al., *energy*, pp. 411-3; Alfred D. Chandler, Jr., and Stephen Salsbury, *Pierre S. duPont and the Making of the Modern Corporation* (New York: Harper and Row, 1971), p. 466; *United States vs. DuPont, General Motors et al.*, Volume VI, *United States District Court Proposed Findings of Fact and Conclusions of Law* (February, 1954) (bound privately by Irénée duPont, EMHL).

18. Williamson et al., *Energy*, pp. 413-4; Boyd, "Pathfinding," p. 189; Harry Mack, "That Day in Dayton," *Ethyl News* (May/June 1959), pp. 37-40; Paul H. Giddens, *Standard Oil Company (Indiana), Oil Pioneer of the Middle West* (New York: Appleton-Century-Crofts, 1955), p. 289.

19. Williamson et al., *Energy*, p. 414; Boyd, "Pathfinding," pp. 189-90.

20. Giddens, *Oil Pioneer*, p. 292.

21. "Sunoco Gas Proves Itself in Special Test," *Our Sun* 2, 8 (May 29, 1925), p. 2.

22. Memo: "Blue Gasoline, 1931:" Memo: "Color of Blue Sunoco," Acc. #1317, EMHL, J. Howard Pew papers, Box 56.

23. Thayer interview; "Refinery Notes, June 1, 1936," Acc. #1317, EMHL, J.H. Pew papers, Box 50, p. 5b.

24. "First Fifty Years," *Our Sun*, p. 6; Thayer interview.

25. Williamson et al., *Energy*, pp. 414-5; Thayer interview.

26. A. Ludlow Clayden to J.H. Pew, January 11, 1927, January 19, 1927; Graham Edgar, Ethyl Corporation, to Clayden, January 17, 1927, Acc. #1317, EMHL, J.H. Pew Papers, Box 55.

27. "Advertising Reflects the Progress of Sunoco," *Our Sun*, 50th Anniversary Issue, 3, 1 (1936), p. 39.

28. "Story of Sun," *Our Sun*, 75 Anniversary Issue, 26, 3–4 (Summer/Autumn 1961), p. 23; "Blue Sunoco 25th Anniversary, 1927–1952," *Our Sun* 17, 2 (Spring 1952), p. 29.

29. "Advertising Reflects," *Our Sun*, p. 39; "Blue Sunoco 25th," *Our Sun*, p. 30.

30. "Advertising Reflects," *Our Sun*, pp. 36–9; "Sun Oil Company Broker Prospectus, 1930," Acc. #1317, EMHL, J.H. Pew papers, Box 58.

31. When it eventually abandoned its no-lead policy, Sun still did not conform to the industry norm. Rather, the firm introduced its custom blending multigrade pump that still dispenses gasoline from Sunoco stations.

32. "Sun Oil," *Fortune* 23, 2 (February 1941), p. 52.

33. Thayer interview; Gustav Egloff and Jacque C. Morrell, "Cracking Spindletop Crude Yields Gasoline of 45-50 Per Cent Benzol Equivalent," *National Petroleum News* 19, 15 (April 17, 1927), pp. 87–90.

34. James J. Flink, *the Car Culture* (Cambridge, Mass.: The MIT Press, 1975), p. 121; Williamson, et al., *Energy*, pp. 414, 605–7.

35. Williamson et al., *Energy*, p. 414; Paul Truesdell, "Widening Sale of Ethyl Gasoline Carries Challenge to Refiners," *National Petroleum News* 18, 38 (September 22, 1926), pp. 21–3.

36. Contract with Standard Oil of California, September 10, 1929, Acc. #1317, EMHL, File 21-A, Box 93; see correspondence between Sun Oil and Atlantic Refining Company, Humble Oil Company, Acc. #1317, EMHL, 21-A, Box 93.

37. "Economic Outline and Data Relating to the Petroleum Industry," U.S. Congress, Temporary National Economic Committee, *Hearings Before the Temporary National Economic Committee Part 14-A, Petroleum Industry*, September 25, 1939 (Washington, D.C.: U.S. Government Printing Office, 1940), pp. 7801, 7805.

38. Williamson et al., *Energy*, pp. 416–7; "The Toledo Refinery Looks Back Over 50 Years," *Our Sun* 2, 3 (July/August 1945) pp. 2–3, 6.

39. "First Fifty Years," *Our Sun*, p. 6; Williamson et al., *Energy*, p. 430.

40. Williamson et al., *Energy*, p. 430; Board Minutes, February 6, 1924, Minute Book #3, p. 234.

41. "First Fifty Years," *Our Sun*, p. 6.

42. "Our Mercury-Made Oils," Memo, 1931, Acc. #1317, EMHL, J.H. Pew Papers, Box 5.

43. "First Fifty Years," *Our Sun*, p. 6; Thayer interview.

44. "Advertising Reflects," *Our Sun*, pp. 36–8, see especially advertisements of February 1921, and August 1924; A.L. Clayden, "Oil in the Internal Combustion Engine," *Our Sun* 2, 7 (April 30, 1925), pp. 6–7.

45. "Advertising Reflects," *Our Sun*, pp. 36–7.

46. R.E. Lamberton to J. Howard Pew, April 4, 1930; Memo, "Regarding Red River Refining Company Suit," Acc. #1317, EMHL, J. Howard Pew Papers, Box 41.

47. Frank Busser to J. Howard Pew, July 1, 1940, Acc. #1317, EMHL, J. Howard Pew Papers, Box 41.

48. "First Fifty Years," *Our Sun*, p. 6; Thayer interview. The main problem with the use of mercury was a loss of vapor in the boilers. When the price of mercury accelerated in the forties, these boilers became a liability.

49. "Broker Prospectus, 1930," Acc. #1317, EMHL, J. Howard Pew Papers, Box 58, p. 7; *Forty Years Growth in the Oil Industry* (Philadelphia: Sun Oil Company, 1926), p. 16.

50. *Forty Years Growth*, Sun Oil Company, p. 16; "In the Oil Fields with Sun," *Our Sun*, 50th Anniversary Issue, 3, 1 (1936), pp. 32–3.

51. "In the Oil Fields with Sun," *Our Sun*, pp. 31–2.

52. "Net Production and Oil Purchased for Account of Philadelphia," July 1925, Acc. #1317, EMHL, File 21-A, Box 13; Cook, Roy C., *Control of the Petroleum Industry by Major Oil Companies,* Monograph No. 39, Temporary National Economic Committee (T.N.E.C.) (Washington, D.C.: U.S. Government Printing Office, 1941), p. 70.

53. Williamson et al., *Energy*, note on p. 382; *Forty Years Growth*, Sun Oil Company, p. 7; "Broker Prospectus, 1930," p. 5.

54. "Tankers Under the Sun Flag," *Our Sun*, 50th Anniversary issue, 3, 1 (1936), p. 20; "Broker Prospectus, 1930," pp. 5-6.

55. Mira Wilkins, *The Maturing of Multinational Enterprise: American Business Abroad From 1914-1970* (Cambridge, Mass.: Harvard University Press, 1970), p. 85.

56. "Golden Anniversary in London," *Our Sun* 24, 2 (Spring 1959), p. 24; "Mineraloelwerke Albrecht & Companie," *Our Sun* 3, 4 (January 30, 1926), pp. 3-4; Board Minutes, December 5, 1919, Minute Book #3, pp. 13-4.

57. Wilkins, *The Maturing of Multinational Enterprise*, pp. 85-6; Craig Thompson, *Since Spindletop: A Human Story of Gulf's First Half-Century* (Pittsburgh: Gulf Oil Corporation, 1951), p. 70; Board of Minutes, October 16, 1925, Minute Book #3, p. 325; Robert Sobel, *The Age of Giant Corporations: A Microeconomic History of American Business, 1914-1970* (Westport, Conn.: Greenwood Press, 1972), p. 125.

58. John L. Kelsey, "A Financial Study of the Sun Oil Company of New Jersey From Its Inception through 1948" (M.B.A. Thesis, Wharton School, University of Pennsylvania, 1950), p. 14.

59. Wilkins, *Maturing of Multinational Enterprise,* pp. 113, 116-7.

60. See file, "Foreign Company," Acc. #1317, EMHL, J. Howard Pew Presidential Papers, Box 57.

61. George Sweet Gibb and Evelyn H. Knowlton, *History of the Standard Oil Company (New Jersey),* Volume II, *The Resurgent Years, 1911-1927* (New York: Harper & Brothers, 1956), pp. 384-6; Edwin Lieuwen, *Petroleum in Venezuela*, Volume 47, University of California Publications in History (Berkeley: University of California Press, 1954), p. 30; Wilkins, *Maturing of Multinational Enterprise,* p. 115. In practice, Gomez awarded concessions to political favorites who then sold them to American companies at large profit. A 15,000-hectare lease (37,065 acres) under the 1918 law brought as much as $30,000 to one of these middlemen. Much of the money flowing into the Venezuelan treasury from the oil companies for leaseholdings and royalties eventually found its way into the pockets of Gomez and his associates.

62. Gibb and Knowlton, *Resurgent Years*, pp. 386-7; Lieuwen, *Venezuela*, pp. 30-2.

63. Board Minutes, Sun Oil Company, January 25, 1921, Minute Book #3, p. 80; Lieuwen, *Venezuela*, p. 44; "With 'The Sun' in Venezuela," *Our Sun* 3, 9 (July 31, 1926), p. 3; Wilkins, *Maturing of Multinational Enterprise*, note 51, p. 507.

64. Liewuen, *Venezuela,* p. 44.

65. *Ibid.*; "With 'The Sun,'" *Our Sun*, pp. 3-7.

66. "Broker Prospectus, 1930," pp. 7-8; *Annual Report of the Sun Oil Company,* 1928, p. 1.

67. "Sun Ship's 30 Years," *Our Sun* 12, 2 (July/August 1946), p. 5; "Tankers Under the Sun Flag," *Our Sun*, p. 35; "Tankers Built for the Sun Oil Company by the Sun Shipbuilding and Drydock Company, 1917-1938," Acc. #1317, EMHL, J. Howard Pew Papers, Box 5.

68. "Tankers Under the Sun Flag," *Our Sun*, p. 35; Folder, "Diesel Engines," Acc. #1317, EMHL, File 21-A, Box 268.

69. "Sun Oil Company, Stock Owned in Subsidiaries, March 6, 1930, Acc. #1317, EMHL, J. Howard Pew Papers, Box 58.

70. *Ibid.*; "Our Company, Its History, Scope and Facilities," *Our Sun* 2, 12 (September 30, 1925), pp. 4-5; Board Minutes, Sun Oil Company, December 31, 1929, Minute Book #4, p. 266.

71. "In the Oil Fields With Sun," *Our Sun*, p. 32; Daniel J. Bergen, "Oklahoma," *Our Sun* 13, 7 (July 1947), p. 4.

72. Giddens, *Oil Pioneer*, p. 525; Leonard M. Fanning, *The Story of the American Petroleum Institute* (New York: Leonard M. Fanning, Editor of *World Petroleum Politics*, 1959), p. 67.

73. *Poor's Manual of Investments, Industrials,* 1915-1920; Board Minutes, Sun Oil Company, December 4, 1919, Minute Book #3, pp. 13-4.

74. Kelsey, "Financial History," p. 13; Williamson et al., *Energy*, p. 305; Board Minutes, Sun Oil Company, January 25, 1921, Minute Book #3, pp. 60-1.

75. Board Minutes, Sun Oil Company, November 7, 1922, Minute Book #3, p. 109; December 5, 1922, Minute Book #3, pp. 115, 122; Franklin Waltman to Robert L. Klaus, September 7, 1961, Memo, *Our Sun* 75th Anniversary Issue, Folder #2, "Sun Oil History," Library File, Sun Oil Company Library, Marcus Hook, Pennsylvania, p. 18.

76. Board Minutes, Sun Oil Company, May 7, 1923, Minute Book #3, pp. 172, 177; June 25, 1923, Minute Book #3, p. 209.

77. "Facts About the Sun Oil Company," *Our Sun* 1, 12 (September 1924), p. 2; Board Minutes, Sun Oil Company, August 30, 1924, Minute Book #3, pp. 266-7; "Sun Oil Company: Statement of Financial Structure," prepared for Temporary National Economic Committee, Acc. #1317, EMHL, J. Howard Pew Papers, Box 45.

78. Frank Cross to J.N. Pew, Jr., June 27, 1925, Acc. #1317, EMHL, File 21-A, Box 269; Board Minutes, Sun Oil Company, May 26, 1925, Minute Book #3, p. 290; June 8, 1925, Minute Book #3, p. 294; "Inside Stories of Great Corporations, Sun Oil," Wall Street News (December 7, 1925), p. 3, clipping in Folder #2, "Sun Oil History," Library File, Marcus Hook; Waltman to Klaus, *Our Sun*, p. 18; "Inside Stories, Sun Oil," *Wall Street News*, 1925, p. 3.

79. Board Minutes, Sun Oil Company, May 16, 1927, Minute Book #4, p. 78; January 28, 1930, Minute Book #5, p. 4.

80. For an excellent analysis of welfare capitalism in this country, see Stuart D. Brandes, *American Welfare Capitalism, 1880-1940* (Chicago: University of Chicago Press, 1976). Another recently published and very useful work is Daniel Nelson, *Managers and Workers: Origins of the New Factory System in the United States, 1880-1920* (Madison: University of Wisconsin Press, 1975).

81. "Proposed Employees Stock Purchase Plan," June 24, 1925, Acc. #1317, EMHL, File 21-A, Box 85; Employees Booklet, "Sun Oil Company Stock Purchase Plan," Acc. #1317, J. Howard Pew Papers, Box 58; "Announce Stock Purchase Plan for Employees," *Our Sun* 3, 8 (June 30, 1926), p. 3; Kelsey, "Financial Study," p. 15.

82. Memo, "Stock Purchase Plan," June 8, 1926, Acc. #1317, EMHL, J. Howard Pew Papers, Box 58.

83. Kelsey, "Financial Study," pp. 88-9; "Holders of Common Stock, May 25, 1928," Acc. #1317, EMHL, J. Howard Pew Papers, Box 58.

84. Board Minutes, Sun Oil Company, December 14, 1927, Minute Book #4, p. 139.

85. See "I Remember When; Marcus Hook Refinery Oldtimers Recall," *Our Sun* 16, 4 (Autumn 1951), p. 19; "Recreation Has its Part in Company Activities," *Our Sun*, 50th Anniversary Issue, 3, 1 (1936), pp. 44-6.

86. "Robert C. Pew Dies at Toledo," *Our Sun* 2, 9 (June 30, 1925), pp. 2-3.

87. "First Fifty Years," *Our Sun*, p. 8; Waltman to Klaus, *Our Sun*, p. 19.

88. Waltman to Klaus, *Our Sun*, p. 19; "Sun Oil," *Fortune,* 1941, p. 53; "Story of Sun," *Our Sun*, p. 23.

89. John G. Pew, "Fifty Years of Industrial Relations," in *Spindletop, Where Oil Became an Industry* (Beaumont, Texas: Spindletop 50th Anniversary Commission, 1951), pp. 121-2; Jno. G. "Jack" Pew, interview held at Dallas, Texas, November 18, 1975.

90. Fanning, *American Petroleum Institute,* pp. 58-60. For a discussion of the origins and

development of corporate public relations during this period, see Richard S. Tedlow, "Keeping the Corporate Image: Public Relations and Business, 1900–1950" (Ph.D. Dissertation, Columbia University, 1976). Also of value is Tedlow's article, "The National Association of Manufacturers and Public Relations During the New Deal," *Business History Review* 50, 1 (Spring 1976), pp. 25–45.

91. Folder, "A.P.I. Public Relations," Acc. #1317, EMHL, File 21-A, Box 20; Fanning, *American Petroleum Institute,* pp. 112–113.

92. Fanning, *American Petroleum Institute,* p. 113.

Figure 1: Joseph Newton Pew, Sr, the founder of the Sun Oil Company (courtesy Sun Company).

Figure 2: J. Howard Pew (courtesy Sun Company).

Figure 3: Joseph Newton Pew, Jr. (courtesy Sun Company).

Figure 4: J. Edgar Pew (courtesy Sun Company).

Figure 5: John G. Pew (Courtesy Sun Company).

Figure 6: Jno. G. "Jack" Pew (courtesy Sun Company).

Figure 7: Arthur E. Pew, Jr. (courtesy Sun Company).

Figure 8: View of Spindletop field looking east (circa 1902–1903). These derricks mushroomed literally over night (courtesy Sun Company).

Figure 9: Marcus Hook refinery (circa 1901–1904). The photograph shows early coal-fired batch stills at left (courtesy Sun Company).

Figure 10: Launching of the *Chester Sun*, the first vessel built at the Sun Shipyard, October 30, 1917 (courtesy Sun Company).

Figure 11: Early Sun service station marketing Blue Sunoco, the "knockless highpowered fuel at regular gas price," in the late 1920s (courtesy Sun Company).

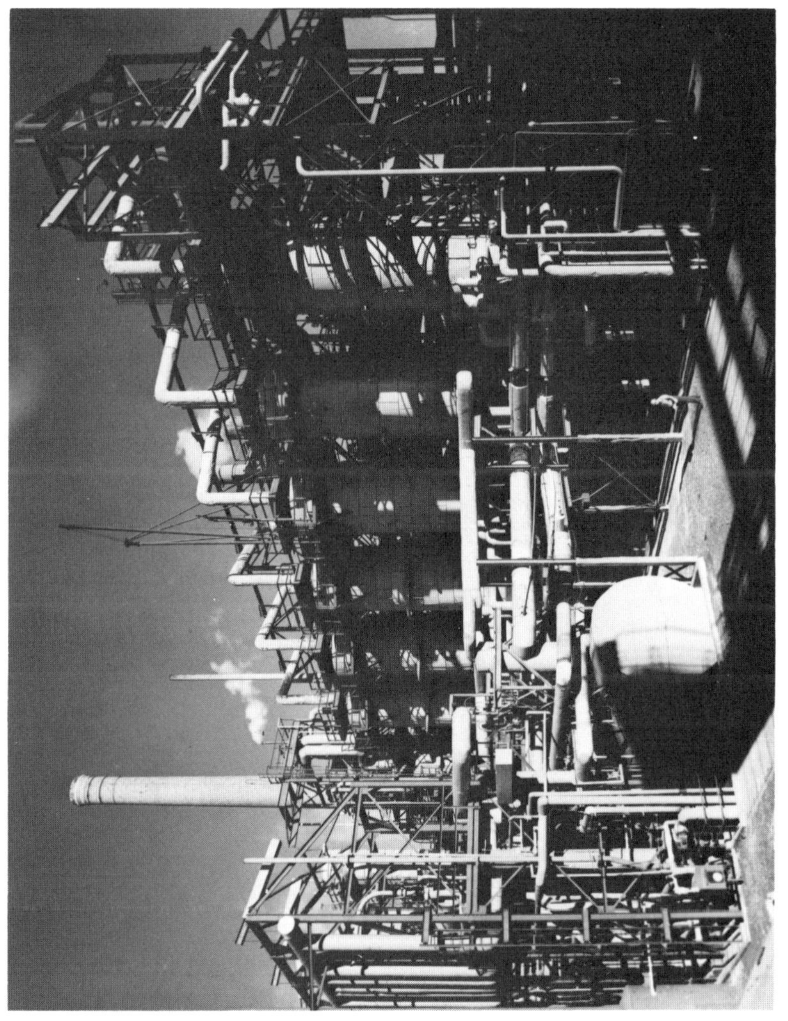

Figure 12: Houdry catalytic cracking plant 11-4, Marcus Hook refinery (courtesy Sun Company).

Figure 13: Dedication of Marcus Hook aviation gasoline plant, October 27, 1943. Lowell Thomas is at the microphone with J. Howard Pew and Harold Ickes at his left. Left inset is of Pew, right is of Ickes (courtesy Sun Company).

Figure 14: Sun Oil Company executives, 1945. Seated left to right are Jno. G. "Jack" Pew, J. Edgar Pew, Frank Cross, and J. Howard Pew; standing from left to right are Robert G. Dunlop, Frank S. Reitzel, Walter C. Pew, Clarence H. Thayer, Samuel B. Eckert, and Joseph N. Pew, Jr. (courtesy Sun Company).

CHAPTER V

Business and Government in the 1920s: Sun Oil and the Emergence of a New Oil Policy

The Sun Oil Company and members of the Pew family were in the forefront of major policy struggles within the petroleum industry in the 1920s. These struggles arose over how the industry and government should respond to problems related to the economics of oil production. The "feast or famine" resulting from the cyclic nature of oil field discoveries had long characterized domestic crude production, but in the post-World War I era, new demand for petroleum products placed a greater emphasis on oil conservation. In Britain, Lord Curzon's frequently repeated statement that "the allies had floated to victory on a sea of oil" summarized what most people now realized. Oil supplies were, and would continue to be, a military necessity as well as an important element in the economic health of nations. The growing popularity of the automobile also created new peacetime demands for petroleum products that placed an even greater strain on known world reserves.

Cooperation was a central element in the main policy debates of the decade. To what extent would oil executives cooperate among themselves and with government in mutual programs of foreign exploration, oil conservation, and industry regulation? The differing positions industry members took on these issues reflected the fact that the petroleum business was composed of many elements. Different groups perceived problems in different ways. Large, fully integrated companies, for example, found production curtailment more palatable than did small independent producers who depended entirely on crude sales for their survival. Similarly, stands on other issues often depended on whether the individuals were strictly domestic producers or oil importers; a small integrated firm or a large integrated firm; an independent producer, refiner, or marketer or an integrated one; a former Standard Oil company or an old-line, anti-monopoly independent. Given its history to 1920, Sun was a highly competitive, conservatively managed firm, very wary of government involvement in the marketplace except to protect individual rights against monopoly control.

Sun's management spoke out on the issues in the 1920s, and one individual in particular, vice-president J. Edgar Pew, played a crucial role as an industry-wide leader. As the industry and government at both state and federal levels thrashed out a new public oil policy in the post-monopoly era, Sun's role not only tells much about its own growth and development but offers insight into the larger industry pattern.

THE CONSERVATION ISSUE

The origins of industry problems in the 1920s lay in the pre-World War I period. Leading geologists and technicians offered gloomy estimates of crude reserves and awakened concerns about scarcity that had persisted since the development of the Appalachian pool in the nineteenth century. In 1907–1908, David T. Day, Director of the Petroleum Division of the United States Geological Survey, estimated recoverable reserves in the United States at a minimum of 10 billion and a maximum of 24.5 billion barrels. He later revised his figures downward to 8 billion minimum and 22 billion maximum. In ensuing years, other experts at the Geological Survey lowered these estimates to minimum figures under 7 billion barrels. Assuming the accuracy of these estimates, depletion of known petroleum reserves would have occurred in ten to twenty years at then-prevailing rates of consumption.[1]

As early as 1909, Secretary of the Interior Richard A. Ballinger warned President Taft of the danger of an oil shortage and suggested the development of a federal program to reduce oil field waste. Ballinger first recommended legislative restrictions on the private entry into publicly owned petroleum reserves.[2] Mark L. Requa, a consulting engineer for the Bureau of Mines, later wartime oil administrator and an important figure throughout the twenties, made a dire prediction in 1916. Requa submitted an influential report to the United States Senate arguing that the United States could meet petroleum demands for only the next five to ten years, and that "in the exhaustion of its oil lands, and with no assured source of domestic supply in sight, the United States is confronted with a national crisis of the first magnitude."[3]

Requa himself was representative of a new class of managerial and scientifically oriented men who were becoming involved in the industry. A Californian and a political disciple of Herbert Hoover, Requa was also an exponent of the theories of scientific management propogated by

Frederick Winslow Taylor. More than any other single person, he became the champion of cooperation in the development of oil conservation legislation.[4] The Pews always distrusted Requa and became leaders of the segment of the industry opposed to cooperation with the government in stabilizing the uneven production that usually characterized oil field development.

The move for stabilization in the pre-war period arose during periods of glut rather than scarcity. Following the large Glenn Pool strikes in Oklahoma during the period 1905–1907, many producers turned to government seeking relief from declining crude prices. The Oklahoma Corporation Commission established a minimum price per barrel of crude and the Oklahoma legislature passed laws restricting pipeline operators. Both responses were early precedents for later regulatory activity at both the state and federal levels.[5]

When the Cushing and Healdton fields brought a price-depressing glut in 1914–1915, the Oklahoma legislature took tentative steps toward prorationing legislation, or mandated limitations of amounts of oil that could be drilled from a given pool. Oklahoma, however, was a unique case since there had always been some degree of governmental involvement because of the leasing of lands on Indian territory. Consequently, the oil men in that state had developed a more cooperative attitude and a greater willingness to deal with government agencies.[6]

The Oklahoma "feast" by 1915 had brought enormous amounts of oil to Sun's Twin State subsidiary, and the firm built its Yale, Oklahoma refinery to utilize these supplies. It was the only plant Sun built in the western oil fields, and it ran this refinery as a relatively inefficient "skimming" plant for as long as the Cushing and Healdton fields kept flowing. But in less than a year following the peak of 1915, "famine" again threatened, and a national shortage scare developed. The growing wartime demand for oil largely accounted for this, but Henry Ford's Model T and increasing sales of domestic gasoline also accelerated the consumption of petroleum products. Mid-Continent crude that sold for forty cents a barrel on August 1, 1915, had risen to $2.05 a barrel by March 15, 1916.[7]

Even in this early period, the paradox of gloomy predictions of known reserves on the one hand and high levels of production in new fields on the other pointed to the unstable nature of petroleum production. Conservationists, military planners, and the technicians of the United States Geologic Survey expressed concern over petroleum reserves. Industry advocates of limited production and other types of regulated oil field activity reflected their economic concern for the general price declines accompanying the discovery of new fields. The movement for oil field efficiency became common ground for advocates of the two positions.

The elimination of wasteful practices and the scientific management of oil pools would help preserve crude for future use and would also serve to stabilize the short-term market for petroleum.[8] This goal became the rallying cry of many oil men during the 1920s.

The oil executives most affected by short-run price declines were independent western producers. These non-integrated oil operators, many of them very small, depended on the price of crude for their livelihood. When prices dropped dramatically, their income was severely reduced. The larger, integrated companies, on the other hand, could more easily weather the storm because they could rely on profits from other branches of their business, including pipelines, refining, and marketing. Most of the larger firms also purchased a large share of their crude needs, and in a period of overproduction, they could obtain the amounts they wanted at bargain prices. Sun Oil had been a large purchaser of crude ever since the Spindletop discovery in 1901 and continued to purchase a large share of its crude in the Texas and Mid-Continent fields.[9]

Changes in the competitive structure of the American petroleum industry in the pre-war era also had important consequences for later developments in conservation. The discovery of new western fields after 1900 had promoted the rapid entry of new firms in the industry and had eroded much of the old Standard Oil domination by the time of the dissolution decree in 1911. In a more competitive environment, "independent" companies were no longer preoccupied with the struggle against monopoly, but were themselves faced with the complex problems of overproduction, scarcity, and price stabilization.

The old antagonisms of the independents like Sun, however, died hard. One long-time observer of the industry, Leonard Fanning, recalled that "many old-time Independents would not be caught dead talking to a Standard Oil man."[10] The fragmented trade associations that existed prior to World War I were strongly independent. Sun Oil belonged to the largest of these, the National Petroleum Association (N.P.A.), a group that spoke primarily for the independent interests and had been aligned against the Standard Oil Company for many years. The *National petroleum News*, edited by the influential Warren C. Platt, began publication in 1909 and had been in the forefront of the attack against Rockefeller. Although this organ had tempered its tone since 1911, it still reflected the views of the smaller firms.[11]

The success of conservation and stabilization programs aimed at eliminating waste and insuring long-term supplies of oil would depend on the degree of cooperation that the oil companies could effect, both among themselves and with government. On the eve of World War I, however, the industry remained a house divided.

THE WARTIME EXPERIENCE

In many respects, Sun Oil's modern era began during and immediately after World War I. The record profits the firm made in sales of lubricants and fuel oils provided it with the capital to buy out the United Gas Improvement interest in Sun and enabled the company to launch its marketing of gasoline and motor oils. The Sun Ship Company was a creature of wartime expansion, and both its business and that of the Sun Oil Company blossomed under the auspices of federal war agencies. Yet the Pews, unlike some oil men during the war, were not comfortable with the regulations and directives emanating from Washington. After experiencing this arrangement, J.N. Pew, Jr. wrote to a friend in 1918 that "Oil [is] doing well; [it is the] only industry that has regulated itself regarding prices and allotment of production. This government regulation is neither desirable or necessary."[12]

Yet, as Gerald Nash argues, the war experience solidified a victory for the "New Nationalism" espoused earlier by Theodore Roosevelt. The tremendous demand for petroleum created by the war brought the industry and government together in a mutual effort to maximize the production of crude and its products.[13] The spirit of cooperation developed during the war spilled over into the twenties, and some individuals continued to advocate increasing cooperation in the post-war era.

Even before the entry of the United States in the war, President Wilson had taken steps toward preparedness by establishing a Council of National Defense in April 1916. A.C. Bedford, Board Chairman of Jersey Standard, became chairman of an Oil Advisory Committee that served under Bernard Baruch of the Council of National Defense. This advisory committee would help coordinate the production and distribution of petroleum and its products. It functioned until July 1917, when its personnel shifted to a new committee under the War Industries Board.[14]

In December 1917, the committee formally became the National Petroleum War Service Committee (N.P.W.S.C.), which functioned throughout the war. J. Howard Pew, while president of Sun Oil, became the youngest member of this group, which eventually had thirty-nine members representing twenty-one companies and eight associations. Meanwhile, A.C. Bedford moved to the vanguard of those advocating cooperation and addressed his theme to various elements of the industry. To the Western Petroleum Refiners Association in Kansas City, a group composed mostly of old opponents of the Standard Oil trust, he urged that "We must keep our eyes on the goal of a still more complete and wholehearted cooperation, of a more perfect coordinated unity of aims and methods."[15]

While the industry moved toward internal cooperation, President Wilson acted to create more formal regulatory authority in Washington. In August 1917, the President created the United States Fuel Administration under authority of the Lever Act, a sweeping emergency law giving the executive branch authority over foodstuffs, fuel, fertilizers, and farm implements. In January 1918, Wilson established an Oil Division of the Fuel Administration with Mark L. Requa at its head.[16]

Requa supervised some of the most significant experimentation in industrial planning of the war. He fostered committees to regulate and stimulate production in the oil regions, encouraged pooling operations and price stabilization, and sponsored the development of scientific drilling, production operations, and refining. Requa gratified the industry by backing liberalization of the oil depletion tax allowance first granted by the Revenue Act of 1913, and by blocking a profiteering investigation by the Federal Trade Commission in 1918.[17]

The exigencies of the war brought many of the industry's leaders around to a spirit of cooperation. A Warren C. Platt editorial in the *National Petroleum News* hailed the "new era" in the oil business, and the National Petroleum Association supported resolutions favoring cooperation over competition at its 1918 meeting.[18] At the time of the Armistice, many felt that they could not go back to the pre-war conditions in the industry. A.C. Bedford articulated this view at a Tulsa meeting: "If it is so easy for us to work together for our country's sake, I cannot see why, after the war is over, we cannot work together for the good of the industry itself."[19]

THE AMERICAN PETROLEUM INSTITUTE

Two aspects of the cooperation issue had emerged during the war. Intra-industry cooperation, embodied in the National Petroleum War Service Committee, proved attractive to the majority of oil men in wartime, and there was support for extending these efforts into the post-war era. Cooperation with government regulatory agencies, however, proved more difficult to swallow in peacetime, and industry executives like the Pews were not interested in preserving or extending government intervention. Others, like Mark Requa, spent the next several years seeking a business-government consensus on the need to stabilize the industry.

Requa and A.C. Bedford became the prime movers in the creation of a new petroleum trade association. They envisioned a permanent body modeled on the American Iron and Steel Institute. On March 24, 1919, the members of the wartime Petroleum War Service Committee voted to abolish that group and to found a new organization, the American Petroleum Institute (A.P.I.). All members of the N.P.W.S.C. and

Mark Requa became founding members and directors of the new association. The first goal—some significant degree of intra-industry cooperation—was realized. J. Howard Pew was a founder and director of the A.P.I. and actively supported its efforts over the next several years.[20]

On the related issue of government regulation, continuity was not so clear. On May 15, 1919, the United States Fuel Administration ceased operations, thus ending the smooth-running controls effected during the war. Requa took a job with the Sinclair Oil Company and continued his crusade for cooperation from within the board of the A.P.I. But tight market conditions in 1919-1920 discouraged the movement for conservation and production controls. Rising crude prices and the absence of new booming fields led many oil men to long for a return to the unregulated, more competitive situation that had existed before the war. For the most part, the government-business cooperation that arose during the war deteriorated rapidly in the immediate post-war era.[21]

THE PEWS: DIFFERING PERCEPTIONS

The two young brothers, J. Howard and J.N. Pew, Jr., had experienced a working relationship with the government during the war and they had not liked it. They were willing to accept emergency mobilization measures, but they believed them to be inefficient, and welcomed a return to the pre-war pattern of unfettered competition.[22] But although they remained skeptical of Jersey Standard's motives and Requa's ambition, they did agree to cooperate with the American Petroleum Institute. This organization rapidly became a spokesman for the larger, integrated firms. From within the A.P.I., however, the Pews represented that group opposed to government regulation of the industry.

Just as significant as their philosophical opposition to business-government cooperation was the Pews' perception of the oil industry. J. Howard and J.N., Jr. were Easterners, largely schooled in the refining and marketing end of the business. They were aware of the unstable nature of crude production, but their experience indicated that the free marketplace would ultimately take care of any short-term economic disequilibrium. As J. Howard later commented: "My father was one of the pioneers in the oil industry. Periodically ever since I was a small boy, there has been an agitation predicting an oil shortage, and always in succeeding years the production has been greater than ever before."[23]

The older cousin, J. Edgar Pew, however, had already developed a different position on the fundamental issue of production controls and regulation. Although he too was skeptical of regulation, J. Edgar was more sensitive to the problems of production because of his greater experience in that facet of the business, and his personal relationships with Texas and Mid-Continent producers.

J. Edgar had only recently rejoined Sun in 1918 after his stint as an independent Oklahoma producer and an executive with the Carter Oil Company, a Jersey Standard subsidiary. While with Carter, he had helped found the Mid-Continent Oil and Gas Association, and he served as its second president in 1917.[24] During the war, he also served as a member of the Advisory Committee for Mid-Continent production under the National Petroleum War Service Committee. In January 1919, Pew served on a committee of the Texas Division of the Mid-Continent Oil and Gas Association investigating the relationship of the Texas Railroad Commission to conservation practices. The committee reported that "entirely too many wells have been drilled too near property lines and entirely too many wells drilled on individual leases."[25] The Texas legislature had given some regulatory power to the Railroad Commission in 1917 and extended its authority to enforce oil conservation practices in 1919.[26] Later, the commission evolved into the powerful state regulatory agency that it is today.

J. Edgar Pew had been one of the pioneers of the Spindletop field and had remained in Texas to head Sun's western division in 1901. His later experiences both outside and within the Sun Oil Company centered in the wide open western oil fields. He had become a Texan rather than a Pennsylvanian, and his vast experience had made him an industry leader in the southwest. An ironic testimony to his high standing came during the fall of 1917 when he was still with Carter Oil in Tulsa. In an unsuccessful effort to strike a blow at the American war effort, members of the "Wobblies" (Industrial Workers of the World) planted a dynamite bomb under the front of Pew's home. The perpetrators testified in court that they could think of no better way to sabotage the oil industry than to eliminate Pew.[27]

Another aspect of J. Edgar's background contributed to making him more a disciple of cooperation than his cousins. J. Edgar Pew is generally acknowledged as the "father of standardization" in oil field equipment. Prior to 1920, there were no standard sizes of pipe, thread sizes differed from firm to firm, the size and height of drilling and pumping derricks varied, and there were no industry-wide standards for wire, rope, and a range of other equipment and machinery. The multiplicity of sizes raised costs and sowed confusion among oil operators.[28] Oil was by no means unique in this regard, for much of American industry was only beginning to become interested in basic standardization programs. What is significant, however, is that J. Edgar Pew's support of standardization suggests that he was closer to the rationalization and scientific management ideas represented by Mark Requa than were his Philadelphia cousins.

J. Edgar Pew conceived a standardization plan, took it first to the Texas group of the Mid-Continent Oil and Gas Association, and then sold it to the Kansas-Oklahoma Division in 1919. He chaired a joint committee of these groups which presented a final report to the A.P.I. meeting in 1922. This report listed all oil field equipment then in use and recommended standardized sizes for each. The A.P.I. adopted the report and named J. Edgar Pew to head a General Committee on Standardization, which he chaired until his death in 1946.[29]

The key to the standardization program was cooperation among oil men and equipment manufacturers. Standardization became an important life's work to J. Edgar, and in the process he naturally developed an appreciation of the problems faced by people in various branches of the industry. His work in standardization helped launch him to the presidency of the A.P.I. in 1924, and prompted the *National Petroleum News* to include him "in the front rank of the progressive element, always eager to advance the technique of the art."[30] Advocates of conservation-related production regulation in the twenties definitely did not think of J. Howard and J.N. Pew, Jr. when they pondered "the progressive element."

FOREIGN EXPLORATION

When the Wilson administration dismantled the apparatus of the war effort in 1919, the direction of government policy toward the oil industry remained generally unclear. There was, however, one area that was an exception to this state of affairs. This was the active encouragement and assistance given by agencies of the federal government to American companies engaged in the exploration of foreign oil fields. These activities grew directly out of the world-wide panic over oil depletion following the war. Moreover, the publication of the Anglo-French San Remo Agreement of April 1920 convinced government and oil industry leaders that the Europeans were attempting to freeze the United States out of promising Middle Eastern reserves.[31]

This agreement was part of a general plan to carve up the Middle East between the victorious European powers. Mark Requa had attempted to introduce the Wilsonian principle of the "Open Door" into oil discussions at the Versailles Conference but had failed. Americans ignored the facts that their country controlled the lion's share of known reserves and that the United States had turned its back on international cooperation by rejecting the Versailles Treaty. Instead, many viewed the Anglo-French policy as naked imperialism and a threat to America's long-term ability to supply its energy needs.[32]

Early in 1919, Requa, Van H. Manning of the Bureau of Mines, and George Otis Smith had urged the formation of a government-backed, world-wide exploration company. In May 1920, Senator James Phelan, Democrat from California who had long represented oil interests, introduced a bill to create the "United States Oil Corporation," a government-sponsored private company to engage in all aspects of the petroleum business on a global basis. Although it did not pass, the bill demonstrated the thinking of many government officials at the time. In early 1921, new Secretary of Commerce Herbert Hoover joined with Secretary of State Hughes in "urging oil companies to acquire all territory in South America and elsewhere before European countries preempted all of it."[33] Hoover's view was particularly significant because he represented an element of continuity between the Wilson and Harding Administrations, having served the previous Democratic president as director of Belgian Relief and of the United States Food Administration.

In August 1920, J. Howard Pew was invited to a dinner for oil men hosted by Requa at the New York Bankers Club. In addition to Pew, Requa had summoned several executives, including Walter Teagle and A.C. Bedford of Jersey Standard, C.W. VanLaw of Sinclair, Henry Doherty of Cities Service, and R.D. Benson of the Tidewater Associated Oil Company, among others. Pew was unable to attend, but he received a copy of Requa's memorandum the following day from Bedford. This document outlined the proposed incorporation of "an exploration and development [company] to operate in foreign oil lands."[34]

J. Howard wrote Requa on August 14 indicating Sun Oil's interest in the project and a willingness to discuss the proposal further. But he also demonstrated caution by voicing an insistence on safeguards for the smaller companies that did not already have prior investments in foreign property. Aware of efforts that Jersey Standard had already made in Mesopotamia (Iraq), Pew suggested that the participants in any joint company not be allowed to obtain independent leases in areas where the group operated. Moreover, he maintained that territory already controlled by individual companies in areas of joint exploration should be turned over to the new corporation or the joint company should confine efforts to virgin territory.[35] Pew's lack of enthusiasm for the project reflected his own skepticism concerning the petroleum shortage and his long-standing suspicion of Jersey Standard's motives. Sun could not ignore the open opportunity to engage in foreign exploration, but its attitude was lukewarm.

Requa's original decision to include Sun, and later even smaller firms, seemed to be an attempt to attract government support through broader participation. A syndicate including smaller units better represented the principle of the "open door" than a consortium controlled by Standard (New Jersey). There is evidence to indicate that this concern played a role in Standard's executives' thinking as well. Jersey still had a tarnished public image and had to be concerned about playing down the monopoly label. In 1921, E.J. Sadler of Jersey wrote to Walter Teagle that in order to gain State Department support of the firm's efforts in Mesopotamia, "it will be necessary to take some other interests with us and a part of whom, at least, should be outside the subsidiaries."[36]

In the fall of 1920, Requa presented his views on foreign exploration in a series of articles written in the *Saturday Evening Post;* at the same time, plans continued for creation of the "Foreign Company."[37] J.N. Pew, Jr. served on the general planning committee, and Sun attorney Francis McIlhenny worked on a legal proposal for the corporation with Ira Jewell Williams, counsel for the Atlantic Refining Company. At this stage there were twenty larger companies, all A.P.I. members, cooperating in the venture. Among the most important issues discussed in committee were potential antitrust problems, the tariff on imported oil, and the method of dividing ownership in the proposed joint firm.[38]

Sun argued for an equal management voice for all parties involved, and opposed a Requa plan to designate one "founders' share" to be divided up among very small companies. Here Sun sided with Jersey Standard and the other large firms in making the venture a project undertaken by the major integrated companies. Sun had long considered itself an independent, anti-monopoly firm, but in the new industry structure following the 1911 dissolution and the war, it was becoming one part of a new oligopoly of about twenty fully integrated companies. This was the group for which the A.P.I. spoke.[39]

As the Pews expressed a growing interest in the foreign company, however, they grew more skeptical of Requa's role in the project. They increasingly began to view the Californian's many schemes as ways to create a job and make money for himself. Moreover, the Pews had a deep distrust for the new managerial class which emerged in the industry following the war, a category into which Requa fit perfectly.

Requa epitomized the scientific managers of the time who were preaching the gospel of rational control of economic expansion and voluntary cooperation to this end. Opposed to the perceived American tradition of *laissez faire* (that most of them surely saw as dead by 1920), the key to their approach was the use of scientific expertise to improve socioeconomic conditions. In addition to this group of scientific managers, there was also another category of business executive rising to the top of the corporate ladder in the twenties. Unlike the entrepreneurial owner-managers of the past, these men had trained in the nation's professional business schools and were rapidly becoming an important new white-collar class in American society. The Pews were suspicious of individuals representing both poles of this new talent pool.[40]

In addition to these more general concerns, J.N. Pew, Jr. related to his brother a growing opposition among the oil companies to Requa's becoming president of the proposed joint exploration company. He cited Requa's connections with the Sinclair Company, a firm "in rather bad repute with the public," and suggested making Requa Board Chairman with a president as real acting head.[41]

The founders suggested many names for the new joint venture, including J. Howard Pew's entry of the "American Oil Mining Company." In the spring of 1921, the planning committee decided on the "American Oil Producers Company," and drew up a proposed certificate of incorporation. The proposal called for a capitalization of 100,000 shares of par $100 common stock, of which each of the twenty founders would subscribe 5,000 shares.[42] The wording of the agreement indicated that Mesopotamia would be the first target of exploration but "thereafter, at the discretion of the Board of Directors, the company will proceed with like activities in other countries."[43] They focused on Mesopotamia for the obvious reason that there were already proven reserves of petroleum there. But the decision also reflected an American business and government interest in breaking the Anglo-French monopoly as well as the desire of Jersey Standard to gain support for negotiations it had already begun independently.

Secretary of Commerce Hoover lent his enthusiastic support to the plan in May 1921, and prospects looked favorable for the project. However, the cooperative spirit of the spring deteriorated over the summer months. Atlantic Refining and other former Standard companies were uneasy about possible antitrust difficulties stemming from the joint venture. There was discussion of a special congressional bill to grant antitrust immunity for the proposal, but Sun and other former "independents opposed this tactic. They were wary of monopoly criticism and concerned that the government might get too deeply involved.[44] Standard Oil of New York adopted the position that only firms with established interests in the area should be included in any syndicate, a view that most benefited them and Jersey Standard. The seven largest firms realized that the "American Oil Producers Company" would most likely aid competitors. In September 1921, E.J. Sadler wrote Walter Teagle of Jersey Standard an internal memo in which he declared that "any success in Mesopotamia would result in bringing all these people into competition with us in the Mediterranean and that the association is highly undesirable except to gain support of the State Department."[45]

Hoover summoned a conference of oil companies in November 1921, at which he learned of the divided industry opinion on the Requa project. The conference put an end to the original joint company venture, and the story of American efforts overseas then shifted to the attempts by Jersey Standard and the State Department to negotiate with the Turkish Petroleum Company and the British Government for concessions. These approaches did succeed, and Jersey ultimately led six other companies into opening the Middle East to American oil interests.[46]

With the failure of the foreign corporation proposal of 1920–1921,

Sun's potential involvement in Middle East oil and politics came to an abrupt end. The company did not pursue independent exploration in the area and did not even begin to purchase Middle Eastern crude until the 1950s, when it became the last major American firm to do so. Initially skeptical of Requa's proposal, the Pews had become willing to participate in a joint venture with a limited financial risk on their part. Although J. Howard and J.N. Pew, Jr. believed the shortage scare probably to be unfounded, they knew that adequate reserves were important for their overall business strategy, and they initially cooperated with the other members of the A.P.I. and the government in the plan. In many respects, Sun had much to gain and very little to lose in the venture. As it turned out, Standard (New Jersey), Standard (New York), Standard (California), Gulf, and Texaco realized this and were mainly responsible for vetoing the proposal.

Despite the support given by Washington at the diplomatic and consular level, Sun discovered that foreign exploration and development was a risky business. The Pews' experience in Venezuela had confirmed this.[47] More significantly, new domestic strikes by 1924 had removed the fear of imminent shortage and the industry had once again gone into a period of glut rather than shortage. But Sun's early negative experiences with the foreign exploration company and with Venezuelan exploration at least partially explain why the firm remained a strictly domestic producer for so long.

Sun had one other contact with the government on the subject of foreign exploration and development. The Bureau of Foreign and Domestic Commerce, a division of Hoover's Commerce Department, took an active interest in encouraging private oil companies to invest abroad during the twenties. The Bureau sent a series of confidential reports compiled from diplomatic and consular information to them in hopes of encouraging foreign exploration. The communications sent to Sun during the decade are filed in the company records "for eyes of J. Howard Pew and J.N. Pew, Jr." One typical letter written in 1922 by J.K. Towles, acting chief of the Bureau of Foreign and Domestic Commerce, contains detailed information on concessions granted, financial arrangements, success of drilling operations, and possibilities for entry of U.S. companies in Venezuelan fields. In one area of business-government cooperation, a consensus was reached very early in the decade.[48]

DOMESTIC OIL PRODUCTION

The new oil strikes of the early twenties did remove the immediate pressure for foreign exploration, but they reawakened the movement for scientific management of pools and elimination of oil field waste. When oil producers refused to adopt conservation measures on a volunteer basis, some individuals advocated that government effect compulsory regulation of oil production. Chief among this group were the ubiquitous Mark Requa, still lobbying for cooperation, and Henry Doherty of Cities Service Oil Company. Among those who became leading spokesmen against compulsory regulation and federal involvement in the industry was J. Edgar Pew.

Like Pew, Doherty had been opposed to the continuation of regulation immediately after the war, but his views changed with the discovery of the Huntington Beach, Signal Hill, and Santa Fe Springs strikes in California beginning in late 1920. Doherty had heavy investments in the California fields and saw the price of crude plummet in the face of unrestricted production. From that point on, Doherty became a crusader for oil field utilization and a federal conservation program.[49]

Unit pool development, or unitization, was a movement to apportion production in a given field according to some rational, pre-arranged plan. By operating each underground pool as a cooperative unit, the wasteful law of capture would be replaced by orderly, systematic production, which would, proponents hoped, not lead to extremely low crude prices. Doherty proposed a plan for compulsory unit production administered by a series of oil districts run under state law. The owner or company holding the majority leaseholding in the pool would head the operation of each unitized pool, and each operator would pump oil according to his percentage of the entire holding.[50]

J. Edgar Pew opposed the plan because it was compulsory. He was not against voluntary agreements to limit production but drew the line at governmental involvement. His cousins, as we have seen, opposed the idea of any artificial restraints on production and preferred that the free market determine price. Small independent producers opposed any plan to restrict production on the grounds that restrictions like unitization favored the largest producers and integrated companies who could afford to hold oil off a depressed market or stop production altogether. These independents would be cutting off their only source of income if they stopped drilling, and oil sold at a low price was better than no sale at all.[54]

In 1923, the A.P.I. appointed a committee to investigate Doherty's plan but rejected it as both impractical and unconstitutional. William S. Farish and Walter Teagle of the Jersey organization lent sympathy if not support to the proposal, but Amos L. Beaty of the Texas Company attacked unitization as an infringement of the legal right of "law of capture."[52]

Frustrated by the industry's failure to support his program in 1923–

1924, Doherty took the bold step of taking his case directly to his friend and fellow Republican, President Calvin Coolidge.[53] He had violated the canons of the industry by appealing to Washington, but he later defended his actions in the following way:

> To bring about stabilization of the oil business, I worked behind closed doors for nearly four years but got no place. I repeatedly warned my associates that if they refused or failed to do something, I would consider it my duty to take the matter up with the Federal Government. This I finally did and I notified all of the officers of the American Petroleum Institute when I did this and each of my fellow committeemen.[54]

At the time of his White House visit in the fall of 1924, however, many of his colleagues considered Doherty's action a betrayal of the industry.

The years 1923 and 1924 were difficult for the oil industry. The LaFollette Senate Committee investigating pricing in the oil industry published its report in 1923. The report generally indicted the industry and concluded that its competitive structure in 1920–1923 differed little from the pre-1911 era of Standard Oil control. The senators pointed in particular to evidence of collusion among former Standard companies, an arbitrary pricing system, extravagant profits, and a great deal of waste. LaFollette had also sided with Gifford Pinchot and other conservationists on the oil reserve dispute, and his probing led to the disclosures of the Teapot Dome and Elks Hill oil scandals.[55]

In October 1923, a Senate subcommittee under Senator Thomas Walsh of Montana began the intensive investigation that ultimately led to the revelations of bribery and political corruption surrounding the transfer of naval oil leases to Edward Doheny and Harry Sinclair.[56] In addition to revealing the sins of the Harding administration, the Teapot Dome scandal put the oil industry on the defensive. It was at this time that Doherty decided to approach Coolidge with his plan for compulsory unitization.

THE FEDERAL OIL CONSERVATION BOARD

Hubert Work, Coolidge's Interior Secretary, had been urging the President to act on oil conservation for some time. Influenced by Doherty, Work, and an awareness of public opinion against the oil industry, Coolidge created the Federal Oil Conservation Board (F.O.C.B.) by executive order on December 18, 1924. A creature of the executive branch, the Board consisted of the Secretaries of Interior, Commerce, War, and Navy. It was empowered to investigate and make recommendations for oil conservation legislation at both the state and federal levels.[57] Secretary of Commerce Hoover later wrote in his memoirs that Coolidge hoped to "pick up the pieces after the Fall-Doheny-Sinclair scandals of the Harding administration" with the establishment of the F.O.C.B.[58] No doubt this was an important consideration, but the Doherty-Requa plea for the end of ruinous cutthroat competition also influenced this first tentative entry of the federal government into the peacetime oil industry.

At this pivotal point in the decade, the A.P.I. held its annual meeting in Fort Worth, Texas, in December 1924, and J. Edgar Pew became a leading participant in the course of events. Doherty's opponents had thwarted him by keeping the compulsory unitization issue off the Fort Worth agenda. Incensed, he hired a hall outside the convention to address the members. Judge Amos L. Beaty, a leading opponent of the plan, countered by renting a local theater to present his views. J. Edgar Pew later related that "each had a chip on his shoulder" at the time. At the A.P.I. general session, J. Edgar Pew introduced a motion that the full A.P.I. Board should meet again in Colorado Springs that January or February for the express purpose of providing Henry Doherty a fair hearing for his plan. The proposal passed and Doherty and Beaty both agreed to call off their extracurricular sessions. The following day, December 10, the convention elected J. Edgar Pew the second president of the American Petroleum Institute.[59] Faced with a serious rift in the organization, the A.P.I. opted for someone who appeared able to fill the role of compromiser.

Pew was very well known because of his ties to the old Pennsylvania and Ohio industry, his pioneering efforts in the Spindletop and Mid-Continent fields, and his leadership of the standardization movement. He appeared to be an advocate of industry cooperation. In 1924, however, J. Edgar feared the expansion of government intervention that compulsory unitization promised. His opposition stemmed not only from the philosophical position he shared with his Philadelphia cousins but from his knowledge that the independent western producers would not cooperate with the program.[60]

When Coolidge announced the formation of the F.O.C.B. on December 18, J. Edgar Pew and many other oil men were taken by surprise. Now the threat of federal intervention seemed imminent. Pew immediately cancelled the planned Colorado Springs meeting and summoned a full A.P.I. Board meeting in Atlantic City for early January. He also issued a public statement that the A.P.I. would appoint a special committee to represent the Institute in cooperative meetings with the new Conservation Board.[61]

The F.O.C.B. was an entirely unknown element, and the A.P.I. moved cautiously in early 1925. The day prior to the Atlantic City meeting, J. Edgar met with A.C. Bedford, Jersey Standard chairman, in New York.

Bedford was still the single most powerful man in the industry and Jersey Standard still the dominant firm. Bedford had been the prime mover in the Petroleum War Service Committee and a co-founder of the A.P.I. with Mark Requa. Pew realized that he would need Bedford's support in Atlantic City.

According to Pew's version of their meeting, he suggested to Bedford that "four first-class resignations from the Board should be had" to insure a sympathetic ear in Washington. He referred to Harry Sinclair, Edward Doheny, J.E. O'Neill, and Robert W. Stewart. Sinclair and Doheny were the prime offenders in the Teapot Dome scandal, and O'Neill and Stewart (who was President of Standard of Indiana) had come under scrutiny following disclosures of related unethical business dealings in 1924. Bedford agreed to use his influence to remove the first three, but Bob Stewart was an old friend and a former Standard colleague.[62] The Pews had wanted to remove Sinclair from the Board for some time. In June 1924, J. Edgar had written J. Howard Pew criticizing Sinclair not only for his connections with the oil scandals but for his reputation as "a horse race man and gambler."[63]

J. Edgar Pew and Bedford met again prior to the Board meeting in Atlantic City, and Bedford agreed to support A.P.I. cooperation with the Conservation Board. At the full Board meeting, Bedford moved a resolution creating a committee to go to Washington to offer its services and compile data for use by the F.O.C.B. The motion named J. Edgar Pew as chairman of a "Committee of Eleven" in his capacity as A.P.I. president and empowered him to appoint ten other members.[64]

Pew offered a committee position to Bedford, but he refused, preferring to stay in the background. The Jersey chairman did suggest that Pew appoint Bob Stewart to the committee, but Pew refused. In the belief that this was an attempt to vindicate Stewart by "the Standard interests," Pew held firm during a two-hour argument that evening and an embarrassing confrontation the next day.[65] This incident was only a further reminder that the old Standard Oil connections still remained years after the court-ordered dissolution of 1911.

At the Atlantic City meeting, Doherty utilized the unlimited time allotted him, and he continued his afternoon presentation well into the evening. The next day Amos Beaty argued his side of the issue, a plea for voluntary rather than compulsory limitations on production, and the Board retired to vote. Doherty's plan was decisively defeated; only two men voted for it. Whether they were voting against compulsory unitization or against Doherty for running to Washington is impossible to tell.[66]

After rejecting the Doherty plan for compulsory unitization, the A.P.I. adopted a wait-and-see attitude. They did not have to wait long, for on January 17, 1925, Interior Secretary Hubert Work, F.O.C.B. chairman, sent questionnaires to leading oil men seeking specific information and opinions on a number of issues. Coolidge had declared that his primary concern had been for national security, but that he hoped the Conservation Board could make recommendations for the conservation and stabilization of the oil industry. Work's initial letter posed queries about future shortages, the extent of United States reserves, the future of the refining industry, petroleum exports and imports, and foreign exploration.[67]

Other questionnaires followed, including a specific series of questions on methods of recovering crude petroleum. Much of the cooperative attitude wilted when oil men began to view these inquiries as improper meddling in their affairs. Replies came in very slowly to the F.O.C.B. and reports circulated in Washington and New York that the industry was no longer cooperating. Secretary Work attempted to smooth over any controversy by stating publicly in February that industry cooperation had been "prompt and energetic."[68]

If Sun Oil was at all typical, and apparently it was, industry leaders went to no great trouble to answer either promptly or completely. J. Howard Pew consulted with J. Edgar, by this time back in Dallas, before submitting responses to the questionnaires. J. Howard deeply resented what he saw as an invasion of privacy and unconstitutional interference from the federal government. J. Edgar Pew replied to one telegram from Howard by charging that the questionnaires came "from a bunch of clerks and not the board. . . . Hoover wrote me that the Board had never had a meeting since their first organization meeting."[69]

The "clerks" J. Edgar referred to were staff members of the F.O.C.B. who handled the day-to-day operations of the agency. For the most part, they were conservation-minded individuals who were preparing detailed reports on waste and inefficiency in oil production. They also reflected the membership of the Board itself. Hoover was on record as favoring a less wasteful scientific management of production pools, a position perfectly consistent with his general concern for efficiency. Hubert Work was a Westerner, a physician from Colorado, who had long been identified with the conservation movement. Work had favored the continuation of naval reserves, and he reflected the Army's and Navy's concern about military supplies of crude. The reports eventually compiled by the F.O.C.B. staff were important documents in crystallizing thinking on the conservation argument.[70]

J. Howard Pew did not give the F.O.C.B. very much help in preparing its reports. He did not send an answer to Work's first request on January 17 until June 18, 1925, and his answers were very brief, included no statistics, and were generally uninformative. In his closing statement, he summed up the *laissez faire* attitude that he and a handful of others were beginning to represent: "It is my firm belief that the oil industry is subject to the same national economic laws of supply and demand as is every other industry. Any effort to combat these economic laws with man-made laws, will destroy the very industry you seek to conserve."[71] For J. Howard, the vision of government-regulated production was anathema.

THE COMMITTEE OF ELEVEN

While the F.O.C.B. pondered the meager data it was able to gather, J. Edgar Pew's Committee of Eleven concentrated on preparing its own report. Led by Pew and A.P.I. general counsel R.L. "Bob" Welch, the committee was composed of many important oil executives, including William Farish (Humble Oil), K.R. Kingsbury (Standard of California), George S. Davison (Gulf), and E.W. Marland (Marland Oil and later Governor of Oklahoma). J. Edgar had stacked the committee against the Doherty proposals and the idea of federal intervention. Davison of Gulf was even more adamant in his opposition to regulation than J. Howard Pew.[72]

In an attempt to head off any federal entry into the industry, the committee hastily assembled its report, *American Petroleum Supply and Demand,* during the spring of 1925. J. Edgar Pew formally submitted the report to the F.O.C.B. on August 6, 1925, but a preliminary version had gone to the A.P.I. Board earlier. The main theme of the report was a denial that there was any serious shortage of petroleum reserves. This optimistic study estimated known recoverable petroleum at over 31 billion barrels, and cited over 1.1 billion acres of major reserves. This latter figure was later popularly referred to as the "billion acre reserve." The report listed coal, shale, and lignite deposits as additional sources of oil and emphasized new techniques of deep drilling and petroleum cracking to further allay fear of shortage. But the most controversial claim in the report was a denial of wasteful practices in the production, refining, and distribution of petroleum and its products.[73]

The Committee of Eleven report encountered much criticism. Many petroleum geologists questioned the optimistic reserve estimates and the refusal to admit to wasteful practices.[74] Both Henry Doherty and Mark Requa openly attacked its findings. Doherty wrote to a Dallas publisher soon after the report appeared that "the makers of the report have tried in every way to create the impression that we have an abundance of ground reserves of petroleum but, realizing that they could not prove this, they have attempted to show that we have unlimited supplies of oil which can be secured by the processing of coal."[75] After seeing a copy of this letter, J. Edgar Pew wrote to J. Howard that Doherty "had gone just a little too far."[76]

A few years later, Doherty actually went much farther in an open letter written to the A.P.I. membership. He described the Committee of Eleven report as "so untrue and misleading as to be nothing less than grotesque."[77] Requa wrote to J. Howard Pew in October 1925, expressing his opinions of both the report and the "*laissez faire*" doctrine represented by the Pews. He cited the report as a splendid compilation of facts, but seriously challenged its conclusions. Moreover, he pressed the argument that only cooperation between industry and government could solve the serious problems facing the industry. Requa drew the battle line as he saw it: "The roads diverge. One leads to cooperation, the other to obstruction of all cooperative effort. The industry must elect which one it will follow. I cannot but think the decision will be of vital importance."[78]

J. Howard Pew defended the committee report in a sarcastic reply to Requa. He argued that the nitrates in the soil would disappear, timber reserves become depleted, and the rivers of the world change their course before petroleum reserves were exhausted. Pew also discussed the possibilities of obtaining oil from shale and cited German success in developing synthetic fuels from the hydrogenation of coal. He concluded by emphasizing his belief that "if the government interferes with the oil industry in any way, those who have spent their lives in the development of the most efficient and effective industry in the world, will no longer have the incentive to continue in the business, and I have no hesitancy in telling you that I will be one of the first to seek other activities."[79] The Sun president then apologized for his frankness, citing his close relations with Requa during the war as his reason for speaking his mind. When J. Edgar Pew saw Howard's reply to Requa, he complimented its "very clean-cut, sarcastic arrangement" and suggested that if Requa "possessed any practical sense, he would admit the situation; he hasn't though."[80]

The Pews were suspicious of Requa's personal motives and blamed him and Doherty for introducing the threat of federal intervention in the form of compulsory unitization and the F.O.C.B. Howard suspected that Requa was again job-hunting, as he had done in the foreign corporation episode, and J. Edgar Pew cited information that Requa was a direct influence on F.O.C.B. Chairman and Interior Secretary Hubert Work.[81]

Requa, perhaps the key figure in oil policy debates throughout this entire period, was involved in every major policy struggle and showed an amazing consistency in the policies he advocated. Requa had enjoyed his powerful position in the wartime Oil Division of the United States Fuel Administration, and he had been much impressed with the efficiency achieved through business-government cooperation. In the twenties, he and Henry Doherty became the leading spokesmen for the handful of people who were deeply committed to regulation of the oil industry.

For the most part, advocates of regulation came from the large integrated companies and opponents from the smaller independent producers. But there were outspoken critics of Requa and government regulation from the major companies as well, such as the Pews, Davison of Gulf, and Beaty of the Texas Company. Others, like Walter Teagle of Jersey Standard, saw merit in Requa's position but doubted its practicability. However, the discovery of even more new oil fields soon won converts to regulation, if not to the personal leadership of Mark L. Requa.

Even sympathetic historians of the petroleum industry generally conclude that J. Edgar Pew's Committee of Eleven Report exaggerated the figures on known reserves, given the information and technology available at the time.[82] Yet the discovery of the Seminole field in Oklahoma in 1927 and the East Texas field in 1929–1930 justified, if perhaps by chance, many of the report's conclusions. J. Howard Pew could later argue in good conscience that "there well may have been at the time some criticism of the report—there always is in such matters; but the fact remains that the Committee of Eleven's Report was the first time that a precise effort along somewhat scientific lines had been made to determine oil reserves and resources, and that its forecasts have been fully sanctioned over the subsequent years."[83]

THE F.O.C.B. REPORT

Althouth Interior Secretary Hubert Work, F.O.C.B. Chairman, adopted a neutral stance in public, he increasingly leaned in the direction of the segment of the industry favoring some form of regulation. On November 16, 1925, he sent a circular letter which stirred up the internal debate even more. Contrary to the Committee of Eleven's conclusions, Work cited evidence of waste that pointed to serious future depletion, and he called for a F.O.C.B.-sponsored hearing to be held in Washington the following February.[84]

The A.P.I. responded by calling an emergency meeting of its Board of Directors in New York. J. Edgar Pew was in Dallas at the time and J. Howard represented Sun at the meeting. Jersey's Walter Teagle suggested that the institute employ Mark Requa to represent it in the upcoming public hearings before the F.O.C.B. Although Teagle had been openly critical of wasteful practices and was leaning toward some form of production regulation, he was primarily motivated by a desire to present the correct image in Washington. J. Edgar Pew and William Farish (of Humble-Jersey Standard) huddled in Texas and conferred with Teagle by telephone. Farish convinced Teagle that he "was on the wrong track" and that Requa's appointment would create great problems with the western contingent of the industry.[85]

J. Edgar Pew then called a meeting of the A.P.I. Board for December 12. At that gathering, he moved to hire Charles Evans Hughes to represent the Institute in Washington. Hughes, the distinguished New York Republican and former Presidential candidate, had recently resigned as Secretary of State in the Coolidge administration. The A.P.I. voted to retain Hughes on December 12.[86] J. Edgar explained his opposition to hiring Requa in the following way:

> all these men who are talking co-operation with the government do not realize what the term means. . . . They do not have any desire for restrictions on the oil industry, and are only hoping to conciliate, to the extent that these restrictions may not come. But if they are not careful they will find they are too late. . . . If they wish to make fools of themselves next year [after Pew's presidency] I have no objections, but I do not want to see them do it this year.[87]

Prior to the F.O.C.B. hearings, the A.P.I. held its sixth annual convention in Los Angeles. In his departing presidential address on January 19, 1926, J. Edgar Pew maintained that petroleum supplies were "inexhaustible" and provided data on deep drilling, oil substitutes, and alternative crude sources that came from the Committee of Eleven Report.[88] Pew's friend William Farish succeeded him as president of the Institute and Edgar left office convinced that the industry had steered a safe course through the treacherous waters of regulation.

Charles Evans Hughes did not present his formal arguments to the Conservation Board until May. At the time, he argued against federal intervention from a constitutional perspective rather than debating the issue of waste and scarcity. Hughes attacked the legality of mandated production curbs and maintained that voluntary agreements could best accomplish any desired production limitations. Henry Doherty once again filed a brief for compulsory unitization, independent producers testified against regulation of production, and other oil men put forth the *laissez faire* position which the Pews advocated.[89]

The long-awaited first F.O.C.B. report appeared in October 1926. It opted for a moderate position somewhere between the extreme industry positions symbolized by Beaty and Doherty. Although the report avoided a direct stand on the scarcity issue, it questioned the Committee of Eleven's "billion acre reserve" and stressed the necessity for proven future supplies of crude. The report urged the development of state regulations when necessary and tentatively supported a growing industry movement for loosening antitrust statutes in order that voluntary agreements on production restriction would not be challenged in the courts. Free-market advocates were moving toward a position that recognized the need for some form of price stabilization in periods of overproduction, and voluntary measures appeared the most palatable. William Farish had adopted this view in 1926.[90]

STATE CONSERVATION MEASURES

A statement Walter Teagle made at the F.O.C.B. hearings soon proved prophetic. Teagle testified against federal intervention, but indicated that some direct action might become necessary if operators discovered a number of rich fields simultaneously, since "a flood of oil is always a possibility."[91] When the vast Seminole field appeared on the heels of the F.O.C.B. report at the end of 1926, the petroleum industry exhibited more confusion and disunity. The major strike at Seminole City on July 26, 1926, was followed by other large discoveries at Searight and Earlsboro. Together known as the Greater Seminole Field, this huge producing area was characterized by much waste, and its tremendous outpouring of crude threatened to disrupt the price structure of the industry. Despite some attempts at voluntary restriction of drilling, the price of crude reached an all-time low of 15 cents a barrel in the Seminole district in February 1927. By July 1927, the field reached a peak production of almost 526,000 barrels a day.[92]

The harsh reality of the Seminole boom and its depressing effect on prices colored the debates held at the A.P.I.'s Tulsa meeting in December 1926. Some representatives still supported the F.O.C.B. warnings of impending shortages, while the Pews and others remained committed to the Committee of Eleven Report. In order to achieve some degree of unity, the A.P.I. adopted a resolution approving the F.O.C.B. report and its admonition against oil field waste.[93] This action reflected a growing industry awareness of the problem of wasteful overproduction and the effect it was again having on price levels for crude.

The A.P.I. named a new Committee of Seven to draft recommendations for curbing production in pools where waste and inefficiency were clearly evident. Chairman E. W. Marland of Oklahoma appointed J. Edgar Pew a member of this group. The committee concentrated on the Seminole problem and explored an entirely new approach to eliminating waste. Marland, himself an independent producer, opposed federal intervention and felt that conservation legislation on the state level would better insure the rights of the independents who had influence in their own states. The key to the Committee of Seven approach was the relationship of natural gas to the drilling and pumping of oil.[94]

A form of this idea was embodied in the Marland Bill introduced into the Oklahoma legislature. The legislation would create a new agency, the

Oil and Gas Conservation Board, to establish and enforce ratios of natural gas production to oil production. From the very beginning, rapid exploitation of abundant fields had meant tremendous waste of natural gas. Oil under gas pressure was pumped off rapidly and gas was simply allowed to go into the atmosphere. Not only did this waste gas, but scientific studies showed that less oil was obtained from the wells that had lost natural gas pressure. By enforcing a gas-petroleum ratio of 500 cubic feet of natural gas recovered to each barrel of oil, more efficient and slower drilling techniques would have to be employed, and the exploitation of the Seminole pool would come under control. All drilling would slow down and wells that could not meet the 500 cubic feet ratio would close down. The gas conservation approach would give the Oklahoma state commission the power to limit indirectly the flow of oil from Seminole.[95]

The most significant aspect of the gas conservation approach was that it developed as a state rather than a federal program. This appealed to the Pews, who feared federal intervention more than state regulation and opposed direct compulsory limitations on drilling. J. Edgar Pew published an article in the *National Petroleum News* in early 1927 supporting the gas conservation remedy. He pointed to the impracticality of agreement on unit production and revision of state and federal antitrust laws to permit voluntary production curbs. Rather, he argued: "the enforcement of the laws of gas and oil conservation, both to the end that gas cannot be wasted, and that the fullest recovery of oil should be had, will accomplish all the results and benefits of a unit operation, and without interfering with the right of each holder to get his proper proportion of oil and gas from under any tract owned by him."[96] J. Howard Pew agreed that gas conservation legislation at the state level was the best way to combat overproduction without endangering the freedom of individual operators.[97]

The Marland proposal was controversial from the beginning and ran into intense opposition in the Oklahoma legislature from independent producers. Marland himself had been a long-time opponent of any form of regulation, but he was persuaded by the chaos of Seminole to change his position. The majority of smaller producers, however, still fought any form of regulation. No relief from the Seminole problem came until after the peak production of July, when the Oklahoma Corporation Commission sanctioned a private, voluntary agreement to limit daily production to 450,000 barrels.[98]

THE COMMITTEE OF NINE

Although no antitrust action resulted from this Oklahoma agreement, the operators realized that the future of voluntary production curbs would greatly depend on the attitude of the state and federal courts. On August 30, 1927, F.O.C.B. Chairman Work proposed that a new committee be formed to investigate and draft appropriate statutes affecting the legal status of the oil industry. The resulting "Committee of Nine" or "Legal Committee," formally organized on December 10, 1927, consisted of three representatives each from the industry, the government, and the American Bar Association (A.B.A.). Past A.P.I. presidents Thomas O'Donnell, J. Edgar Pew, and William S. Farish represented the industry; all three men were spokesmen for the interests of the large, integrated firms. Government members were Walter F. Brown, Assistant Secretary of Commerce; Edward C. Finney, Assistant Secretary of Interior; and Abram F. Myer of the Federal Trade Commission. The legal representatives all came from the A.B.A. Mineral Section Committee, but reflected a bias toward the larger integrated companies. They were Professor Henry M. Bates, University of Michigan Law School; Warren Olney, Jr., of San Francisco; and James A. Veasey of the Carter Oil Company (Standard New Jersey), a friend of Mark Requa.[99]

Before the Committee of Nine formally met, the A.P.I. Board of Directors adopted a resolution supporting state and federal legislation authorizing voluntary production agreements. An A.P.I. legal committee chaired by Amos L. Beaty of the Texas Company, long-time supporter of voluntary agreements, sponsored the resolution. J. Edgar Pew, William S. Farish, and Thomas O'Donnell also gave their backing. Secretary of Commerce Hoover had already publicly supported a revision of the Sherman Act allowing voluntary agreements.[100]

After nearly a year, the Committee of Nine released its report in November 1928. It recommended that oil companies be exempted from the antitrust laws in order to effect voluntary limits on production during peak periods. The committee suggested that any agreements be subject to approval by the Federal Oil Conservation Board. According to "inside information" J. Edgar Pew later supplied to R. C. Holmes of the Texas Company, Hoover had delayed the release of the controversial report until after the 1928 political campaign.[101]

Although J. Edgar Pew had always favored voluntary measures over compulsory oil field regulation, his support of antitrust law modification appears inconsistent with the traditional Pew philosophy of opposition to monopoly. He explained his position at the time with the following statement: "In any event, and without knowing anything about the law, I have always believed, as a practical question, that if the administration was willing to give us this relief, such action would remove many of the excuses of those in the industry which were being put forward and which were blocking the fullest cooperation."[102]

The Seminole panic had accomplished what Requa and Doherty

had failed to do in the past. J. Edgar Pew, William Farish, Amos L. Beaty, E.W. Marland, and a majority of the A.P.I. leadership were now committed to some form of cooperative effort leading to a more scientific management of producing pools. Conflicts over what form this cooperative effort would take and battles with individual independent producers continued for some time. Some oil men, like J. Howard Pew, would never feel comfortable with restriction programs.

Encouraged by the apparent government blessing bestowed on the Committee of Nine report, the A.P.I. created a "Committee on World Production and Consumption of Petroleum and its Products" in February 1929. Chaired by R.C. Holmes of the Texas Company, the committee drafted a proposal for voluntary production limitation which it presented to the full Board in March. The heart of the proposal was a plan to limit 1929 United States production to the daily rate of production of 1928. The grandiose scheme loosely coincided with a global plan to restrict production being fostered at the time by Sir Henry Deterding of Royal Dutch Shell. The A.P.I. Board adopted the Holmes plan and resolved to send it to the F.O.C.B. for approval.[103]

On April 2, President Hoover dropped a bombshell by declaring that the proposal violated the Sherman Act and suggesting that the industry's goal was really price-fixing. Hoover's decision reflected a great deal of behind-the-scenes discussion with his key advisors. Particularly important was the legal opinion of Attorney General William D. Mitchell that the F.O.C.B. had no authority to grant antitrust immunity. The industry was once again in a quandary. On April 8, Ray Lyman Wilbur, Hoover's Interior Secretary and new Chairman of the F.O.C.B., wrote to R.C. Holmes, virtually killing the A.P.I. proposal. Wilbur emphasized that legal authority resided in the states and that state governments might enter into voluntary compacts limiting production and then obtain congressional authorization. Here was the germ of another idea now supported by Hoover—the interstate compact.[104]

THE COLORADO SPRINGS CONFERENCE

J. Edgar Pew, an active participant in most important industry debates since 1924, was noticeably absent from the center of activitiy at this time. His physical condition largely accounts for his unusually low profile. He had been suffering from acute attacks of sciatica which were to plague him in later life. But Pew's appetite for battle was whetted by the entry of his old foe, Mark Requa, into the lists. The indomitable Requa had successfully managed Hoover's presidential campaign in California and now enjoyed more influence in Washington than he had in either the Harding or Coolidge administrations.[105]

On April 27, 1929, Requa issued a press statement in support of Secretary Wilbur's interstate compact and announced a personal trip west to confer with the Governors of Oklahoma, Texas, and California. Under pressure from Wilbur, the A.P.I. adopted a resolution supporting the interstate compact, and Hoover announced plans for a special oil conference to be held in Colorado Springs on June 10. Although Requa had great influence with Hoover, his status was unofficial until May 28, when the President appointed him chairman of the conference. Requa extended invitations to the governors of eleven states, the A.P.I., and all other major trade associations, including groups representing independent producers.[106]

In early May, the Interior Department drew up an interstate compact proposal for discussion at Colorado Springs. The core of the plan was an Interstate Conservation Committee empowered with authority to control output, including the authority to suspend state antitrust statutes. The proposal defined waste to include "economic waste," formalizing for the first time what had really been the driving force behind the adoption of conservation laws.[107]

Prior to the Colorado Springs meeting, both J. Edgar Pew and J. Howard Pew continued to express reservations about Requa's role in the affair. The participation of Congress in approving an interstate compact seemed to once again raise the specter of federal intervention. J. Edgar wrote to R.C. Holmes of the Texas Company that "I think it would be a tragedy for the oil industry to admit that the public has some interest in our industry that would necessitate the placing of our industry under the political control which I fear would result if the plan of Mr. Requa is carried out 100 percent."[108] It is interesting that Edgar only objected to adopting the plan "100 percent"; he now appeared to embrace some form of production restriction.

J. Howard Pew, however, reacted strongly against the entire concept of the interstate compact. He pointed to the dangers of federal intervention, industry price-fixing of refined products as well as crude, and the "political voters" of the western oil-producing states whom he considered too radical and opposed to the interests of the major firms. Howard's specific reference to the price-fixing of refined products also reflected his concern about Blue Sunoco gasoline marketing. Any mandated price system for gasoline would probably establish a minimum price for "regular" and "premium" gasoline. Sun sold Blue Sunoco at a price slightly higher than regular gasoline, but three to five cents below most other premium or Ethyl grades. Because of Sun's unique one-brand marketing, J. Howard Pew carried on a continuous fight against the price-fixing of refined products.[109]

Any form of artificial production restriction was a form of indirect price-fixing, a fact that the oil companies did not like to admit. It was of course much easier to place the price-fixing of crude within the context of conservation, efficiency, and the elimination of waste. J. Howard Pew still thought like a refiner. He was frankly not very keen about having an oil conference at all because he believed that the crude market would rebound as it had always done. J. Edgar had adopted a more moderate position, one that also reflected his greater empathy for the problems of crude production.

In reply to J. Howard's strong denunciation of the interstate compact, J. Edgar offered a more balanced view. He first complimented the work of Holmes's committee on World Production, and opted for some compromise with the Hoover administration. He cited the administration's commitment to efficiency and its "general tendency to undertake a solution of all problems." Given this situation, J. Edgar urged that something had to be done about the overproduction problems. Furthermore, to ignore the President's request for a conference at this time "would merely aggravate the situation and make the administration more determined to press its own plan."[110] Its own plan seemed to promise some form of federal regulation.

The Colorado Springs conference was a complete failure for Mark Requa. He began on the worst possible note in his opening address when he hinted at federal coercion if the states failed to adopt a cooperative agreement restricting production. This put the delegates on the defensive, and open controversy then erupted over two specific issues. The western Governors opposed a recent Hoover ban on new drilling permits on public lands, and the independent producers backed a high tariff on imported oil rather than the production curbs favored by the integrated firms. Many of the large integrated companies now had significant investments in foreign production, and they opposed this method of supporting the domestic price structure.[111]

The A.P.I., representing the larger integrated firms, issued a statement supporting the government in any "practical" conservation measure, but specifically rejecting the joint commission representing several states, the heart of the interstate compact. J. Edgar Pew was pleased with the results at Colorado Springs and expressed the belief that the best answer to overproduction lay in compulsory state gas conservation laws and the possible use of voluntary production agreements immune from antitrust.[112] The former wartime oil administrator had lost again, but as J. Edgar Pew wrote, "Requa is very adaptable, however, and expresses himself as quite satisfied with the results of the conference. This, of course, is nothing more than a gesture, unless he has lost all his sense of humor, if he ever had any."[113]

In reality, Requa was upset with what he perceived as a lack of direction from Hoover, and he resigned as chairman of the conference in July. Hoover himself later charged that the major oil companies opposed the interstate compact because they really wanted federal control! He called their "demand for collectivism" "stupid." With some justification, he became bitter when the interstate compact, his proposal, came into operation during the New Deal. Hoover argued, however, that oil executives only backed this after they tasted the collectivism of the National Recovery Administration and rejected it.[114] The issues in 1929 were more complex than Hoover either understood or cared to admit. The intra-industry split between the independent producers and the major integrated firms remained an obstacle to a policy consensus for the next several years. But even though they disagreed on the exact form, the majors, including Sun, now advocated some form of production curtailment to combat "economic waste."

THE NEW ERA

As if to symbolize the changes that had occurred since 1920, J. Edgar Pew spearheaded a movement in west-central Texas in 1929, which some oil men consider to be the first fully unitized field. Sun, Humble, Pure Oil, Texas, and Shell all held leases in the Van Pool in Van Zandt County, Texas. They agreed to operate the pool as a unit and let the largest operator, Pure Oil, administer the production. A Texas state conservation bill had passed in 1929, but without a provision for unitization. The five companies therefore proceeded in full risk of antitrust prosecution. The experiment proved very successful, and the controlled exploitation of the field resulted in a high percentage yield of oil.[115]

There was much truth in an open letter that the old champion of unitization and scientific production, Henry Doherty, wrote to the A.P.I. directors in 1930: "I don't think there is now much difference between my views and recommendations and those of my former opponents."[116] One year later, J. Edgar Pew praised his old opponent, Doherty, as the founder of unitization in a speech delivered in Dallas entitled "The New Conception of Oil Production."[117] The discovery in 1930 of the East Texas field, perhaps the biggest strike of all, plus the general effects of the deepening depression pushed J. Edgar Pew and other industry leaders even closer to the concept of oil field regulation which Doherty and Requa had advocated a decade earlier.

NOTES

1. Harold F. Williamson, Ralph L. Andreano, Arnold R. Daum and Gilbert C. Klose, *The American Petroleum Industry,* Volume II, *The Age of Energy, 1899-1959* (Evanston: Northwestern University Press, 1963), pp. 47-8.

2. Gerald D. Nash, *United States Oil Policy, 1890–1964* (Pittsburgh: University of Pittsburgh Press, 1968), p. 10.

3. J. Edgar Pew, "United States Petroleum Resources," *Testimony before the Special Committee to Investigate Petroleum Resources of the United States Senate* (New York: American Petroleum Institute, 1945), p. 7.

4. Nash, *Oil Policy*, pp. 30–1.

5. Williamson et al., *energy*, pp. 49–50.

6. *Ibid.*, pp. 50–4,

7. Daniel J. Bergen, "Oklahoma," *Our Sun* 13, 7 (July 1947), p. 3; Fanning, Leonard M., *The Story of the American Petroleum Institute* (New York: Leonard M. Fanning, Editor of *World Petroleum Politics*, 1959), p. 15.

8. Nash, *Oil Policy*, pp. 15–6.

9. "In the Oil Fields with Sun," *Our Sun*, 50th Anniversary Issue 3, 1 (1936), pp. 31–2; Roy C. Cook, *Control of the Petroleum Industry by Major Oil Companies*, Monograph No. 39, Temporary National Economic Committee, (Washington, D.C.: U.S. Government Printing Office, 1941), p. 70.

10. Fanning, *American Petroleum Institute*, p. 4.

11. *Ibid.*, p. 6.

12. J.N. Pew, Jr. to Lt. Page A. Watson, September 14, 1918, Accession #1317, Eleutherian Mills Historical Library, File 21-B, Papers of Joseph Newton Pew, Jr., Box. 1.

13. Nash, *Oil Policy*, p. 23.

14. Ibid., pp. 24–5; Williamson et al., *Energy*, pp. 268–9; Fanning, *American Petroleum Institute*, p. 4.

15. Nash, *Oil Policy*, pp. 26, 28–9; Norman Emmanuel Nordhauser, "The Quest for Stability: Domestic Oil Policy, 1919–1935" (Ph.D. dissertation, Stanford University, 1970), pp. 1–4; Fanning, *American Petroleum Institute*, p. 21; Williamson et al., *Energy*, p. 270.

16. Nash, *Oil Policy*, pp. 29–30; Williamson et al., *Energy*, pp. 269–70.

17. Nash, *Oil Policy*, pp. 34–35; Williamson et al., *Energy*, pp. 270–92, passim.

18. Nash, *Oil Policy*, p. 26.

19. Nordhauser, "Quest," p. 3.

20. *Ibid.*, pp. 6–7; Nash, *Oil Policy*, pp. 39–41.

21. Nordhauser, "Quest," pp. 4, 10–1.

22. J.N. Pew, Jr. to Samuel Eckert, June 3, 1918, Acc. #1317, EMHL, File 21-B, Box 1.

23. J. Howard Pew to Mark L. Requa, November 20, 1925, Acc. #1317, EMHL, File 21-A, Administrative, Box 23.

24. "Notes Made by J. Edgar Pew, 1942," Folder #1, "J. Edgar Pew," Library File, Sun Oil Company Library, Marcus Hood, Pennsylvania, p. A-7.

25. *J. Edgar Pew, His Life and Times, 1870–1946* (Philadelphia: Sun Oil Company, 1947), p. 11 (Reprint from *American Petroleum Institute Quarterly* [April 1947], p. 1).

26. Nash, *Oil Policy*, p. 113.

27. *J. Edgar Pew, Life and Times* p. 5.

28. "A.P.I. Past Living Presidents," *National Petroleum News* 38, 44 (October 30, 1946), p. 53.

29. *Ibid.;* J. Edgar Pew, Life and Times, pp. 8–9.

30. "A.P.I. Past Living Presidents," p. 53.

31. Nash, *Oil Policy*, p. 47.

32. Williamson et al., *Energy*, pp. 517–8; George Sweet Gibb and Evelyn H. Knowlton, *History of Standard Oil Company (New Jersey)*, Volume II, *The Resurgent Years, 1911–1927* (New York: Harper and Brothers, 1956), pp. 283–4; Fanning, Leonard M., *American Oil Operations Abroad* (New York: McGraw Hill, 1947), pp. 2–3; Nash, *Oil Policy*, p. 47. In 1919, the United States produced 378.4 million barrels of crude out of a world total of 555.9

million barrels. Middle Eastern production amounted to only 10.1 million barrels. (See Table 8:1 in Williamson et al., *Energy,* p. 262.)

33. Nordhauser, "Quest," p. 9; Fanning, *Oil Abroad*, p. 4; Herbert Hoover, *The Memoirs of Herbert Hoover,* Volume II, *The Cabinet and the Presidency, 1920–1933* (New York: MacMillan, 1953), p. 69.

34. Mark L. Requa to J. Howard Pew, August 12, 1920, Acc. #1317, EMHL, J. Howard Pew Papers, Box 57; Nordhauser, "Quest," pp. 10–12; Nash, *Oil Policy,* p. 44; "Memorandum Having to do with the Incorporation of An Exploration Company to Operate in Foreign Oil Fields," Acc. #1317, EMHL, J. Howard Pew Papers, Box 57; see folder "Foreign Company," in J. Howard Pew Papers, Box 57.

35. J. Howard Pew to Mark L. Requa, August 14, 1920, Acc. #1317, EMHL, J. Howard Pew Papers, Box 57.

36. Gibb and Knowlton, *Resurgent Years,* p. 292.

37. Mark L. Requa to J.H. Pew, August 12, 1920, Acc. #1317, EMHL, J. Howard Pew Papers, Box 57.

38. See folder: "Foreign Company," Acc. #1317, EMHL, J. Howard Pew Papers, Box 57; Memorandum, J.N. Pew, Jr. to J. Howard Pew, December 30, 1920, Acc. #1317, EMHL, J. Howard Pew Papers, Box 57.

39. *Ibid.*

40. For an excellent discussion of the engineering mind-set as applied to socioeconomic problems, see ch. II, "The Engineer as Progressive," in Joan Hoff Wilson, *Herbert Hoover: Forgotten Progressive* (Boston: Little Brown, 1975), pp. 31–53. More detailed discussion of the engineering profession and the broader implications of Taylorism may be found in Edwin T. Layton, Jr., *the Revolt of the Engineers: Social Responsibility and the American Engineering Profession* Cleveland: The Press of Case Western Reserve University, 1971) and Monte A. Calvert, *The Mechanical Engineer in America, 1830–1910* (Baltimore: the Johns Hopkins Press, 1967). The best discussion of the new managerial class that emerges in the 1920s may still be found in the classic study by Adolf A. Berle, and Gardiner C. Means, *the Modern Corporation and Private Property* (New York: MacMillan, 1933). A more recent discussion of the managerial revolution is found in the important synthesis by Alfred D. Chandler, Jr., *The Visible Hand: The Managerial Revolution in American Business* (Cambridge, Mass.: Harvard University Press, 1977), especially ch. 14, "the Maturing of Modern Business Enterprise," pp. 455–83.

41. Memorandum, J.N. Pew, Jr. to J. Howard Pew, December 30, 1920, Acc. #1317, EMHL, J. Howard Pew Papers, Box 57.

42. Telegram, Ira Jewell Williams, Atlantic Refining Company, to M.L. Requa (Copy), March 4, 1921; "Certificate of Incorporation, American Oil Producers Company," Acc. #1317, EMHL, J. Howard Pew Papers, Box 57.

43. "Certificate of Incorporation," Acc. #1317, EMHL, J. Howard Pew Papers, Box 57.

44. Francis McIlhenny to J. Howard Pew, May 9, 1921, Acc. #1317, EMHL, J. Howard Pew Papers, Box 57.

45. Gibb and Knowlton, *Resurgent Years,* p. 292.

46. J. Edgar Pew to J. Howard Pew, June 3, 1929, Acc. #1317, EMHL, File 21–A, Box 42; Nash, *Oil Policy,* p. 56; for a full discussion of Jersey Standard Operations in the 1920s, see Gibb and Knowlton, *Resurgent Years*, Chapter 50, "The Quest for Crude Oil, 1919–1928."

47. Edwin Lieuwen, *Petroleum In Venezuela*, Volume 47, University of California Publications in History (Berkeley: University of California Press, 1954), p. 44. Sun's Venezuelan operations are discussed in chapter IV.

48. J.K. Towles, Bureau of Foreign and Domestic Commerce, U.S. Department of Commerce, to J. Howard Pew, August 12, 1922, Acc. #1317, EMHL, J. Howard Pew

Papers, Box 56. This evidence supports in part the contentions of William Appleman Williams and his followers that American isolationism in the 1920s was largely a legend. The active support and assistance given by the United States government to American companies demonstrates that there were close connections between economic interests and an activist foreign policy. Where American oil companies did succeed in foreign operations, United States diplomatic and consular support proved a definite asset. (See William Appleman Williams, *The Tragedy of American Diplomacy* [New York: Delta, 1962], especially Chapter 4, "The Legend of Isolationism," and Carl P. Parrini, *Heir to Empire: United States Economic Diplomacy, 1916-1923* [Pittsburgh: University of Pittsburgh Press, 1969]).

49. John Ise, *The United States Oil Policy* (New Haven: Yale University Press, 1926), Chapter 12; Nordhauser, "Quest," pp. 14-15; Nash, *Oil Policy,* pp. 82-84.

50. Nordhauser, "Quest," pp. 17-18; Gibb and Knowlton, *Resurgent Years,* pp. 432-3; Nash, *Oil Policy,* p. 83.

51. Gibb and Knowlton, *Resurgent Years,* p. 430.

52. Henrietta M. Larson, and Kenneth Wiggins Porter, *History of Humble Oil and Refining Company: A Study in Industrial Growth* (New York: Harper Brothers, 1959), pp. 251-3; Gibb and Knowlton, *Resurgent Years,* p. 432; Nordhauser, "Quest," pp. 19-23.

53. Nash, *Oil Policy,* p. 83.

54. Henry Doherty to A.P.I. Board of Directors (Copy), November 1, 1930, Acc. #1317, EMHL, File 21-A, Box 52.

55. Williamson et al., *Energy,* pp. 307-8; Gibb and Knowlton, *Resurgent Years,* pp. 383-4; Nash, *Oil Policy,* pp. 74, 77-8.

56. *Ibid.*, pp. 75-6; for the best discussion of the Teapot Dome, see Burl Noggle, *Teapot Dome: Oil and Politics in the 1920's* (Baton Rouge: Louisiana State University Press, 1962).

57. Nash, *Oil Policy,* p. 85.

58. Hoover, *Memoirs,* Volume II, p. 237.

59. "J. Edgar Pew Notes, 1942," pp. A-7, A-10; *J. Edgar Pew, Life and Times,* pp. 10-1; see also Folder, A.P.I., 1924, Acc. #1317, EMHL, File 21-A, Box 16.

60. "J. Edgar Pew Notes, 1942," p. A-8.

61. *National Petroleum News* (December 24, 1924).

62. "J. Edgar Pew Notes, 1942," p. A-10; Henrietta Larson, Evelyn H. Knowlton and Charles S. Popple, *New Horizons, 1927-1950*, Volume III, *History of Standard Oil Company (New Jersey)* (New York: Harper and Row, 1971), p. 48; Paul H. Giddens, *Standard Oil Company (Indiana): Oil Pioneer of the Middle West* (New York: Appleton-Century-Crofts, 1955), pp. 361-5; Noggle, *Teapot Dome,* pp. 180-2.

63. J. Edgar Pew to J. Howard Pew, June 3, 1924, Acc. #1317, EMHL, File 21-A, Box 22.

64. "J. Edgar Pew Notes, 1942," pp. A-11, A-12.

65. *Ibid.*

66. *Ibid.;* J. Edgar Pew, Life and Times, p. 12.

67. Hubert Work, F.O.C.B., to J. Howard Pew, January 17, 1925, Acc. #1317, EMHL, File 21-A, Box 22.

68. *National Petroleum News* (February 11, 1925), p. 3940.

69. J. Edgar Pew to J.H. Pew, June 13, 1925, Acc. #1317, EMHL, File 21-A, Box 22.

70. Nash, *Oil Policy,* pp. 84, 87.

71. J. Howard Pew to Hubert Work, F.O.C.B., June 18, 1925, Acc. #1317, EMHL, File 21-A, Box 22.

72. R.L. Welch, A.P.I., to J. Howard Pew, January 23, 1925, Acc. #1317, EMHL, File 21-A, Box 20; Nash, *Oil Polciy,* p. 89.

73. *New York Times* (August 7, 1925), p. 22; "Digest of the Several Reports Prepared

Under Direction etc.," Acc. #1317, EMHL, File 21-A, Box 20; Williamson et al., *Energy,* pp. 317-8; Nash, *Oil Policy,* p. 88; Nordhauser, "Quest," pp. 32-5; for entire Committee of Eleven Report, see American Petroleum Institute, *American Petroleum Supply and Demand* (New York: McGraw Hill, 1925).

74. Nash, *Oil Policy,* pp. 88-9.

75. Henry Doherty to J.W. Mahon, R., Managing Editor, *Dallas Times* (Copy), August 19, 1925, Acc. #1317, EMHL, File 21-A, Box 20.

76. J. Edgar Pew to J. Howard Pew, September 3, 1925, Acc. #1317, EMHL, File 21-A, Box 20.

77. Henry Doherty to President and Board of Directors, A.P.I., November 1, 1930, Acc. #1317, EMHL, File 21-A, Box 52.

78. M.L. Requa to J. Howard Pew, October 12, 1925, Acc. #1317, EMHL, File 21-A, Box 23.

79. J. Howard Pew to M.L. Requa, November 20, 1925, Acc. #1317, EMHL, File 21-A, Box 23.

80. J. Edgar Pew to J. Howard Pew, November 4, 1925, November 25, 1925, Acc. #1317, EMHL, File 21-A, Box 23.

81. J. Howard Pew to J. Edgar Pew, November 21, 1925; J. Edgar Pew to J. Howard Pew, November 25, 1925; "Result of Interview with 'H'," Acc. #1317, EMHL, File 21-A, Box 20.

82. Gibb and Knowlton, *Resurgent Years,* p. 433; Williamson et al., *Energy,* pp. 318-9.

83. Fanning, *American Petroleum Institute,* p. 115.

84. Hubert Work to J. Howard Pew, November 16, 1925, Acc. #1317, EMHL, File 21-A, Box 20; Nordhauser, "Quest," p. 36.

85. J. Edgar Pew to J. Howard Pew, November 25, 1925, Acc. #1317, EMHL, File 21-A, Box 20.

86. Nash, *Oil Policy,* p. 89.

87. J. Edgar Pew to J. Howard Pew, November 25, 1925, Acc. #1317, EMHL, File 21-A, Box 20.

88. *New York Times* (January 20, 1926), p. 8.

89. Nash, *Oil Policy,* pp. 89-90; Nordhauser, "Quest," p. 37; see *Complete Record of Public Hearings Before the F.O.C.B.,* February 10, 11, 1926 (Washington, D.C.: U.S. Government Printing Office, 1926).

90. Nordhauser, "Quest," pp. 38-40; Nash, *Oil Policy,* p. 90; Gibb and Knowlton, *Resurgent Years,* p. 434.

91. Quoted in Nordhauser, "Quest," p. 37.

92. Williamson et al., *Energy,* pp. 322-5.

93. *New York Times* (December 10, 1926), p. 23; Nash, *Oil Policy,* p. 91.

94. *New York Times* (December 10, 1926), p. 23; Nash, *Oil Policy,* p. 96.

95. J. Edgar Pew, "Saving and Re-Utilization of All Gas Proposed Over-Production Remedy," *Our Sun* 4, 6 (June 30, 1926) (reprint from *National Petroleum News*), pp. 16-17; Nordhauser, "Quest," pp. 41-2; Nash, *Oil Policy,* pp. 96-7.

96. Pew, "Saving and Re-Utilization," p. 17.

97. J. Howard Pew to J. Edgar Pew, June 17, 1929, Acc. #1317, EMHL, File 21-A, Box 52.

98. Nash, *Oil Policy,* pp. 96-7; Nordhauser, "Quest," pp. 41-43; Williamson et al., *Energy,* p. 323.

99. *National Petroleum News* (December 14, 1927), pp. 21-22.

100. *National Petroleum News* (December 7, 1927), pp. 17-18; *Oil and Gas Journal* (December 9, 1926), p. 164.

101. Nordhauser, "Quest," p. 52; J. Edgar Pew to R.C. Holmes, Texas Company, June 3, 1929, Acc. #1317, EMHL, File 21-A, Box 42.

102. *Ibid.*

103. R.C. Holmes to E.B. Reeser, A.P.I. (Copy), March 27, 1929, Acc. #1317, EMHL, File 21-A, Box 42; Nordhauser, "Quest," pp. 53-4; Nash, *Oil Policy,* p. 102; R.C. Holmes to E.B. Reeser (Copy), March 27, 1929, Acc. #1317, EMHL, File 21-A, Box 42.

104. Nash, *Oil Polciy,* p. 102; Nordhauser, "Quest," pp. 57-8; William D. Mitchell to Ray Lyman Wilbur (Copy), March 29, 1929, Acc. #1317, EMHL, File 21-A, Box 42; William Starr Myers and Walter H. Newton, *The Hoover Administration* (New York: Charles Scribner's Sons, 1936), p. 375; Ray Lyman Wilbur, and Arthur M. Hyde, *The Hoover Policies* (New York: Charles Scribner's Sons, 1937), pp. 391-2; Ray Lyman Wilbur to R.C. Holmes (Copy) April 8, 1929, Acc. #1317, EMHL, File 21-A, Box 42.

105. J. Howard Pew to J. Edgar Pew, May 29, 1929; J. Edgar Pew to J. Howard Pew, June 3, 1929; J. Edgar Pew to R.C. Holmes, June 3, 1929, Acc. #1317, EMHL, File 21-A, Box 42; Hoover, *Memoirs,* Volume II, p. 56.

106. "Statement of Mark L. Requa to Associated Press," April 27, 1929, Acc. #1317, EMHL, File 21-A, Box 42; "American Petroleum Institute Press Release," May 27, 1929, Acc. #1317, EMHL, File 21-A, Box 42; "Department of Interior Press Release," May 27, 1929, Acc. #1317, EMHL, File 21-A, Box 42.

107. J. Edgar Pew to R.C. Holmes, June 3, 1929, Acc. #1317, EMHL, File 21-A, Box 42.

108. *Ibid.*

109. J. Howard Pew to J. Edgar Pew, May 29, 1929, Acc. #1317, EMHL, File 21-A, Box 42; Melvin G. de Chazeau and Alfred E. Kahn, *Integration and Competition in the Petroleum Industry* (New Haven: Yale University Press, 1959), p. 94.

110. J. Edgar Pew to J. Howard Pew, June 3, 1929, Acc. #1317, EMHL, File 21-A, Box 42.

111. J. Edgar Pew to J. Howard Pew, June 14, 1929, Acc. #1317, EMHL, File 21-A, Box 42, Nash, *Oil Policy,* pp. 104-5.

112. J. Edgar Pew to J. Howard Pew, June 14, 1929.

113. *Ibid.*

114. Nash, *Oil Policy,* p. 105; Hoover, *Memoirs,* Volume II, pp. 237-9.

115. *J. Edgar Pew, Life and Times,* p. 12; Larson, Porter, and Popple, *New Horizons,* p. 88.

116. Henry Doherty to E.B. Reeser (Copy), September 2, 1930, Acc. #1317, EMHL, File 21-A, Box 52.

117. J. Edgar Pew, "The New Conception of Oil Production," speech delivered at the A.P.I. meeting, Division of Production, (Dallas, 1931), Folder #1, "J. Edgar Pew," Library File, Sun Oil Company Library, Marcus Hook.

CHAPTER VI

Growth in the Depression Decade: Sun, 1930–1940

Although the Great Depression did not affect the large integrated oil companies and other oligopolistic firms so adversely as it did most other units in the economy, Sun's gross income declined from over $98,000,000 in 1930 to only $69,000,000 in 1931. During the same period, net income dropped from nearly $8,000,000 to slightly over $3,000,000. In the following year, however, fortunes turned upward and, with the exception of a downturn in 1938 following the nation-wide recession, Sun exhibited steady growth and increased profits for the remainder of the decade.[1]

Petroleum, chemicals, automobiles, steel, aircraft, food production, and other large-scale oligopolistic industries turned to technological advances and efficiencies to maintain their profit margins and shares of the market. An important key to recovery and continued growth in these industries was their ability for self-financing. The major petroleum companies were strong and possessed very healthy cash flows on the eve of the Depression. Then, despite the general malaise of the economy, the demand for petroleum products continued to expand during the 1930s. The number of passenger cars and trucks in service increased steadily after 1932 and the rapid growth of the aircraft industry created important new markets for aviation fuel. This increased gasoline demand, and higher octane requirements for airplanes and higher compression automobile engines stimulated the development of new refining processes. In 1937, refineries employed $38,700 of capital per worker as compared with only $3,700 for all manufacturing industries.[2]

The oil industry had been divided for years over the issue of regulated crude oil production. Under the duress of economic depression and increasing pressures from government, however, a basic structure of planned production evolved (this is discussed in detail in chapter VIII).

The net result of government-supervised rather than voluntary elimination of waste was to increase the profit margins of the major integrated firms. Furthermore, despite initial decline in the early thirties, the price of gasoline showed moderate steady increases in order to offset the high capital costs of developing new plant facilities. Thus, although the 1930s was a genuine period of serious economic maladjustment, for certain large oligopolistic industries, it was a time of surprising growth and expansion.[3]

Sun's history closely followed this pattern, and the Pews invested heavily in new pipelines and refining processes. Because Sun had followed conservative financial policies in previous decades and was not burdened by debt, the firm was in a relatively good position to generate recovery through reinvestment of earnings. By the end of the 1930s, Sun had achieved a cheap and efficient transportation system, a position of industry leadership in petroleum cracking, and a strong share of the gasoline and aviation fuel markets. J. Howard Pew later reflected that "the depression was in many ways the best period we ever had."[4]

PRODUCT PIPELINES

In order to expand their marketing of Blue Sunoco in western Pennsylvania, Ohio, and New York, Sun's managers embarked on a major project, the construction of one of the first gasoline product pipelines in the United States. In 1929, Jersey Standard had begun to use its old Tuscarora pipeline to pump gasoline inland from its Bayway refinery to a new territory serviced by its marketing affiliate, Standard Oil of Pennsylvania. The cost savings Standard enjoyed with this innovative product pipeline seriously threatened Sun's home business in Pennsylvania. Accordingly, J.N. Pew, Jr. approached the railroads and requested a rate cut on Sun products to meet the challenge of the Tuscarora pipeline competition. When they refused, Sun planned construction of its own product pipeline from Marcus Hook west and north to Pittsburgh and Cleveland, with a branch line north to Syracuse.[5]

This was one of the first pipelines built expressly for transportation of refined products. Either Sun or the Phillips Petroleum Company deserves the credit for this "first," since both firms constructed similar lines in different parts of the country at the same time. Sun formed a company to begin construction of the line in 1930, and preliminary work began on clearing a right of way. Joseph N. Pew, Jr. was the moving force behind the project from the start, and he organized the work necessary prior to construction.[6]

The right of way required 1,058 permits to cross highways, 161 to cross railways, 33 for electric lines, 37 federal permits for navigable stream crossings, and 671 state permits for the crossing of smaller streams. Sun needed permission from hundreds of property owners, and eventually obtained 3,300 easements. J.N. Pew, Jr.'s experience dealing with government in clearing the way for the Susquehanna Pipeline whetted his appetite for politics, an avocation that he pursued with ever-increasing intensity.[7]

The physical obstacles were also enormous. The construction crew had to clear the Allegheny mountains and cross the Susquehanna River three times. Work began in April 1930, and by June 21, 1931, Sun began pumping gasoline from Marcus Hook to the western terminus in Cleveland. Shortly after, gasoline moved through the northern branch line to Syracuse. When completed, the line consisted of 664 miles of six- and eight-inch pipe and 68 miles of branch lines of smaller size, a total of 732 miles.[8]

Under the pipeline amendment to the Hepburn Act of 1906, all interstate pipeline operators had to function as federally regulated common carriers and file tariffs with the Interstate Commerce Commission. One of the managerial responses to this legislation designed to minimize its impact was to operate sections of the pipeline system as private lines. These lines would not have to operate as common carriers nor file tariffs with the ICC. This strategy, first employed by Standard Oil early in the century, was finally negated by a series of court decisions culminating in 1914.[9] However, the industry saw the product pipelines as an entirely new ballgame, and Sun apparently was following this line of action in the organization of the Susquehanna line. Sun established the Sun Pipe Line Company of Delaware to subscribe all the capital stock of the new system, but organized three separate operating companies: the Susquehanna Pipe Line Company in Pennsylvania, the Sun Pipe Line Inc. in New York, and the Sun Oil Line Company in Ohio. The Susquehanna Pipe Line Company carried gasoline to the borders of both New York and Ohio and filed rates with the ICC. The other companies joined at the Pennsylvania border to pipe gasoline to New York and Ohio respectively. Sun's position was that these companies were private lines operating intrastate only.[10]

Another tactic used after 1906 to circumvent the goals of the Hepburn Amendment was the requirement of minimum "tender" regulations. This was the smallest amount of crude that a prospective shipper would have to offer to send to a designation before the pipeline would accept his order. By establishing high tender requirements, the major pipeline operators could easily freeze out the smaller operators whom common carrier legislation was designed to help. The Susquehanna Pipe Line Company established a minimum tender of 75,000 barrels, and no gasoline was carried for outsiders during the first six months of operation. The full legal status of product pipelines under the Hepburn Act was still not settled in the 1930s.[11]

In addition to the legal questions surrounding the product lines, there were technical problems that had to be overcome. If these lines did operate as common carriers, there was concern for mixing differing grades of gasoline during transit. New federal government specifications for liquid fuels and improved methods for testing them partially alleviated this problem. In actual practice, the contamination problem proved to be minimal. When Sun did ship outside gasoline in the Susquehanna line, it used dye markers to indicate the end points of runs in an effort to avoid mixing leaded gasoline with Blue Sunoco. The introduction of improved seamless and welded pipe and the use of electric welding of pipe joints by the end of the 1920s had also solved the problem of leakage.[12]

Because of the volatility of gasoline as compared with crude oil, it was also desirable to eliminate tank storage of the product at pump stations. To do this, you needed an automatic system to control the flow of gasoline and safety devices to shut down operations when necessary. Some success in this regard had been achieved in the operation of an intrastate gasoline pipeline in California in 1929, and the Susquehanna Pipeline employed a new version of this technology in moving its refined products westward. Rather than using the customary diesel-driven positive displacement pumps, Sun employed electrically-driven centrifugal pumps. These devices allowed continuous operation, with the pumping at each station synchronized with that at each other station. Later pipelines copied this system pioneered by Sun.[13]

As a response to Jersey Standard's and Sun's challenge, the Atlantic Refining Company built its Keystone Pipeline in 1931. This line more or less paralleled the Susquehanna line from the Delaware River to Pittsburgh. Atlantic later used Sun's completed line to Cleveland and Sun used an Atlantic line north to Buffalo and Rochester. In 1934, Sun built an extension through New Jersey with a terminus in Newark. This branch was operated by the Middlesex Pipeline Company (New Jersey).[14]

While many industries retrenched during the early years of the Depression, Sun invested more than $9,000,000 in the construction of these new product pipelines, lines that gave them a cost advantage in the transportation of gasoline to markets in the east and midwest. By 1939, Sun ranked fifteenth in net investment in pipelines as a result of its construction of product lines. The Cleveland line allowed Sun to distribute in Canada, and, along with the output of the Toledo refinery, in Ohio and Michigan. The Syracuse branch linked up with the New York barge canal system and aided distribution throughout upper New York.

The line from Marcus Hook to Newark, New Jersey provided gasoline for the New York City marketing area and for shipping up the Hudson River by barge to Albany or to New England by coastal tank ship. By 1940, through the use of its own pipelines and those of others, Sun enjoyed an almost unbroken flow of water and pipeline transportation from its oil wells to the filling station pump. At the end of the decade, Sun moved most of its gasoline by cheaper transportation than it had used in 1930. By 1940, twenty-six percent of total barrel-miles was by water, sixty-four percent by pipeline, and only four percent by rail, five percent by truck and one percent by outside bulk haulers.[15]

Sun's innovations in product pipeline development demonstrated its constant search for independence and self-sufficiency. J.N. Pew, Jr. originally conceived the Susquehanna pipeline as a necessary competitive tool against Jersey Standard's expansion into the Pennsylvania marketing area. It evolved into a fully integrated system which allowed Sun to have control over its own cheap transportation. In previous years, Pew's father had relied on tank ships carrying Texas crude to circumvent the old Standard monopoly of railroad transportation. Sun's fully integrated tanker and pipeline shipments of crude and refined product now insulated the firm even more effectively in competition with the industry giants.

SALES AND MARKETING

Sun was no longer a small company. In 1932, the firm accounted for 2.8 percent of all gasoline sold nationally, and by 1935 this figure had risen to 3.4 percent, making Sun the ninth largest retailer of gasoline in the United States.[16] Sun accomplished this with efficient marketing operations. Samuel B. Eckert, vice-president in charge of marketing, believed that Sun's success was largely due to its entering gasoline marketing late. In his words, Sun entered the field "after most other companies had made about all the mistakes possible."[17] But more important than this were Sun's innovations in pipeline distribution for high-volume marketing.

The company avoided supplying hard-to-reach locations and concentrated in high density areas near its product supply lines. Sun also used strict, some said high-handed, regulations in supplying its retailers. It made deliveries only once a week, and its stations had to fill up then or wait for the next visit of a Sunoco tank truck. Furthermore, Sun had a relatively small number of gasoline stations which it either owned and operated outright or leased to others. In 1928, Sun had 276 such stations, and the number increased gradually over the decade until, in 1938, there were 682 of these installations. But even though this figure had risen to well over 1,500 by 1940, it represented only about twenty percent of the more than 9,000 retail outlets selling Blue Sunoco gasoline.[18]

Sun did not invest heavily in its own service stations during the thirties for a number of reasons. The industry had generally discovered that it was more profitable to lease the sale of gasoline through independent station owners because the profit margin on retailed gasoline was slim and the large companies could get better return on investment in other operations. The individual gas station operator made his profits on the sales of oil and other automotive accessories, not on the sale of gasoline, where he operated on a dealer margin of less than two cents a gallon. The threat of heavy taxation and the anti-chain store legislation of the thirties also discouraged the direct operation of retail service stations. The Temporary National Economic Committee later charged that the integrated companies dropped their stations to avoid paying social security taxes to their employees, an accusation that J. Howard Pew directly denied. Of the stations it did own, Sun leased the majority to independent operators, but retained some which were operated for product research and the training of personnel.[19]

In the 1930s, the older practice of "split stations," service stations that sold more than one brand of gasoline, began to fade away. Even though many of the stations were not operated or leased by major companies, the large firms pressured these stations to handle one company's products exclusively. Sun, for example, charged its split station customers one-half cent more per gallon than it did for its one hundred percent dealers. The company justified this on the grounds that these stations sold a higher volume on the average and that Sun's delivery trucks did not have to make as many stops to keep the stations' underground tanks filled with Blue Sunoco. Critics of the major integrated companies argued that this was a discriminatory practice designed to compel independent retailers to sell one brand of gasoline.[20]

The best advantage Sun had in the distribution of its gasoline in the competitive eastern and midwestern market was the popularity of Blue Sunoco. Sun's exclusive dealerships had an average sales volume of 8,000 to 9,000 gallons of gasoline per month, a figure well above the industry average of 6,500 gallons per month in the period 1933–1934.[21] This high volume per pump was partially due to the relative scarcity of Sun stations, but it was also a genuine measure of consumer acceptance of the firm's brand. Sun's advertisements stressed the "premium high test" and "premium knockless" qualities of its Blue Sunoco in 1930 and 1931 newspaper and magazine spots. A November 1931 example advertised Blue Sunoco as having a 72 octane rating and characterized it as "modeled after fighting grade aviation gasoline." These advertisements continued to stress the idea of premium gasoline at regular prices, at least three cents below the premium fuels of competitors.[22]

In 1930, Sun began to supplement these campaigns with radio advertising on "The Sunoco Show," an entertainment variety program featuring a live orchestra. When these shows came to an end in 1932, Sun sponsored a program that continued for many years as an important launching pad for Sun products, the Lowell Thomas News of the World Program, formally sponsored by the *Literary Digest.* Carried on the NBC network, Lowell Thomas was one of the most popular news commentators on the airwaves, and his fifteen-minute evening program achieved excellent ratings for many years.[23]

These advertising campaigns also reflected the increasing competition that Blue Sunoco was encountering in the gasoline market. Sun's aggressive policy of selling one grade of antiknock gasoline at a "regular" price three cents below most "premium" grades brought competitive retaliation in the form of other "third grade" fuels dyed blue, increased advertising by the Ethyl Corporation, and other tactics by the competition. However, the greatest challenge to Blue Sunoco came from the increasing octane demands of higher compression automobile engines. Unless Sun opted to use tetraethyl lead in its thermally cracked gasoline, it would be unable to meet the high octane rating of the premium gasoline sold by competitors. It was this problem that thrust Sun into making one of its most important long-range decisions in 1933—the adoption of the Houdry Process for the catalytic cracking of petroleum. This step enabled Sun to emerge by the end of the decade as a leader in cracking technology and a potentially major producer of high octane aviation fuel on the eve of World War II, a story discussed in detail in the following chapter.

CRUDE PRODUCTION AND TRANSPORTATION

During the 1930s, Sun expanded its domestic production, but continued to follow a policy of purchasing the largest share of its total crude needs. The firm's leaseholdings increased from just over 2,346,000 acres in 1930 to 4,524,000 acres in 1938, and the number of wells it owned or operated jumped from 800 to almost 2,000. Sun's 1930 crude production of 8,387,000 barrels exceeded its purchases of 7,366,000 barrels that year, the only exception to an otherwise consistent pattern of purchases greater than its own production. In 1938, for example, the firm's production had risen to 14,550,000 barrels, but the Pews bought an additional 24,112,000 barrels on the open market, a ratio more typical of the company's policies.[24] Sun's own 1938 production represented 1.11 percent of the total United States effort, a percentage that declined only slightly to 0.98 in 1939 and 0.91 in 1940.[25]

Sun's tactic of purchasing a large share of its crude needs demonstrated the Pews' overall emphasis on the refining and marketing sides of their business as well as an abundance of cheap oil in the 1930s. Indeed, the instability of the crude markets in the Depression led J. Howard Pew to the conclusion that Sun's production department was a drain on the company's business. After first testifying at the T.N.E.C. hearings that the Sun production department had been running in the red for several years, Howard commented that "there is nothing that would give me so much satisfaction as to be able to buy all the crude oil that our company refines."[26]

The oil producing business had been in the doldrums ever since the giant East Texas field was discovered in 1930–1931. Thus, the general problem of price decline in the Depression years was further complicated by the continued overproduction, which confronted independent and integrated companies alike. A major thrust of public policy at the state and federal level during the New Deal period focused on various remedies to control production under the banner of conservation and industrial cooperation, a subject discussed below.

Within Sun Oil company management, there were differences between J. Howard and J.N. Pew, Jr. on one hand and their influential cousin, J. Edgar Pew, who ran Sun's production department on the other. J. Edgar had moved into the forefront of the prorationing movement during the decade as an advocate of the scientific limitation of production in flush fields. Never content with the concept of artificial limitations on the market, J. Howard and J.N., Jr. publicly expressed their main desires to obtain crude at cheap prices so that they could sell gasoline cheaply. Howard believed that if the market were left alone to run its course, there would always be oil to purchase. In his support of conservation legislation, however, J. Edgar had a better long-term understanding that crude production would ultimately be the most important phase of the oil business.[27]

The views of Sun's Philadelphia management were the object of displeasure of many independent oil producers, who found themselves the weak element in an industry increasingly dominated by the major firms. One Texas independent operator expressed this outlook when writing to Jack Pew in Dallas concerning Howard's T.N.E.C. testimony:

> However, when you get away from the part of the Sun Oil Company that you and your father control, which is this end of it down here, it is my opinion there is not too much admiration for your company on the part of a vast majority of the oil

> men, . . . some of them feel that the Philadelphia office of your company was one of the toughest hurdles and that Mr. Pew [J. Howard] was one of the last converts to an increase in price.[28]

Despite J. Howard's views, Sun continued to acquire extensive producing acreage in the thirties. The Pews realized that an integrated production arm was an essential weapon in competition with other integrated firms and insurance against future shortages in crude production. During the period of peak production in East Texas, Sun had a five percent interest in the field, with some 950 producing wells.[29] In May 1931, the firm made plans to construct a new 10-inch crude pipeline from the center of the East Texas field to Sun Station on the Gulf coast, 200 miles away. The Sun Pipeline Company of Texas built the line in conjunction with the Yount-Lee Oil Company, a Texas firm with East Texas holdings and a terminal at Smith's Bluff. With the use of high speed ditching machines and other mechanized equipment, Sun completed the line in thirty-nine days. In addition to the Sun-Yount-Lee line, the firm built up a system of over 150 miles of gathering lines in the East Texas field.[30]

The firm also completed crude pipelines comprising approximately 116 miles from its Yale, Oklahoma refinery to operating leases in Payne and Creek counties. Sun invested heavily in the Conroe and Anahuac fields in Texas, and discovered the Guerra and Sun fields in that state and the Chacohoula field in Louisiana. Sun also obtained increased producing acreage in Michigan, Indiana, and Illinois. A total production of 6.7 million barrels of crude in 1929 had nearly doubled to 12.3 million barrels by 1939.[31]

FINANCE

During the Depression decade, Sun invested heavily in its product pipeline, the Houdry Process for catalytic cracking, and expansion of its marketing system; all were moves that had important long-range effects on the firm's growth. The primary financial tool that Sun used was its traditional policy of a high reinvestment of earnings. Although Sun used this technique more than other oil companies, this was the dominant form of financing in the industry.

Sun's cash dividends on common stock as a percentage of net income were significantly lower than the petroleum industry average (for major companies), as reflected in the figures compiled by the Temporary National Economic Committee. The average for the seventeen companies that submitted data to the T.N.E.C. for the period 1929–1938 was 71.1 percent, while Sun paid only 27.2 percent of its net income out as cash dividends. Jersey Standard paid out 72.2 percent, Texaco 91.0, Atlantic

52.2, and Standard of Indiana 72.1 percent. Gulf Oil, which, like Sun, was largely owned by one family, (the Mellons), did not submit data to the T.N.E.C. Sun had an advantage in the close family ownership of the company. There was little pressure from stockholders for a higher cash dividend since the family owned such a preponderance of shares. In 1928, members of the Pew family owned eighty percent of the outstanding shares in Sun common stock, and by 1938, this figure had slipped to only seventy-one percent.[32]

Sun's policy of paying a low dividend on its common stock later had distinct tax advantages for the Pew family. Given the higher personal income taxes levied in the late thirties and afterward, family members would have been forced to pay high taxes on dividend incomes. Rather, they preferred to reinvest profits in the firm, thus enhancing the company and its value on the market. If and when a member of the family decided to sell stock, the capital gains taxes paid would be substantially lower than personal income tax rates. Although the policy of high reinvestment had been maintained for many years and was part of a well-integrated financial philosophy of the firm, one cannot discount the tax advantage argument in explaining why the Pews continued this policy over other possible methods of financing.[33]

In 1930, Sun increased its authorized common stock from 1,600,000 shares to 1,800,000 and approved issuance of 50,000 additional shares of preferred $100 par stock which had been authorized in 1927. These funds, along with earnings, were used primarily to fund improvements at the Toledo refinery.[34]

On July 1, 1931, Sun also issued $4,000,000 of two-year, 6 percent gold notes to provide funds for plant improvement obligations. The gasoline product pipeline constructed in 1930–1931 required a capital investment of $9,000,000. The firm borrowed $2,000,000 in short-term loans, raised $5,000,000 by a private placement of bonds with Metropolitan Life, the Equitable Assurance Society, and Prudential, and raised the other $2,000,000 with an equity stock offering in the Sun Pipe Line of Delaware. The depressed market conditions of the thirties increasingly found firms bypassing the investment banker and privately placing securities with institutional investors. Insurance companies needed an investment outlet for their collected funds, and Sun and other companies needed a market for their securities.[35]

The $2,000,000 of Sun Pipe Line of Delaware stock was broken down to fifty-five percent Class A and forty-five percent Class B stock. The Pew family subscribed to all of the Class A stock, which had a six percent earnings equity clause, a provision to limit the personal incomes of the family. The Sun Oil Company of New Jersey bought all of the Class B stock, which received all earnings above the six percent going to the

Pews. Thus, the Pew family retained control of the entire product pipeline system through their ownership in the Sun Pipe Line Company of Delaware and their majority ownership in Sun Oil.[36]

Once Sun had made an initial commitment to invest in, and later adopt, the Houdry Process, the company required large amounts of capital. Costs included the initial license purchase from Houdry, the purchase of Houdry Process Corporation stock, Sun's own research and development, and the erection of plants at both Marcus Hook and Toledo. In addition to reinvesting earnings, Sun sold new issues of common stock, made public bond offerings, and placed additional bonds with institutional investors.

In March 1934, Sun increased the value of its no-par common stock from 1,800,000 to 2,300,000 shares, and in August of that year made a semi-public offering of $6,500,000 in five-year 3¾ percent bonds. The firm placed $2,900,000 of this issue privately with two insurance companies and sold the rest through six investment banks in Philadelphia and Boston.[37]

In January 1937, the firm privately placed $9,000,000 in 2¾ percent serial debentures with Prudential Insurance, and in March increased its common stock authorization to 2,500,000 shares. By 1938, the construction of Houdry cracking units at Toledo and Marcus Hook demanded more capital, and in March, the board increased the common stock authorization from 2,500,000 to 3,000,000 shares. In December, Sun also placed $12,000,000 of ten-year, 2⅞ percent debentures with the Equitable Life Insurance Society. Most of this capital went to build Houdry units, but funds also went to construct crude pipe lines from the Michigan and Illinois fields to Toledo, to extend the product line in Ohio, to build crude lines in Stan County, Texas, and to enlarge storage facilities at Toledo.[38] Thus, like most of the other integrated oil companies, Sun invested heavily in plant expansion, new technologies, and improved transportation systems in the 1930s.

MANAGEMENT

In 1930, J. Edgar Pew, vice-president in charge of production, moved from Dallas to Philadelphia, where he worked until his death in 1946. Thus, Sun's decision-making process became even more centralized. J. Howard, J.N., Jr., and J. Edgar had offices near one another on the same floor in the Philadelphia office and they worked well together. There was no official executive committee until a managerial reorganization in 1947, and Sun's management made top level business decisions on a very informal basis. No minutes exist from these meetings, and many of them took place on the elevator or in the hall.[39]

Family management was not limited to the three top men. John G. Pew, J. Edgar's twin brother, was entrenched as the president of Sun Ship in Chester, Sun's largest subsidiary, and Arthur E. Pew, Jr. was vice-president in charge of refining. Some problems did exist at this level. J.N. Pew, Jr., for example, had difficulty keeping his hands off Sun Ship, a business that he had organized in 1916. As chairman of the Sun Ship Board, he often meddled in the daily operations of the firm and sometimes came into conflict with John G.[40] Similarly, J. Howard Pew was a refiner all of his life and found it difficult to grant Arthur, Jr. autonomy in this end of the business.

Conducting business this way enabled the firm to act quickly, but generally, it seems that too many routine operating decisions were made by top level management. For example, in reply to a 1932 *Fortune* questionnaire on the role of corporate purchasing, J. Howard Pew replied that "I meet with our purchasing agent at 8:30 every morning and spend about 15 minutes with him checking over every purchase which he proposes to make for the day, all of which he has carefully tabulated before him."[41] This procedure insured that Pew had his finger on the pulse of the company, but it was probably not the best use of his time. Although the various functional divisions had a degree of autonomy, all final decisions had to be approved by the Philadelphia management.

Sun's highly centralized management structure was quite typical of most larger firms in the late nineteenth and first half of the twentieth centuries. The firm's organization centered around functionally designated operating divisions—production, transportation, refining, and marketing—and the involvement of Pew family members at the head of each insured tight control. Managers funnelled reports into Philadelphia and all major decisions emanated from the central office. After World War I, however, new forms of corporate organization began to develop in American industry in response to the particular needs of firms that had grown extremely large or had diversified into related production areas.

In his pioneering study, *Strategy and Structure*, Alfred D. Chandler, Jr. traced the emergence of these new forms of business organization. Large industrial enterprises adopted new organizational structures as a response to demands that followed from the market strategies that they followed. Chandler focused on the evolution of the new, but today more common, forms of the multidivisional, decentralized business firm. Opposed to the highly centralized, functionally organized businesses that preceeded them, a handful of United States corporations developed independently, in response to particular needs, a structure consisting of four clearly separate levels: the general office, the central office of each division, the departmental headquarters, and individual field units. For his analysis, Chandler used four examples of firms that pioneered this pattern in the first three decades of the twentieth century—DuPont, General Motors, Standard Oil of New Jersey, and Sears Roebuck.[42]

The Standard case is most relevant to an analysis of Sun Oil. Chandler argues that beginning in 1925 and developing over a period of years, Standard Oil became the first major oil company to rationally develop a multidivisional, decentralized organization. Jersey had "fashioned for the first time a structure that formally defined the relationship between three basic administrative units—the functional department, the regionally defined multifunctional affiliate, and the general office which coordinated, appraised, and set policy for the autonomous divisions in the interest of the enterprise as a whole."[43] In this view, Jersey's new organization freed the general office from daily operating decisions and enabled Standard's top managers to concentrate on long-range planning and overall company policy.

The development of the multidivisional structure was not as simply conceived as this brief discussion suggests, and Chandler may have overstated the substantive changes Standard Oil actually made in the period studied, but the experience of Standard and the other firms he analyzed did pave the way for a new business institution. These organizational changes followed strategic responses to population growth and shifts, the vicissitudes of the business cycle, and the increasing pace of technological change. Sun Oil has felt the need to adopt this pattern only in the very recent past; it undertook a major company reorganization in 1975.[44]

Sun's integration under Pew management was organized vertically and functionally. Unlike another famous Pennsylvania oil family, the Mellons of Gulf Oil, the Pews did not diversify into other lines. Even their most distinct integrated step, the Sun Shipbuilding and Drydock Company, was in an enterprise directly tied to oil industry integration through oil tankers. The autonomous divisions identified by Chandler developed around either a distinct product line, as in General Motors' Chevrolet and Pontiac divisions or DuPont's paint and chemical divisions, or ultimately around geographic regions, as in the case of Jersey Standard. Sun Oil did not diversify to any degree and was not a large enough national, let alone multinational, company to require a radical reorganization. Rather, Sun remained a highly centralized, functionally organized company, small enough to operate effectively, but just large enough to run into managerial problems, particularly in the matter of long-range planning.

Sun had grown tremendously by the end of the 1930s, but in many respects, its management acted much the way that it had under J.N. Pew, Sr. when the firm was a small, specialized unit. The Pews had invested heavily in pipelines and refining processes, but moved quickly to reduce their debt as soon as possible and carefully reinvested a large part of their own profits into these projects. J.N. Pew, Jr. later commented that he wanted to go into stockholders' meetings confident that he "had the votes of his brother and sisters in his pocket,"[45] an attitude that reflected the personal, entrepreneurial view that he had of the Sun Oil Company. The upper management of the firm was clearly in the hands of the family, with members dominating each of the main functional divisions: production, transportation, refining, marketing, and shipbuilding. Because of this, the Pews shaped Sun Oil in very personal and particular ways.

LABOR RELATIONS

If family control reigned in the upper levels of Sun management, paternalism dominated the relations between Sun and its employees. In the post-New Deal world of industrial unionism, paternalism is a pejorative word, but in the older context of nineteenth- and early twentiety-century employer-employee relations, it described a working arrangement based in part on what the entrepreneur saw as his responsibilities to his workers. As long as wages and employee benefits satisfied the work force, they were happy with this relationship and resisted joining labor unions.

Sun had institutionalized this paternalistic outlook by adopting a policy of welfare capitalism in the twenties through its employee stock purchase plan, adoption of a life insurance scheme, and the sponsorship of a full-scale program of employee social and recreational activities. The Pews actively cultivated the idea of management and employees as one happy family, and this approach continued into the more difficult Depression years of the thirties.

Implicit in this relationship, however, was the assumption that management rights superseded employee rights, a view the Pews embraced. J. Howard Pew was a true patriarchal figure by virtue of his fifty-year tenure at the head of the Sun Oil Company. Employees felt that they could bring their problems to J. Howard, and the firm's family character helped it avoid major labor disputes through most of its history. It is significant that the first protracted strike at the Marcus Hook refinery did not occur until 1973, soon after Howard's death. At that time, the Oil, Chemical, and Atomic Workers Union, A.F. of L./C.I.O. organized the refinery. Up to that time, a combination of Sun paternalism and active resistance to organized labor had kept the unions out.[46]

Another significant measure of Sun's employee relations during the Depression lies in its excellent employment record. Unlike the general industrial pattern of lay-offs and wage cuts, Sun maintained a reasonably steady rate of employment throughout the 1930s. During the darkest days of the Depression in the fall of 1932, Sun was able to keep all of its employees on the payroll. Two-thirds of Sun's employees were working a five-day week and one-third a six-day week.[47] In the summer of 1933, the average Sun refinery worker worked forty hours a week and received a wage of $25.48. This compared favorably with the industry average of $26.46 in the refining sector. The average hourly wage at the large Marcus Hook refinery held constant, never falling below the 1929 rate of 58 cents an hour and increasing steadily after 1933, reaching 96 cents an hour by 1938.[48]

The Pews did balk at various schemes to reduce employee hours and hire more unemployed men. In correspondence with both Walter Teagle of the industry's "Share the Work Movement" and the American Petroleum Institute, J. Howard Pew expressed the view that this action would be self-defeating. By reducing the wages of present employees, there would be demoralization as well as real decline in their purchasing power. Pew also pointed out the cost involved in hiring many men to do the work of the present force.[49]

J. Howard Pew believed that the fundamental cause of the Depression lay in the maldistribution of wages in the twenties and the failure of purchasing power to maintain itself. He articulated this view in the Sun Oil Company Annual Report in 1932:

> Even before the depression it was plain that continued prosperity demanded the maintenance of a broad-based buying power such as could be assured only by a general program of liberal wages and salaries. If a larger share of prosperity's profit had gone to wages, there would have been more consumption and less speculation.[50]

This view was not unique among certain "enlightened" businessmen at the time, but it is interesting to note that the "conservative" Pew adopted the same position. More importantly, Pew became convinced that this was the correct view and actively moved to maintain wages in his firm and influence others to follow suit.[51]

Various academicians and public officials working on solutions to the unemployment problem contacted business leaders, including J. Howard Pew, on the unemployment question in the early days of the Depression. There was an interesting correspondence in the spring of 1931 between Pew and Alvin H. Hansen, professor of economics at Minnesota. Hansen would soon leave for Harvard, where he contributed, independently of John Maynard Keynes, to the body of macroeconomic theory explaining the business cycle which some economists later termed the Keynes-Hansen thesis. At this particular time, Hansen worked as a consultant for the Pennsylvania Unemployment Committee and had solicited Pew's views on the problem. Pew cited his own firm's record of maintaining 1929 levels of employment and surprisingly advocated a program of unemployment insurance that "would compensate workers when out of employment when such unemployment is brought about by conditions entirely beyond their control."[52] This position coincided with Pew's own emphasis on the consumption function as the key to understanding the business decline.

J. Howard Pew maintained a consistent posture on this issue. After conferring with Pew in June 1932, Dr. Joseph H. Willits of the Wharton School of Finance and Commerce sent data to "illustrate your theory that the depression has been created, or at least deepened by the failure of wages to keep pace with production and profits."[53] When, a year later, the National Recovery Act was being ironed out in the Congress, J. Howard Pew wrote to the president of the National Association of Manufacturers that "the time has come for the Government, in cooperation with industry and workers, to adjust wages." This, of course, was the period of the business "honeymoon" with Roosevelt and the New Deal, a period that was exceedingly brief for the Pews. But, although J. Howard Pew became disillusioned with N.R.A. and particularly with its collective bargaining provisions, he firmly supported the N.R.A. effort to increase wages. Pew later referred to the "chiseling of wages" by corporations which resulted in lower wages than was "clearly the spirit and intent of the N.R.A. and the respective codes."[54]

The Pews drew the line, however, at provision 7a of the N.I.R.A., and later at the Wagner Act which replaced the N.I.R.A. collective bargaining clause in 1935. It was perfectly logical to J. Howard Pew that wages should be raised, but this should be done at the discretion of the employer, who had a duty to do so. It should not occur as a result of the threat of industrial action. On this issue, Pew conservatism and paternalism held forth. After passage of the N.R.A., Sun Oil complied with section 7a by adopting an employee representation scheme or company union, the "Sunoco Employee and Management Counsel."[55] This company-dominated union remained the designated employee bargaining unit at the Sun Oil refineries until 1973 when the A.F. of L./C.I.O. gained representation.

Although the International Association of Oil Field, Gas Well and Refinery Workers had some organizational success during the N.I.R.A. era, most major oil companies had organized either company unions or employee representation schemes by 1935. In addition to Sun, Phillips, Continental, Texaco, Magnolia, Gulf, Pure, Standard of California, and Union also organized company unions. In 1935, the International changed its name to the Oil Workers International Union and affiliated with the renegade C.I.O. It also had very limited success, however, its best organizational results coming in the refining sector.[56]

There were some attempts by Sun workers to organize with "outside" unions during the thirties, particularly in areas removed from Marcus Hook and Philadelphia. But even here, Sun management kept out the International by matching any settlements the unions achieved and maintaining good relations with their men. There were movements to organize workers in the Oklahoma fields, but Twin State's general manager assured J. Edgar Pew in 1933 that "you can depend on our employees being 100 percent loyal." He went on to say "I assured the boys that we would take care of them as we had always done in the past and that I did not think it would be wise for them to join any organization."[57] One can only guess what was meant by its not being wise, but the managers of Sun's Oklahoma production affiliate had apparently made it clear that they were unhappy with the International's organizational efforts in the Healdton pool.

When union contracts were signed in Sun's midwest marketing territory in 1934, the firm tried to reach agreements with its employees for similar wage settlements. Walter C. Pew wrote to Howard informing him of his and vice-president Eckert's advice that Sun follow the pattern of union negotiated contracts for service station employees, truck drivers, warehousemen, and maintenance men.[58]

At Sun Oil's subsidiary, Sun Shipbuilding and Drydock, there were more serious efforts at outside organization beginning in 1937. After years of bitter struggle with charges and countercharges of abuse on both sides, the Industrial Union of Marine and Shipbuilding workers/ C.I.O./ won a National Labor Relations Board representation election in 1943. Among the charges leveled at Sun were that they had maintained their "creature union" by threats, physical violence by anti-labor goons, and by firing union organizers.[59]

In 1940, J.N. Pew, Jr., by this time well known in national political as well as petroleum circles, commented on unions in the following manner: "Proper unionism puts an intelligent floor under wages on a craft basis without, as the government does, putting a ceiling on wages too."[60] But for Joe Pew, proper unionism meant company-dominated unions. As vice-president of Sun Oil and chairman of the board of Sun Ship, he fought hard to keep things that way. The Pews's reaction to the compulsory unionization and collective bargaining made possible by the New Deal was typical of that of many businessmen. They saw this particular aspect of "federal encroachment" as a threat to their fundamental management rights. This issue was important in turning a significant segment of the business community against the New Deal and Roosevelt. The New Deal onslaught had made a dent in the paternalistic labor-management policies at Sun, but it had not destroyed them. By granting good wage scales and benefits, Sun kept outside unions out for a very long time afterward.

SUBSIDIARIES

Sun made some important changes in the organization of various subsidiaries during the Depression decade. In 1932, the British Sun Company, a firm that traced its history back to the founder, J.N. Pew, Sr., was reorganized as British Sun Oil Ltd. The long association with William Smellie in Liverpool came to an end, and the company's office moved to London with Arthur Stephens at its head. Although the complexion of Sun's European business had changed somewhat since the days of Sun Red lubricating oils, Sun still marketed primarily specialty items like lubricants, waxes, and industrial oils.[61]

Although it had not gained control of the German affiliate, Mineraloelwerke Albrecht & Company, Sun obtained a financial interest in this Hamburg firm during the twenties. Long a distributor of Sun products on the continent, Albrecht had run into severe financial difficulties in the Depression and sought to liquidate in 1931. At this time, Sun had $707,000 in the form of oil advances, stock interest, and loans in the German firm. The Pews and the Albrechts had been friends for many years and they reached an agreement in 1931 for Sun to become a full partner in the firm to forestall liquidation.[62]

Things went downhill, however, and Ernst and Walter Albrecht came to Philadelphia in 1934 to renegotiate the entire agreement. In the summer of 1934, Sun and the Albrechts agreed to liquidate the firm. Once funds earmarked for Sun had been placed in German banks, however, it became difficult to get them out of Germany under Nazi law. To alleviate this problem, Sun purchased six-inch steel pipe manufactured in Germany and had the pipe shipped to Philadelphia. The company originally intended the pipe for the Middlesex product pipeline, but it eventually ended up in Texas, where it was used in crude lines.[63]

With the influx of East Texas oil in 1931 and the gradual diminishing of the older Oklahoma fields, Sun began to phase out its Oklahoma production business. The Twin State Oil Company, Sun's Oklahoma subsidiary, had become a financial liability by 1936, and in November of that year, Sun's board voted to liquidate the company. The parent firm absorbed both Twin State and the Sun Company of Delaware (which operated the Yale refinery). In order to do this, Sun Oil (New Jersey) domesticated itself in the states of Oklahoma and Kansas, and arranged for another subsidiary to assume the operation of some small Twin State holdings in California.[64]

In December 1938, Sun Oil consolidated its holdings in its product pipeline by liquidating and dissolving the Sun Pipe Line Company. It transferred all assets of the company including the outstanding stock of the Susquehanna Pipe Line, Sun Pipe Line, Sun Oil Line Company, and Middlesex Pipe Line to the Sun Oil Company. The parent company and the Pew family now had direct control over the four operating companies that comprised the product pipeline system.[65]

The Sun Shipbuilding and Drydock Company fell on lean years in the interwar period. Unable to keep its ways busy with ship contracts, the yard kept alive with ship repair work, steel plate work, machinery construction, and fabricating refining equipment. However, as war began to threaten in the late thirties, Sun Ship began to land several highly profitable ship contracts. In the earlier years of the decade, Sun's refining and marketing profits made up for the losses that Sun Ship displayed in the corporation's consolidated profit figures. In 1938, however, a bad year for Sun Oil, the income from the Yard and Sun's pipe lines helped offset the short-term losses that the firm showed in its production, refining, and marketing activities. In 1939, the unconsolidated income account reported to the Securities Exchange Commission showed that the parent Sun Oil Company earned $1,000,000 and Sun Ship $3,600,000. In addition to this, crude pipelines brought in $950,000 and the gasoline product lines $1,100,000.[66]

Sun's flexible, functional integration allowed the corporation to offset losses in one area with earnings in another. This was one of the main charges leveled at the major oil companies during the various monopoly hearings held during the thirties. When the Second World War came, Sun Ship was in a position not only to create big earnings for the entire Pew interests, but to provide the United States with badly needed oil tankers. By the end of the war, the Chester concern had become the largest privately owned shipyard in the world.[67]

GROWTH SUMMARY

Sun had performed well during a most difficult decade and emerged at the end much stronger than it had been at the beginning. In two key areas, catalytic cracking and shipbuilding, investment risks and faith in the future soon paid dividends as war created unprecedented demands for aviation gasoline and oil tankers. Investments in pipeline development also strengthened Sun's transportation system and gasoline marketing. If Sun deserves criticism for its resistance to the march of organized labor, it merits praise for maintaining high employment levels within the firm during the Depression. Sun achieved industry prominence through a blend of cautious conservatism on one hand and daring innovation on the other.

Despite the gains the company made on various fronts, it still had not grown to the extent that it really threatened the giants of the industry. One may trace much of Sun's size limitation to the firm's financial policies and the Pews' desire to keep control in the family. However, their philosophy went beyond their views on finance. J. Howard Pew stated in 1939 that "I very definitely believe that there is in every corporation the seeds of destruction, and that inevitably when a corporation gets too large it is going to go broke. I don't believe it is possible for a company to continue growing indefinitely."[68] This attitude was reflected in the way the Pews managed Sun.

This is certainly not to say that Sun had no important impact on the industry. Its gasoline marketing, pipeline developments, and aggressive competitive stance made it a very significant industrial unit. Without identifying the firm by name, economists Edwin G. Nourse and Horace B. Drury discussed Sun's impact in a study of price policies published by the Brookings Institution in 1938. They described "a type of management which consciously subscribes to the philosophy of striving vigorously for the lowest possible costs, and then extending its field of operations by the lowering of prices—at least to whatever point may be necessary to market its expanding volume by output."[69] Nourse and Drury proceeded to discuss the competitive tactics Sun displayed in marketing, refining, pipeline development, and the management of their company, which led it to be recognized as one of the most efficiently organized in the country. More important, they argued, was the fact that the firm had passed on savings achieved through technological process to the consumer.[70]

Sun had remained independent and competitive in a decade when cooperation and economic stabilization were the keynote of the day. There is no better example of this independence than the firm's tenacious adherence to the policy of exclusively marketing lead-free gasoline that led the firm into a major program of technological innovation in the 1930s.

NOTES

1. *Annual Report of the Sun Oil Company, 1930, 1931, 1938.*
2. Robert Sobel, *The Age of Giant Corporations: A Microeconomic History of American Business 1914-1970* (Westport, Connecticut: Greenwood Press, Inc., 1972), pp. 122-8.
3. Ibid., pp. 127-8, 152.
4. "Sun Oil," *Fortune* 23, 12 (February 1941), p. 114; *Annual Report of the Sun Oil Company, 1936.*

5. Arthur M. Johnson, *Petroleum Pipelines and Public Policy, 1906–1959* (Cambridge: Harvard University Press, 1967), p. 255; Henrietta M. Larson, Evelyn H. Knowlton and Charles S. Popple, *New Horizons, 1927–1950*, Volume III, *History of Standard Oil (New Jersey)* (New York: Harper & Row, 1971), p. 233; "550 Mile Pipe Line for Gasoline to be Built," *Our Sun* 7, 2 (February 10, 1930), p. 3.

6. Melvin deChazeau and Alfred E. Kahn, *Integration and Competition in the Petroleum Industry* (New Haven: Yale University Press, 1959), p. 292; Porter Howard, "History of the Sun Oil Company" (unpublished manuscript, 1957(?)), Folder #1, "Sun Oil History," Library File, Sun Company Library, Marcus Hook, Pennsylvania, p. 27; "J.N. Pew, Jr., Marks Half a Century of Service With Sun," *Our Sun* 23, 3 (Summer 1958), p. 12.

7. "History and Growth of Pipe Lines," *Our Sun*, 50th Anniversary Issue, 3, 1 (1936), p. 17; Howard, "History of the Sun Oil Company," p. 27; "Sun Oil," *Fortune*, p. 112; Robert G. Dunlop, Interview held at St. Davids, Pennsylvania, October 31, 1975.

8. "History and Growth of Pipelines," *Our Sun*, pp. 17–18; "550 Mile Pipe Line," *Our Sun*, p. 3; Howard, "History of the Sun Oil Company," pp. 27–8.

9. Johnson, *Petroleum Pipelines*, pp. 32–3, 70–81.

10. *Ibid.*, pp. 255, 262; see folder, "Susquehanna Pipeline," Acc. #1317, Eleutherian Mills Historical Library, File 21-A, Administrative, Box 62; John L. Kelsey, "A Financial Study of the Sun Oil Company of New Jersey From its Inception through 1948" (M.B.A. Thesis, Wharton School, 1959), p. 18.

11. Harold F. Williamson, Ralph L. Andreano, Arnold R. Daum, and Gilbert C. Klose, *The American Petroleum Industry*, Volume II, *The Age of Energy, 1899–1959* (Evanston: Northwestern University Press, 1959), pp. 108–9; *Ibid.*, pp. 255, 263.

12. Johnson, *Pipelines*, p. 254; Kelsey, "Financial Study," p. 112.

13. Johnson, *Pipelines*, p. 261; "History and Growth of Pipe Lines," *Our Sun*, p. 18.

14. Johnson, *Pipelines*, p. 255; "Sun Oil," *Fortune*, p. 112.

15. Williamson et al., *Energy*, pp. 590–1; "Sun Oil," *Fortune*, p. 114.

16. de Chazeau and Kahn, *Integration and Competition*, note, p. 101.

17. "Sun Oil," *Fortune*, p. 114.

18. *Ibid.*; "Sun Oil Company Domestic Service Stations, July 11, 1939," Acc. #1317, EMHL, J. Howard Pews Papers, Box 49.

19. "Sun Oil," *Fortune*, p. 114; "Testimony of J. Howard Pew," U.S. Congress, Temporary National Economic Committee, *Hearings Before the Temporary National Economic Committee of the United States* (T.N.E.C.), *Part 14, Petroleum Industry* (Washington, D.C.: U.S. Government Printing Office, 1940), pp. 7192–3, 7214; Kelsey, "Financial Study," p. 36; *New York Times* (February 24, 1941), p. 25.

20. "J. Howard Pew T.N.E.C. Testimony," pp. 7207–13; "Testimony of J. Howard Pew," *Pennsylvania Oil Investigation*, Book 4 (1937), p. 940, Acc. #1317, EMHL, J. Howard Pew Papers, Box 48; "Blue Sunoco 25th Anniversary, 1927–1952," *Our Sun* 17, 2 (Spring 1952), pp. 31–2.

21. Williamson et al., *Energy*, Table 19:4, p. 681.

22. "Advertising Reflects the Progress of Sunoco," *Our Sun*, 50th Anniversary Issue, 3, 1 (1936), p. 38.

23. *Ibid.*, p. 37.

24. Roy C. Cook, *Control of the Petroleum Industry by Major Oil Companies*, Monograph No. 39, Temporary National Economic Committee (Washington, D.C.: U.S. Government Printing Office, 1941), pp. 64, 68, 70.

25. "Sun Oil Company Ratios, 1937–1947, Table 1-a," May 10, 1948, Acc. #1317, EMHL, J. Howard Pew Papers, Box 58.

26. "J Howard Pew T.N.E.C. Testimony," p. 7206.

27. *Ibid.*, pp. 7206–7; Dunlop interview.

28. J.S. Bridwell, Bridwell Oil Company, to Jno. J. "Jack" Pew, February 17, 1944, Accession #1317, EMHL, J. Edgar Pew Production File #71, Box 88.
29. "The Story of Sun," *Our Sun,* 75th Anniversary Issue 26, 3–4 (Summer/Autumn 1961), p. 29.
30. *Annual Report, Sun Oil,* 1931; "History and Growth of Pipe Lines," *Our Sun,* 50th Anniversary Issue, 3, 1 (1936), pp. 16–7.
31. "History and Growth of Pipelines," *Our Sun,* p. 17; "Story of Sun," *Our Sun,* p. 29.
32. "Holders of Common Stock ,May 25, 1938," Acc. #1317, EMHL, J. Howard Pew Papers, Box 50; "Analysis of Common Stockholdings of 100 Largest Stockholders," A.P.I. study of T.N.E.C. Questionnaire, Question 3, Table IV: "Cash Dividends on Common Stock as a Percentage of Net Income," A.P.I. study of T.N.E.C. Questionnaire, Question 9, Table I; Acc. #1317, EMHL, J. Howard Pew Papers, Box 45; see A. R. Koch, *The Financing of Large Corporations* (Washington, D.C.: National Bureau of Economic Research, 1943), p. 83.
33. Kelsey, "Financial Study," pp. 76–7.
34. Board Minutes, Sun Oil Company, January 28, 1930, Minute Book #5, pp. 4–5.
35. "Statement Prepared for T.N.E.C.." p. 2, Acc. #1317, EMHL, J. Howard Pew Papers, Box 45; Kelsey, "Financial Study," p. 19.
36. Kelsey, "Financial Study," pp. 18–9; "Sun Oil," *Fortune,* p. 51.
37. Board Minutes, Sun Oil Company, March 13, 1934, Minute Book #5, p. 150; August 3, 1934, Minute Book #5, p. 178; "Sun Oil Company Sale of Securities and Long Term Debt," Acc. #1317, EMHL, J. Howard Pew Papers, Box 45.
38. Board Minutes, Sun Oil Company, January 14, 1937, Minute Book #6, pp. 58–9; March 9, 1937, Minute Book #6, p. 140; December 13, 1938, Minute Book #6, p. 176; December 38, 1938, Minute Book #6, p. 181.
39. Dunlop interview; "Sun Oil," *Fortune,* p. 52.
40. "Jack" Pew interview; Dunlop interview.
41. *Fortune* to Howard Pew, July 7, 1932; Pew to *Fortune,* August 11, 1932, Acc. #1317, EMHL, File 21-A, Box 67.
42. See Alfred D. Chandler, Jr., *Strategy and Structure: Chapters in the History of the American Industrial Enterprise* (Cambridge: M.I.T. Press, 1962), introduction, pp. 1–17.
43. *Ibid.,* p. 224.
44. *Ibid.,* p. 25. For a discussion of Sun's recent decentralization, see *Sun Oil: Building Flexibility for the Future* (St. Davids, Pennsylvania: Sun Oil Company, 1975).
45. Dunlop interview.
46. Dan Rottenberg, "the Sun Gods," *Philadelphia Magazine* (September 1975), p. 195.
47. J. Howard Pew to Walter Teagle, "Share the Work Movement," September 24, 1932, Acc. #1317, EMHL, File 21-A, Box 70.
48. J. Howard Pew to W.R. Boyd, Jr., A.P.I., June 8, 1933, Acc. #1317, EMHL, File 21-A, Box 73; "Sun Oil Company, Marcus Hook Average Hourly Wage, 1913–1938," Acc. #1317, EMHL, J. Howard Pew Papers, Box 45.
49. J. Howard Pew to Teagle, September 24, 1932; Teagle to Pew, September 28, 1932, Acc. #1317, EMHL, File 21-A, Box 70; J. Howard Pew to W.R. Boyd, Jr., June 8, 1933, File 21-A, Box 73; J. Howard Pew to Amos L. Beaty, A.P.I. President, May 28, 1932, Acc. #1317, EMHL, File 21-A, Box 65.
50. *Annual Report, Sun Oil,* 1932.
51. J. Howard Pew to C.A. Musselman, Chilton Class Journals, April 3, 1933, Acc. #1317, EMHL, File 21-A, Box 74; J. Howard Pew to R.L. Lund, N.A.M., June 5, 1933, File 21-A, Box 76; J. Howard Pew to William Smellie, January 38, 1931, File 21-A, Box 62.
52. Hansen to Pew, April 15, 1931; Pew to Hansen, April 20, 1931, Acc. #1317, EMHL,

File 21-A, Box 63. For a discussion of Hansen's contributions to the Keynesian synthesis, see Robert Lekachman, *The Age of Keynes* (New York: Vintage, 1966), pp. 126–37.

53. Willits to J. Howard Pew, June 4, 1932; Pew to Willits, June 8, 1932, Acc. #1317, EMHL, File 21-A, Box 70.

54. J. Howard Pew to J.S. Crutchfield, American Fruit Growers, Inc., November 6, 1933, Acc. #1317, EMHL, J. Howard Pew Papers, Box 68.

55. "Sunoco Employee and Management Council Handbook"; W.D. Mason to J. Howard Pew, June 30, 1933, Acc. #1317, EMHL, J. Howard Pew Papers, Box 61; "The Employee Representation Plan," *Our Sun* 10, 3 (August 1933), p. 2. For a discussion of the causal relationship between N.I.R.A. and the growth of company-dominated unions, see Stuart D. Brandes, *American Welfare Capitalism, 1880–1940* (Chicago: University of Chicago Press, 1970), pp. 142–4.

56. Herbert Werner, "Labor Organization in the American Petroleum Industry," Appendix; Williamson et al., *Energy,* pp. 835–6.

57. E.J. Gorman, Twin State Oil, to J. Edgar Pew, July 11, 1933, Acc. #1317, EMHL, File #71, Box 79.

58. Walter C. Pew to J. Howard Pew, April 14, 1934, Acc. #1317, EMHL, J. Howard Pew Papers, Box 68.

59. "It's C.I.O. at Sun," *Business Week* (July 17, 1943), p. 92; "Rebuke by N.L.R.B.," *Business Week* (June 21, 1941), p. 52.

60. Marquis Childs, "Pennsylvania's Boss Pew, a G.O.P. Power Out of Grim Hate of New Deal," *St. Louis Post Dispatch* (May 26, 1940), Section C., p. 1.

61. William Smellie to J. Howard Pew, December 10, 1931; folder, "British Sun," Acc. #1317, EMHL, J. Howard Pew Papers, Box 59; "Golden Anniversary in London," *Our Sun* 24, 2 (Spring 1959), p. 24.

62. J. Howard Pew to Ernst Albrecht, June 12, 1931, Acc. #1317, EMHL, J. Howard Pew Papers, Box 56.

63. J. Howard Pew to Ernst Albrecht, March 23, 1934, Acc. #1317, EMHL, J. Howard Pew Papers, Box 56; Howard, "History of Sun Oil Company," p. 31.

64. Board Minutes, Sun Oil Company, November 17, 1936, Minute Book #6, pp. 40–1; T.L. Foster to Frank Cross, October 28, 1936; J. Edgar Pew to R.W. Pack, Beaumont, November 30, 1936; J. Edgar Pew to F.S. Reitzel, Philadelphia, November 30, 1936, Acc. #1317, EMHL, File #71, Box 79.

65. Board Minutes, Sun Oil Company, December 13, 1938, Minute Book #6, p. 176; Kelsey, "Financial Study," p. 22.

66. "Sun Ship's 30 Years," *Our Sun* 12, 2 (July/August 1946), p. 5; "Sun Oil," *Fortune,* p. 52; *Annual Report, Sun Oil,* 1931–1939.

67. "Sun Ship's 30 Years," *Our Sun,* p. 4.

68. "J. Howard Pew T.N.E.C. Testimony," p. 7231.

69. Edwin G. Nourse and Horace B. Drury, *Industrial Price Policies and Economic Progress* (Washington, D.C.: The Brookings Institution), pp. 195–6. See "Excerpts From Industrial Price Policies and Economic Progress by Edwin G. Nourse and Horace B. Drury"; Folder #1, "Sun Oil History," Library File, Sun Company Library, Marcus Hook Pennsylvania, pp. 1–3.

70. *Ibid.*

CHAPTER VII

The Ethyl Challenge and Sun's Technological Response: The Houdry Process

Gasoline had become the most important commodity sold by the Sun Oil Company in the 1930s. By 1937, it represented 58.6 percent of all product sales and remained in the range of 56 percent for the rest of the decade.[1] Long noted in former years as a company of lubricant specialists, the firm was now in the mainstream of the industry. Accompanying its increased sales of gasoline in the Depression decade were large investments in petroleum refining. Faced by the pressures of vigorous competition with Blue Sunoco in Sun's marketing territory and rising octane requirements to prevent knocking in the automobile industry's higher compression engines, the firm turned to technological innovation for a solution to its problems.

The literature on technological innovation addresses the subject from many disciplinary perspectives—economics, business management, industrial systems and engineering, sociology, and anthropology.[2] Many recent studies, however, have challenged some of the conventional wisdom concerning the relationship of innovation to firm size, value systems, institutionalized research and development, the process of invention, economic demand, and the role of government.[3] As a result of these studies, one realizes that a simple model to explain the process of innovation does disservice to what is a highly complex phenomenon. The one common denominator that does exist at the start of all innovations is the presence of a perceived need or want, whether this be a strictly economic need or not. In the case of Sun Oil, there were strong economic reasons for wanting to continue marketing Blue Sunoco. Yet, Sun could have licensed tetraethyl lead from the Ethyl Corporation, abandoned its no-lead policy, and probably thrived in the continually growing gasoline markets. Therefore, one can argue that there were other perceived wants that played an active part in Sun's corporate decision-making, among them the continued desire of the Pews to retain their independence and a sincere belief that their gasoline was superior to leaded fuels.

Initially, the firm concentrated on improving its Cross-licensed thermal cracking stills to upgrade the quality of its gasoline. By further increasing the operating pressure and temperature of this equipment, Sun was able to independently obtain the high octane base stock necessary to blend Blue Sunoco. In 1933, however, new circumstances led the company to adopt a totally new technology brought in from the outside, the Houdry fixed-bed process of catalytic cracking. This pioneering process was developed by an independent inventor, Eugene Houdry, in the 1920s, and Sun only became involved after the rest of the petroleum industry had rejected it.

Previous research into cracking had concentrated on the problem of obtaining increased yields of gasoline from each barrel of crude. This had been the initial thrust of Houdry's work also, but in the glutted market following the East Texas strike of 1931, it was no longer a pressing concern for the industry. For Sun, however, another feature of Houdry-cracked gasoline was most attractive—its extremely high octane rating (88 to 92). Thus the unique marketing strategy launched by the firm in 1927 is the key to understanding Sun's leadership in refinery innovation in the 1930s.

THE ETHYL CHALLENGE

As long as Sun could maintain a high octane gasoline without using tetraethyl lead, it enjoyed many marketing advantages. In addition to selling a motor fuel at three to five cents below the "premium" brands of its competitors, Sun saved money because it required fewer bulk-plant tanks, marketing trucks, pumps for gasoline stations, and less labor and bookkeeping expense.[4]

Blue Sunoco competed with all "premium" gasolines as well as "regular" grades, but increasingly, Ethyl gasoline sold by many competitors became the firm's main target. Sales of Ethyl Fluid had steadily increased since its introduction. In 1927, the inaugural year of Blue Sunoco, only two percent of all gasolines sold nationally contained Ethyl. This doubled to four percent in 1928, and again to eight percent in 1929. The Ethyl Corporation hoped to expand its sales to twelve or fourteen percent of the total in 1930.[5]

While Sun advertised the premium qualities of its gasoline sold at regular price, other companies also used improved thermal cracking to upgrade their non-leaded gasolines. These companies usually sold this fuel as a third grade, however, and not as a regular gasoline. The Atlantic Refining Company, for example, introduced its "White Flash" brand with a 65 to 70 octane rating, and sales of Ethyl dropped from sixteen percent to ten percent of the gas sold in Atlantic's marketing area. When Standard of Ohio brought out its similar unleaded "X-70" brand in 1931 at 68-70 octane, Ethyl's percentage dropped dramatically from forty to twenty-four percent.[6]

The Ethyl Corporation (and General Motors, Jersey Standard, and DuPont) realized that by limiting the sale of Ethyl fluid to only premium grades, they were stimulating the refining industry to solve the problem of engine knock in other ways. Thus the competition presented by Blue Sunoco, White Flash, X-70, and other non-leaded, antiknock gasolines pushed the Ethyl Corporation into a major marketing change. In April 1933, the Ethyl Corporation for the first time offered a lead additive to be put into regular gasoline. The new substance, "Q Fluid," was slightly different in composition from "Ethyl Fluid," but it was a similar compound of tetraethyl lead. In its promotional campaign, the Ethyl Corporation stressed the cheaper costs and greater flexibility achieved by using lead rather than extreme thermal cracking to upgrade the quality of gasoline.[7]

There was yet another form of leverage that Ethyl could use to upgrade its sales. The compression ratios of internal combustion engines were increasing incrementally in the early thirties, and the Ethyl Corporation believed that the broadening usage of TEL "would become a yardstick by which the motor manufacturers could design with respect to compression ratios."[8] Ethyl, half owned by General Motors, was in a particularly good position to influence this development. The success of Sun and other lead-free marketers caused a great deal of consternation in the board room of Ethyl Corporation. In 1931, E.W. Webb, Ethyl's president, wrote to Irénée duPont, a member of the board, that

> the competition of the standard brands of Sun Oil Company, Atlantic Refining, and Standard of Ohio, during the period they have been marketing 65 75 octane gasoline has been most distressing to Ethyl. The Ethyl sales have decreased very substantially in such areas. Sun is not a customer of ours and does not sell a premium gasoline, but because of the good antiknock quality of their Blue Sunoco they have made it very difficult for their competitors' brands of regular gasoline as well as Ethyl.[9]

Webb suggested the implementation of a new strategy, one that soon became evident with the introduction of new General Motors automobile models. He argued that "Ethyl sales are going to be hurt unless higher compression engines are brought out," and suggested engines with at least a 78 octane requirement.[10]

Whether one brands the interlocking interests of General Motors and Jersey Standard in the Ethyl Corporation as unfair collusion or industrial efficiency, there is no doubt that an increasing cooperation developed between engine designers and motor fuel manufacturers. A certain amount of cooperation was of course essential, since engine development could not proceed if adequate fuels were not available in the marketplace. Furthermore, the public responded well to the performance benefits of improved engines. One still wonders, however, whether the rate of increase in engine compression ratios was really warranted. By 1938, the octane rating of regular gasoline was greater than that of premium grades in 1930.[11] During this period, the Ethyl Corporation became both an octane and price leader in the petroleum industry and most of the major firms fell into line behind it. All, that is, except Sun Oil.

INCREASED COMPETITION

This independence brought competitive retaliation in the marketplace. In April 1931, the Standard Oil Company of Indiana, the first major licensee of Ethyl Fluid, took actions apparently aimed at Sun. Standard began to market an inferior third grade of gasoline also colored blue. Blue Sunoco's success in Michigan and the prospect of extension of this product into Indiana with the new Sun product pipeline prompted Standard to embark on this campaign, apparently designed to discredit the concept of blue gasoline. Representatives of Jersey Standard observed the Indiana success and moved shortly thereafter to market a third grade of "fighting" blue gasoline in the Philadelphia stations of its subsidiary, the Standard Oil Company of Pennsylvania. When Sun vice-president S.B. Eckert telephoned his counterpart at Pennsylvania Standard, he was informed that all of the Jersey Standard Companies had agreed to put out a fighting quality gasoline as a third grade and colored blue.[12]

Sun's repeated protests and the threats of bad publicity resulting from the old monopoly company attacking an aggressive "independent" caused Jersey's Walter Teagle to halt this practice in August 1931 (Standard then dyed its third grade green). However, the tactic appeared in other marketing areas where Sunoco gasoline began to eat into the sales of competing companies. In 1932, the Continental Oil Company introduced a blue gasoline, "Blue Conoco," an apparent attempt at copying Sunoco, but it stopped in September 1932, following protests from Sun's legal department. Sun did not have a patent on blue gasoline, even though the firm had tried to obtain one. In the case of "Conoco," Sun's lawyers exerted pressure by claiming infringement of the "Blue Sunoco" trademark and the blue color schemes of their advertising and gas stations.[13]

These challenges to Blue Sunoco's place in the market reflected the very real inroads that the Sun product had made into the sale of premium gasolines, particularly Ethyl fuels containing lead. As a result, several firms began to object to the Sun claim that "Blue Sunoco contains more premium qualities than many extra price gasolines." In October 1932, someone, apparently representing a Sun competitor, filed a complaint with the National Better Business Bureau against the Philadelphia firm's advertising campaign in the Detroit marketing area.[14] The bureau ran tests of Sunoco gasoline and other competitors' products, administered by a University of Michigan professor, and charged that Sun could not back up its claim.[15] Sun replied to this group in November that there were more premium qualities than antiknock alone. It cited low sulphur, low gum, quick acceleration, ease of starting, low vapor pressures, and "absence of harmful lead" as premium characteristics. Sun went on to argue that

> In brief, we cannot recognize that the Ethyl Gasoline Corporation holds any monopoly on premium gasolines nor can they set up arbitrary standards and constitute themselves as judge and jury in deciding what a premium product is, eliminating the entire field of premium competition merely by adding a little more lead in order to keep up their theoretical non-detonating value a trifle ahead of competition.[16]

Sun took a further, bolder step in response to the Better Business complaint. The firm commissioned the same Michigan professor, George Granger Brown, to run more comparative tests, and it then developed a full-page national advertisement of the results. Before running the ad, Sun wrote letters to six companies in the Michigan marketing region, Esso (Jersey), Atlantic, Texaco, Tydol, Shell, and Gulf, apologizing in advance for the comparative ad, but claiming that they "were forced" into it by the Better Business Bureau. Sun's sales manager sent an advance copy of the advertisement with each letter on December 29, with the information that it would run the following Saturday.[17]

The advertisement reported the tests of Blue Sunoco and twenty other "premium" grade gasolines. Of the twenty, Sunoco had more premium qualities than thirteen, the same number as five, and a lesser number than only two of the others. The premium qualities in the test were defined as ease of starting, acceleration in responsiveness to throttle, knock rating, gum content, chemical stability, sulphur content, purity (defined as freedom from foreign chemical compounds including lead), freedom from vapor lock tendency, and the number of heat units contained in each gallon of fuel. No names of any of the other twenty products appeared in the advertisement. In the letter written to the six companies, Sun said it would not run the advertisement if the Better Business Bureau dropped its complaint.[18]

No reply came from the Bureau, and the controversial piece ran in several December 31, 1932 newspapers. The other oil companies were not amused, and R.C. Holmes, president of Texaco, wrote J. Howard Pew in January that Sun should discontinue this type of advertising. He indicated that his company had accumulated quite a file on these tactics, going back to 1928 and 1929. Max Leister, sales manager, replied for Sun, maintaining that proof lay behind Sun's claims of better performance.[19]

In fact, Sun was finding it increasingly difficult to match the higher octane ratings that competitors were achieving with the combination of lead and thermally cracked gasoline. Originally, Sun's refiners had been able to meet the octane requirements of new engines rather easily, but increases in performance demands were putting them in a squeeze. The limits of thermal cracking forced the company to consider tetraethyl lead, but J. Howard Pew and his brother, J.N., Jr., were reluctant to surrender to the Ethyl Corporation.

SUN FIGHTS BACK

J. Howard Pew wrote to Sun's patent attorney, Frank S. Busser, in 1930, exploring the possibility of by-passing the Ethyl Corporation's patents on TEL by adding lead outside the United States. At the time, Sun was contemplating construction of a Canadian refinery because of the high duties charged on manufactured gasoline going into that country. Pew first asked clarification on whether Ethyl's patents were good in Canada and whether Sun could freely add lead there. Second, he requested an opinion on Sun's adding lead to crude "at some island in the West Indies," then taking it to Canada for final distillation without incurring damages from the Ethyl Corporation. Busser's reply indicated the plan's futility, since Ethyl's patents were not on a process, but on a fuel comprising gasoline and tetraethyl lead. Once Sun sold the fuel in an area where the patent applied, and he was quite sure Canada was covered, the gasoline itself would represent an infringement of patent.[20]

When asked in 1937 at the Pennsylvania Oil Investigation hearings why Sun had not entered the leaded field, J. Howard Pew recited the belief that Blue Sunoco provided better performance without the use of tetraethyl lead. He also stated that Sun's technical people believed that spark plugs and engine cylinders remained in better condition without TEL. But by this time, the Ethyl Corporation had greatly benefited from the movement to higher compression engines. It could boast that seventy percent of all gasoline consumed in the United States contained some quantity of TEL, and that seventy-seven percent of all gasoline marketed by its customers contained some quantity of either "Ethyl" or "Q" fluid, customers who together marketed ninety percent of all the gasoline sold in the United States.[21]

Sun had meanwhile indirectly benefited from attacks aimed at the Ethyl Corporation and Standard Oil (New Jersey). Organizations of independent oil marketers saw the "Ethyl club" as another way for the majors to control the price of gasoline and keep independents out of the oil business. One organ of this group was the newsletter, *Interesting Oil Facts and Rumors,* which originated in Chicago. This publication attacked the dangers of poisonous tetraethyl lead and labeled the Ethyl Corporation as a monopoly intent on squeezing the public dry. On the other hand, a typical 1933 issue asked readers to "check up on a station that is selling clean antiknock gas and notice some of your old customers who, afraid of Ethyl, are now buying all of their needs from a dealer who does not handle poison."[22]

This was indeed a touchy subject for the Pews. Although they had already irritated many competitors with their comparative advertising, they could further benefit from exploiting the public's fear of TEL. However, they were also members of the American Petroleum Institute, and were attempting to solve industry-wide problems cooperatively in the period 1931–1933.

Sun vice-president S.B. Eckert sent a memo to J.N. Pew, Jr. in August 1933, accompanying an advertisement received from the American All-safe Company, a manufacturer of gas masks. The company was selling these masks for refinery use in the blending of TEL into gasoline. In the memo, Eckert suggested that "this may be of interest to you in your campaign against Ethyl." Handwritten across the advertisement in pencil was another note from Arthur E. Pew, Jr.: "J.N. Pew—What do you think of this—'Use Blue Sunoco and discard your gas mask—Blue Sunoco contains no Ethyl lead!'" There is no evidence that Sun ever used the idea.[23]

At the same time that Sun's top management was exploring its options in regard to the use of tetraethyl lead, the firm did not stand idle in its research laboratories and refineries. After joining Sun in 1926, chief engineer Clarence Thayer had obtained excellent results by increasing the operating temperature and pressure of the Cross cracking stills installed at Marcus Hook. By 1930, Sun was operating this equipment at pressures of 1,200 lbs. per sq. inch and producing superior high octane gasoline. Moreover, in the laboratory, Thayer had been experimenting with a pilot plant running at 2,000 lbs. per sq. inch and found that further improvements were obtained. In order to implement these higher pressures in a full-sized plant, however, Sun would have to install a new heat exchanger, pumps, tubes, and piping able to withstand this extremely high pressure.[24]

Thayer's report was immediately sent to J. Howard Pew through his nephew, Arthur E. Pew, Jr., and a decision was made within hours to proceed with the necessary modifications. Eight weeks later, Sun engineers had accomplished the work and the firm was producing gasoline with the new equipment in 1931. The Pews's quick action was further evidence of some of the advantages of a tightly-run family concern. Encouraged by the success of this first unit, Sun designed and constructed a new 25,000-barrel-a-day unit in 1933. Because of its unique needs for high octane gasoline stock without using lead, Sun had abandoned the older Cross system and had developed its own cracking technology.[25]

Sun's success at Marcus Hook had removed some of the urgency surrounding its need for high octane gasoline. However, the Pews realized that they were reaching the limits of what they could achieve with thermal cracking. Moreover, in addition to the danger involved in operating these units at such high pressure, there were serious maintenance problems with the equipment.

Sun briefly considered one other alternative method of improving octane rating—the blending of alcohol with its Blue Sunoco gasoline. This was not a new idea, and there had been experiments with alcohol-gasoline blends as early as 1906. Agitation for using alcohol for this purpose coincided with periods of petroleum scarcity and farm discontent over low prices. Concerned with a potential wartime fuel shortage in 1915, Henry Ford had announced a plan to extract alcohol from grain and garbage to power his new Fordson tractor, a program consistent with his strong views on wedding industrial science to American agriculture.[26]

The shortage scare immediately after World War I further stimulated the alcohol movement. *The Journal of Industrial and Engineering Chemistry* reported in 1920 that it was inevitable that alcohol would replace gasoline in the internal combustion engine, and it cited a number of advantages to using alcohol-gasoline blends in the short run, including increased mileage, greater power, and the elimination of knock.[27] Large crude oil strikes in California and the Mid-Continent fields in the 1920s had taken the wind out of the sails of the alcohol-fuel movement, but the severe agricultural crisis of the Great Depression once again resurrected the idea.

A coalition of farmers and legislators from the agricultural midwest, alcohol distillers, and industrial chemists from the emerging farm chemurgy movement united to sell the idea of blending between five percent and twenty-five percent ethanol (alcohol made from farm products) to the government and the consumer.[28] Very soon this group met organized resistance from the oil industry and various political and consumer groups who opposed the passage of preferential legislation supporting "power alcohol." The economic arguments were basic. In the oil-glutted market of 1933, alcohol blending would add several cents to the price of a gallon of fuel. In addition to this economic argument, there were technical problems with using alcohol. Unless the alcohol was anhydrous (with very little water absorbed), it would separate from gasoline and cause serious problems in storage as well as use in the engine. Critics also charged that alcohol caused engine wear and that carburetor adjustments had to be made before the blend could even be used.[29]

The American Petroleum Institute launched an organized lobbying effort against the alcohol movement that stressed the costliness of power alcohol as a farm relief program as well as the mechanical difficulties of running an automobile on the mixture. Sun cooperated with the A.P.I. on both the national and state level, and on his radio broadcast of April 25, 1933, Lowell Thomas read a message from the firm attacking power alcohol as a "quack harmful panacea."[30]

The Pews' opposition to necessary state or federal subsidies to make alcohol blends competitive with gasoline was consistent with their general attitude toward the role of government in the marketplace. It was also logical to oppose the power alcohol movement on economic and competitive grounds. Yet there was one characteristic of alcohol blends that was intriguing—the claim that it increased the octane of gasoline.

At the same time that Sun was fully participating in the anti-alcohol lobbying of the oil industry, it was running tests on a mixture of ninety percent Blue Sunoco and ten percent absolute alcohol. These tests, conducted in the spring of 1933, indicated that the addition of alcohol did indeed raise the octane of most of Sun's gasoline blends (Blue Sunoco was a blend of synthetic or cracked gasoline and selected natural gasolines), but that other deficiencies were evident. The presence of water in amounts greater than 0.2–0.3 percent caused turbidity at 60 degrees Fahrenheit that induced separation of the alcohol from the gasoline. It would be too costly and impractical to produce anhydrous alcohol of sufficient quality to eliminate this problem. Sun technicians also reported negative results from other laboratory tests, including high silicone and copper dish gum tests, and lower vapor pressures.[31]

The lack of a clear-cut technical or economical advantage in using alcohol rather than TEL or advanced thermal cracking coupled with the Pews' opposition to federal aid to the farm sector turned Sun away from further consideration of alcohol in 1933. In reply to a complaint about

Sun's position as stated by Lowell Thomas, J. Howard Pew stated that "to mix this alcohol with gasoline violates the age-old American principle, that no one industry should be burdened in order to advantage another."[32]

Sun still faced the increasing octane requirements of the automobile industry, but the Pews chose to attack the problem by continued research into cracking technology. Sun's continued reluctance to use tetraethyl lead and its decision not to experiment further with alcohol led the firm to its most important technological innovation: adoption of the Houdry Process for the catalytic cracking of petroleum.

THE HOUDRY PROCESS

In 1931, a Frenchman, Eugene Houdry, and his associates had established research facilities across the river from Marcus Hook on the refining property of the Vacuum Oil Company in Paulsboro, New Jersey. Although there were rumors about what was happening at the Vacuum refinery, the company had kept the main goal of the project secret. Houdry was trying to develop a commercial process for the catalytic cracking of petroleum. By introducing a catalyst into the cracking of oil at high temperature and pressure, Houdry had found that he could alter the chemistry of petroleum hydrocarbons and produce a product of very high quality. Specifically, gasoline obtained from catalytic cracking was of a much higher octane than that yielded from normal thermal cracking. It was this characteristic that ultimately attracted the Sun Oil Company.[33]

Eugene Houdry was the son of a wealthy French steel manufacturer and a graduate of a Parisian engineering school. He went to work with his father, but interrupted this phase of his career to serve with distinction in the French artillery in World War I. In the post-war years, his new-found hobby of automobile racing turned Houdry to motor fuel research, the endeavor that would take up much of the rest of his adult life.[34]

In 1922, he learned of a chemist in Nice who was making gasoline from lignite or "brown coal," a resource that France had in abundance. At this time, many nations were concerned about a potential petroleum scarcity, and Houdry's motives stemmed from nationalistic as well as economic interest. He remained a French patriot until his death.[35]

The Nice inventor, E.A. Prudhomme, produced water gas from lignite, and then passed this gas over a nickel-cobalt catalyst to produce small quantities of gasoline. There were many others attempting to produce gasoline and other petroleum-type hydrocarbons from coal, most noticeably German efforts in developing a hydrogenation process to produce a form of oil by introducing powdered coal to the action of hydrogen under high temperature and pressure. Catalysts were also not new, and petroleum refiners had experimented with them for many years. But the Prudhomme experiments were very promising, and Houdry entered into an agreement with the Nice chemist to form a corporation for continued research. The firm soon moved from lignite to coal research and, in 1925, to crude oil that France imported from the Middle East.[36]

Work focused on two crucial areas: the development of an efficient catalyst and the regeneration of the catalyst, which quickly became contaminated with coke deposits. This latter problem was the critical obstacle to developing a commercial process.

In the mid-twenties, Houdry severed his relations with Prudhomme after the latter had used questionable judgment in dealings with a Belgian syndicate. In attempting to extract gasoline from coal tar, Prudhomme had added some gasoline to speed the reaction along, a questionable scientific practice smacking of charlatanism. The Belgians backed out, and Houdry was forced to repair his own damaged reputation.[37]

Houdry, with tentative backing from the French government, built a factory to extract oil from lignite. Although the plant produced gasoline, it was unprofitable and the French government withheld a promised subsidy. However, as early as 1926, Houdry had been encouraged by another project. Using a new aluminum silicate catalyst, he had successfully cracked heavy Iraqi crude and produced a clear, apparently high octane gasoline. He tested this gas in his Bugatti racing car, and it demonstrated a high performance. Laboratory research also showed that he could successfully regenerate the contaminated catalyst by blowing air on it under pressure to burn off excess carbon.[38]

Even before the French government had deserted him on the lignite project, Houdry attempted to sell his crude oil process to Anglo-Persian (British), Royal Dutch Shell (Dutch-British), and Jersey Standard. But Anglo-Persian lacked research facilities, and both Shell and Jersey opted to invest in the hydrogenation processes under the control of the giant I.G. Farben German combine. At the time, Farben's Badische coal hydrogenation process looked very promising as a method for turning very poor crude and even tars into gasoline, and its development was more advanced than the Houdry process.[39]

In 1930, Houdry successfully interested Harold Sheets of the Vacuum Oil Company in his work. Vacuum was a petroleum refiner with limited crude reserves of its own and was attracted by the promise of extracting higher gasoline yields per barrel of crude. Houdry signed an agreement with Vacuum in October of that year to come to the United States and construct a ten-ton-a-day commercial unit for the catalytic cracking of petroleum. Vacuum invested $100,000, and Houdry himself would "supervise the construction, erection, and operation of the apparatus." In exchange for its investment, Vacuum received one-third of the shares in a new venture, the Houdry Process Corporation, and Houdry and a group of French investors retained the rest.[40]

Houdry constructed several pilot plants at Paulsboro, and was refining 200 barrels of gasoline a day by 1933. But by this time, Vacuum had merged with Standard Oil of New York (Socony), a firm with extensive production properties, and the flush East Texas field had caused a radical drop in the price of gasoline. The combination of the changed economic environment and some corporate reshuffling after the merger caused Socony-Vacuum management to lose interest in the catalytic experiment. Socony gave Houdry permission to demonstrate his process to others, and he proceeded to negotiate unsuccessfully with eleven oil companies. It was at this point that Sun became involved.[41]

SUN OIL AND THE HOUDRY PROCESS

In the spring of 1933, upon the advice of John G. Pew and other employees of Sun Ship, Sun Oil first expressed interest in the Houdry process. The Sun Shipbuilding and Drydock Company had by this time gone into the business of fabricating oil refinery equipment to offset a drop in business effected by the Depression. The subsidiary had done a great deal of work for Sun Oil, and the shipyard had built the heavy steel equipment that Houdry needed at Paulsboro. The parent firm obtained permission from Socony to talk with Houdry, tested samples of Houdry gasoline, and sent chief engineer Clarence Thayer to Paulsboro to investigate.[42]

Thayer was immediately struck by two apparent weaknesses. First, the catalyst cases were rectangular in shape and riven with a maze of tubes. This design was not the best for maximizing the strength of the case. Second, the long cycles that Houdry was running (up to seven hours) coked the catalyst heavily. Thayer returned to Marcus Hook with a gasoline sample and reported to J. Howard Pew and Arthur E. Pew, Jr. the next morning. The octane rating of this sample was an impressive 81, and Howard became convinced that Sun simply "had to have the process." Thayer had prepared a drawing using two cylindrical catalyst cases. The cylindrical design provided added strength, and by having two cases instead of one, the flow of oil could be diverted to one case while the other was being purged of impurities. The catalyst would be regenerated by burning off the carbon with forced air. This simple drawing became the basis for Sun's commercial process.[43]

Sun arranged a meeting with Eugene Houdry and moved fast to secure a formal working agreement. Sun's quick action illustrates important points about its management. J. Howard Pew, president of the firm, was actively involved in all phases of Sun's business. In a vast corporation like Jersey Standard, it was more likely that something of this nature would go through much committee investigation and discussion before reaching top management. Second, once the issue reached top management, J. Howard Pew, really an owner-manager or entrepreneur in the nineteenth-century mold, was able to ack quickly in a potentially risky situation, confident that he and his family had complete control of the corporation. Finally, in the poor economic environment of 1933, Sun plunged ahead because of the strong conviction that the Pews had in the promise of continued sales for unleaded gasoline. From an industry-wide, purely economic viewpoint, investment in Houdry Process Corporation was perhaps not the wisest course in 1933, but Sun's unique product demands and the independent stance of its management led the firm to proceed with an innovation that would reap future rewards. One representative of a large oil company was quoted in a 1941 *Fortune* article as saying: "Ask those boys how long it took them to decide on adopting the Houdry Process after it had kicked around for weeks in our committees."[44]

Sun did not tip its hand to Houdry at first. In hopes of making a better deal for themselves, the Pews did not wish to make their enthusiasm known. They arranged the first formal meeting in Arthur E. Pew's office on June 6, 1933. Pew did not show up for the gathering, and after cooling their heels for a time, Houdry and his associates left. Pew apologized by letter and arranged a subsequent meeting for June 13. On that occasion, Houdry proposed to grant Sun a license for producing Houdry gasoline in a commercial plant at Marcus Hook. Sun would purchase shares of Houdry stock and provide the capital for the construction of the facility.[45]

Following this initial meeting, Sun explored options with its patent attorneys. This correspondence makes it clear that the controlling concern was the avoidance of using lead. Sun had to decide whether to enter into an agreement with Houdry or to proceed independently with what it already had learned. The Pews made the decision to enter into a short-term agreement with an option to invest more heavily in the Houdry Process Corporation (H.P.C.) later. Among other advantages, this would give Sun access to Houdry patent files and would allow the Pews to decide whether they could proceed independently without patent infringement.[46]

In July, Sun entered into an agreement with Houdry Process Corporation to build a test plant to manufacture gasoline with the process. Sun would provide the investment capital necessary for this project, since H.P.C. had used up most of the funds originally obtained from Vacuum.

Sun Oil also received an option for an exclusive license, and rights to investigate H.P.C.'s patent files, including patents pending at the time. Sun did not at this time buy into H.P.C. There were 300,000 shares of outstanding stock, of which Socony-Vacuum held one-third or 100,000 shares and Eugene Houdry and his French associates the rest. For all practical purposes, Socony was out of the research picture for the next few years.[47]

Sun deferred the decision to construct a commercial facility until it could perfect the process, particularly in regard to the catalyst regeneration phase. Continued testing had demonstrated a consistently higher octane gasoline yield from the process, although the percentage of gasoline obtained from cracking stock, twenty-three to twenty-four percent, was lower than what Sun was getting from existing thermal cracking plants. Eugene Houdry was anxious for Sun to exercise its option for a paid-up license, or even better, to invest in H.P.C. directly. He was desperate for capital and realized that future success would depend on H.P.C. obtaining financing.[48]

In September, Eugene Houdry made a direct offer to J. Howard Pew to sell 41,000 shares of H.P.C. stock to Sun at $10 a share. Thirty thousand of these would be from a new H.P.C. authorization and the additional 11,000 from private individuals. Moreover, Houdry indicated that in his forthcoming trip to France, he would be able to make available an additional 9,000 shares from among the 600 individual French investors in the corporation. The Pews cautiously demurred for the time, but expressed an interest in becoming a part owner in the Houdry Process Corporation.[49]

In February 1934, Sun Oil signed a formal licensing and option agreement with the Houdry Process Corporation. The firm paid $400,000 for a non-exclusive paid-up license to use any and all patents H.P.C. had as of July 21, 1934, and for the next ten years. In return, Sun gave H.P.C. use of its refining patents with the exception of the mercury vapor process and its work on fatty acids.[50] After completing the agreement, H.P.C. vacated its Paulsboro site and moved to Marcus Hook. Houdry and his chief associates, Raymond Lassiat and W. F. Farragher, were then to work closely with Sun's laboratory and research staff. For a short time, Houdry himself lived in the home of Arthur E. Pew, Jr., and Pew's wife served as French-English translator for the planning of the first large pilot plant.[51]

SUN CONTRIBUTIONS TO HOUDRY TECHNOLOGY

To build this first plant, Sun utilized two of its cylindrical Cross cracking stills for catalyst cases. Marcus Hook engineers filled these cases with the alumina-silicate catalyst manufactured by H.P.C. from activated fullers earth purchased from the Filtrol Company. H.P.C.'s Raymond Lassiat supervised the placement of water cooling tubes throughout the catalyst cases, and Sun's Clarence Thayer designed the operating cycles. The charging stock entered the first case where cracking occurred, then the flow switched over to the second case while the catalyst in the first was regenerated. As it evolved, this Sun-H.P.C. process became identified as a fixed-bed, semi-continuous operation. The catalyst remained static while the charge flowed through it. Sun ran heavy oil through the system but encountered difficulty removing residue from this oil before the catalyst had been regenerated. First Thayer's team tried J. Howard Pew's suggestion of a vacuum purge, lessening the pressure so that the oil could more easily vaporize, and later they turned to a steam cycle to solve this problem. Sun engineers also found that it was more efficient to use three catalyst cases rather than two; while one case was being regenerated, cracking took place in the other two.[52]

These operations placed great stress on the valves and tubes in the system. First the pressure had to be reduced, then steam admitted and released, air introduced to burn off the carbon, then steam again to force the air out—a separate cycle for each. Much experimentation went into improved valves and tubes, and eventually into automatic timing devices for operating the cycles. Chief engineer Thayer supervised these equipment modifications while the laboratory ran experiments to determine the optimum cycle time for the catalyst. First Sun used a cracking cycle of forty-five minutes with effective results. Later experiments demonstrated that after fifteen minutes the coke buildup had become heavy, and Sun switched to a fifteen-minute cycle. Soon, by altering pressure and temperature, Sun engineers found that they could raise gasoline yields from heavy oil charging stock from twenty-three to forty-three percent.[53]

One other important breakthrough remained. The cost of pumping air under pressure for the crucial regeneration cycle was very high. Sun sent two experts to Switzerland to inspect a new development, the turbo-compressor, an early prototype of the turbine engine. A combustion gas turbine may be simple defined as a turbine activated by a steady flow of expanding gases from combustion. The main advantage of any turbine is that it produces rotary movement without the addition of pistons, cranks, and other mechanical devices. Manufactured by the Swiss firm of Brown, Boveri, the unit installed at Marcus Hook used refinery gases as an energy supply. The gaseous products of catalyst regeneration were piped into the turbo-compressor to provide the power to pump the air to regenerate the next case. These gases resulting from the burning of the carbon off the coke in the regeneration phase were at a temperature between 800 and 850 degrees Fahrenheit and at a pressure of more than 40 lbs. per sq. in. This was sufficient power to compress, in an axial compressor attached to the shaft of the turbine, enough air from the atmosphere to accomplish regeneration in the next catalyst case. When the rotary motion of the turbine was adapted to generate its own electric power, the process became even more economical to operate.[54]

In a continuing effort to obtain the best economies, Sun worked at cracking the heaviest charges of fuel oil and low grade gas oil, and it soon succeeded with all but the last ten percent of East Texas crude. Thus, by the end of 1935, there was a commercially feasible H.P.C.-Sun design for a fixed bed catalytic cracking process. The two key improvements over the 1933 prototype at Paulsboro were the shorter cracking cycle and the turbo-compressor system.[55] By the end of 1934 alone, Sun had spent $1,341,000 in development of the commercial process. A large part of that amount was used to acquire almost 79,000 shares of H.P.C. stock from the French syndicate. By the end of 1935, Sun's investment had risen to $2,000,000.[56]

Sun had achieved a great deal with its technological improvements, but the Pews realized that further financing would be necessary. Armed with the results of their success thus far, Sun and H.P.C. moved to get Socony-Vacuum actively interested in the process again. Socony was a big company, a large retailer of gasoline in the United States with a well-staffed research arm. It ranked second to Jersey Standard in refining capacity of all major companies, had the second largest marketing territory, and was the fourth largest producer of gasoline in the United States.[57] Arthur E. Pew, Jr. and Thayer went to New York to try to interest the Socony executives in the Sun-H.P.C. process. At this time, A.E. Pew, Jr. became a vice-president and member of the board of H.P.C., in addition to being vice-president in charge of manufacturing for the Sun Oil Company.[58]

Sun had increased its holdings in H.P.C. stock so that it now owned approximately one-third of the total authorization of 333,000 shares of par $10 stock. The acquisition of one-third ownership of Houdry's company had been achieved with full approval of Socony. Now three separate interests owned H.P.C.: Sun, Socony, and Eugene Houdry and his French associates.[59]

On December 23, 1935, Socony concluded a new agreement with H.P.C. whereby it paid $1,900,000 for an unlimited capacity license to use the new process. The parties determined the agreed amount by two negotiated ratios. The first was Socony's greater refining capacity (in 1935, Socony was the fourth largest manufacturer of gasoline among all majors, and Sun thirteenth), and the second was the developmental expense Sun Oil had laid out since 1933. With the influx of this money,

H.P.C. voted a $5 stock dividend for 1935 for a total of $1,422,364. This represented the first profits that Houdry and the French interests had earned from their H.P.C. investment.[60]

Both Sun and Socony began construction of full commercial plants in 1936, Socony in Paulsboro and Sun at Marcus Hook. Socony completed a 2,000-barrel-a-day, three case water-cooled plant in April 1936. It designed the plant to crack light gas oil, a different task than that proposed by Sun, but one which enabled the refinery to begin operating more quickly. This was the first commercial Houdry fixed bed plant. Sun did not complete its more elaborate 12,000-barrel-a-day, heavy gas-oil plant until April of 1937, at a cost of $3,250,000.[61]

A major problem with these first plants involved their cooling systems. During cracking, the catalyst became intensely hot, and water was introduced through a series of tubes to effect cooling. These extreme temperature changes weakened tubes, joints, and valves, and all of the early units installed by Sun and Socony experienced this trouble. It was not until a new system of using a third medium, salt, to transfer the heat from the catalyst was developed by Socony in 1935–1939 that H.P.C. solved this problem. Prior to the adoption of this system, the refiners had to replace tubes, valves, and other equipment after one or two years.[62]

In 1939, Sun Oil completed two large continuous units, one at Marcus Hook and one at Toledo. These facilities processed crude oil and sent the heavy fuel oil charging stock directly to the Houdry fixed-bed bases. Although they were originally designed to process 28,000 barrels of crude a day, Sun quickly expanded these units to 40,000 barrel capacity. Both new plants incorporated the salt case transfer unit, and Sun also converted its original 12,000-barrel-a-day unit to salt.[63]

In 1937, Sun's success with both Houdry cracking and Blue Sunoco sales prompted the firm to consider building a refinery in Texas for the first time. Because of Sun's earlier U.G.I. connection, J.N. Pew, Sr. had not followed the pattern of Gulf, Texaco, and others in building a refining capacity in Texas following Spindletop. Now the East Texas fields had once again provided huge amounts of crude, and Sun contemplated the advantages of lower refinery costs and cheaper freight rates in hauling gasoline to markets. The Pews decided against doing so, however, because of the tremendous cost in building an efficient organization and the relatively unfavorable price structure for gasoline sales.[64]

EXPANSION OF H.P.C. LICENSES

At the time of the 1938 annual meeting of the American Petroleum Institute in November, there were only three Houdry units in operation: the first two at Paulsboro and Marcus Hook, and a third plant owned by

Socony in Italy. On the final day of the conference, Arthur E. Pew read a technical paper, "Catalytic Processing of Petroleum Hydrocarbons by the Houdry Process," which sent shockwaves through the industry. Most insiders knew a little of what had been going on with Houdry and his partners, but now it was totally in the open. As presented by Pew, the Houdry Process was really not one, but six separate catalytic operations. In addition to cracking crude or heavy oil for gasoline, one could reduce the viscosity of residues or tars, catalytically treat gasoline to upgrade it, catalytically desulphurize gases, polymerize butenes in liquid phase, and produce light gravity gas oils or diesel oils by catalytic cracking.[65] The technical paper was also an official announcement that H.P.C. was in the business of licensing the process to others and manufacturing and supplying its customers with the all-important catalyst.

Prior to this meeting, however, Arthur E. Pew, Jr. and Eugene Houdry had been actively trying to sell the process to others. In particular, they were trying to land the biggest fish of all, Standard Oil of New Jersey, the company that had inherited the legacy of its ancester, the old Rockefeller monopoly, and had reasserted itself as the dominant firm in the industry. Standard had expressed interest in the Houdry Process in France in the late twenties, but at the time had opted for the hydrogenation process.

In 1927, Standard Development Company (S.O.D.) made an agreement with I.G. Farben to undertake research into the hydrogenation of heavy oil and to exchange all information on the subject. In 1929, it paid $35,000,000 for the rights to hydrogenation and several other processes. In 1930, Jersey formed the Hydro Patents Company to license the process in the United States. Although the hydrogenation plants built in this country performed satisfactorily, they were not economically practical, particularly in an era of an overabundance of crude and depressed prices.[66]

Late in 1935, Frank Howard, president of Standard Oil Development Company, Frank W. Abrams, president of Standard Oil of New Jersey, and Harry C. Wiess, president of Humble Oil (Standard's subsidiary) opened discussions for obtaining a Houdry license. The original proposal offered by A. E. Pew, Jr. called for Standard to purchase a license to use the Houdry fixed-bed process with an amount of money determined by the operating capacity of the company. Standard would not be a partner in the Houdry patents and therefore would not participate in any royalties obtained from other refiners. The original sum A.E. Pew, Jr. sought from Jersey was $100,000,000, a figure based on Jersey's large world-wide refining capacity.[67]

Assuming that there would be an agreement on a cash payment, Frank Howard proposed a deal based on three general principles. First he insisted that Standard Oil Development remain free to work out a competitive process with rights to license to Jersey units and others. The second point was that Standard and its licensing partner, I.G. Farbenindustrie, be admitted as participants in the Houdry agreement. They would receive a percentage of Houdry profits, and after two years, allow Houdry Process Corporation non-exclusive rights to license certain S.O.D. and I.G. patents, excluding hydrogenation. Finally, in exchange for I.G. participation in Houdry profits, H.P.C. and Sun Oil would come into the U.S. Hydro Patents Company, or in some other way participate in the Standard I.G. hydrogenation patent deal. Socony was already a licensee of Hydro. Sun had earlier shown an interest in hydrogenation and had contacted both German and American licensors as early as 1930–1932.[68]

The parties remained far apart on money, Arthur E. Pew, Jr. holding out for $100,000,000 and Standard offering $14,000,000. But the real obstacle was the issue of Standard participation in the Houdry patents.[69] Sun had never directly become a member of the "patent club" in the twenties, a series of interlocking agreements and cross-licenses among the majors. More importantly, Eugene Houdry and his French backers were only now on the brink of reaping profits from his invention, and they were reluctant to let anyone else in the door. Houdry revealed some of his bitterness in a handwritten letter to Arthur E. Pew, Jr. from Boca Raton, Florida, in February 1938:

> you will have a very tough job to carry this deal home if you do not give a participation because all of them want it, and *are willing to pay for it.* They cannot conceive that somebody in this world (and maybe the next world) would turn down N.J. as a partner. Well I understand such a principle.[70]

Negotiations intensified in the winter and early spring of 1938. Arthur E. Pew, Jr. and others made frequent trips to New York, and a technical team from the Humble Company (a Jersey Standard subsidiary) camped out in Clarence Thayer's office for two months studying every characteristic of the fixed-bed process. In contrast to Houdry's desire to remain independent of Standard, Jersey based its bargaining position on the hard reality of their large share of the refining industry. In 1938, Jersey owned 9.9 percent of total domestic daily crude oil refining capacity, the largest single percentage, while Socony-Vacuum was in second place with 7.4 percent. Sun, on the other hand, ranked fifteenth with 2.0 percent.[71] Writing to J.A. Brown, president of Socony, in April, Jersey Board Chairman William Farish articulated this view:

> We understand that there is not, and cannot be, any truly basic patent position in this field. If this is true, the permanent position of the Jersey interests in this

> development will necessarily be roughly commensurate with their position in the industry—i.e., Jersey will be one of the largest factors in catalytic cracking, technically and from an improvement patent standpoint.[72]

Despite personal meetings among Farish, J. Howard Pew, and Socony president John Brown, Jersey held firm to its demand for participation in the Houdry patents. Clarence Thayer remembers urging Arthur Pew, Jr. to settle for a flat sum of $14,000,000 for allowing Jersey participation, but Pew stood fast, largely pressured by Eugene Houdry's refusal to consider sharing his patents. There was also the concern that if one firm, Jersey, were granted an exclusive deal, then other refiners would insist on similar arrangements, extending the "patent club." Negotiations finally terminated on May 4, 1938, when J.A. Brown informed Farish by telephone that Jersey's offer was unacceptable to the Board of H.P.C.[73] J.A. Brown wrote to J. Howard Pew on May 4, informing him of this event, and the letter ended on what proved to be a rather ironic note:

> I cannot help but share your feeling that they may be a long time finding out what they can do about catalytic operation. If it were such an easy matter that their Technical Department could guarantee at any time, starting from nothing, to put them in a position of their own within two years, it is strange indeed that they did not do it long ago.[74]

This was exactly what Jersey Standard proceeded to do. In a crash research program aided considerably by the war, it developed a competitive catalytic process which eventually proved superior to the Houdry fixed-bed. In October 1938, representatives of Jersey Standard, the M.W. Kellogg Company, Standard-I.G. Company, Standard Oil (Indiana), and I.G. Farbenindustrie met in London to organize Catalytic Research Associates. Soon after, they were joined by Anglo-Iranian, Royal-Dutch Shell, the Texas Company, and Universal Oil Products Company. The goal of this technical consortium was to develop an independent catalytic process that would not infringe on the Houdry patents.[75] The Houdry group now found itself competing against what John T. Enos has called "probably the greatest scientific effort directed at a single project, . . . surpassed only by the development of the atomic bomb."[76]

The companies involved spent over $15,000,000 in four years and developed a successful process that began commercial operation in 1942. Unlike the more individualistic triumphs of William Burton and Eugene Houdry, Enos labels this project an example of truly modern institutionalized process innovation. This new process, fluid catalytic cracking, used a powdered catalyst that functioned within the charging stock itself. These "suspensoid" particles acted as a fluid. After cracking had completed, the particles were separated from the petroleum and regenerated outside the cracking chamber. The big advantage of the process was that it handled a much larger volume of charging stock. The Houdry fixed-bed process continued to compete for a few more years, but soon Socony and H.P.C. adopted their own moving catalyst process to compete with fluid cracking. Had Standard (New Jersey) decided to license from H.P.C. in 1938, the course of refining development might have been different, but in either case, Sun Oil's leadership was probably destined to wane in the shadow of the giants in the industry.[77]

The Houdry fixed-bed units had, however, firmly established Sun and Socony with an important head start in catalytic cracking. In July 1938, H.P.C. signed an agreement with E.B. Badger & Sons of Boston, New York, Philadelphia, San Francisco, and London to be licensors of the process. Badger was a well-established firm of engineers and contractors for petroleum distillation and cracking equipment. In 1938, H.P.C. sold a fixed-bed license to the Tidewater Associated Oil Company, the first contract negotiated outside the Sun-Socony-Houdry group. Soon after, Houdry reached an agreement with the Standard Oil Company of California in February 1939.[78]

California Standard had enjoyed close relations with Sun over the years, the Pews having purchased large quantities of high quality California gasoline to blend into their Blue Sunoco. California had been a member of the "patent club" as a result of their purchase of Universal Oil Products patents in 1931 for $5,000,000. They had also later entered into agreement with the M.W. Kellogg Company, which had ties with S.I.G. (Jersey Standard and I.G. Farben), and the Hydro Patents Company, the Jersey-Farben group which licensed hydrogenation processes in the United States.[79]

Following its agreement with Houdry, California Standard signed an agreement with S.I.G. in which it agreed to purchase one-quarter of a barrel from the Jersey Standard firm for every other barrel of capacity purchased from others. This agreement prompted Frank Howard of S.I.G. to boast that S.O.C.A.L. was paying his firm for the privilege of using the Houdry process, a charge which California protested to A.E. Pew, Jr. was inaccurate. In effect, what S.O.C.A.L. was doing was making payments to I.G. Farben on any capacity, even their own, that they installed other than I.G.'s. Although not pleased by the arrangement, Arthur Pew, Jr. was forced to accept the explanation offered by Standard of California. The important thing was that the Houdry group had sold a license to another firm in the industry. Although H.P.C. had only limited success in this regard, by 1942 there were a total of fourteen fixed-bed units operated by Sun, Socony, Standard of California, and Tidewater.[80]

In 1938, 1939, and 1940, H.P.C. corresponded with firms in France, Britain, Italy, Germany, Rumania, and Sweden concerning licensing of Houdry fixed-bed units. One of the most interesting episodes involved negotiations with Japan in 1938 and 1939. In April 1938, representatives of the Japanese government, the Army and Navy, and Nippon Soda, the largest refiner in Japan (forty-three percent of capacity), contacted H.P.C. with the aim of obtaining a license. The Japanese were very interested in the use of Houdry gasoline for producing 100 octane aviation fuel. Although the octane of Houdry fuel was very high, it had to be supplemented with tetraethyl lead and iso-octane (pure 100 octane) to bring it up to desirable levels.[81]

H.P.C. authorized Badger & Sons to furnish general information to the Japanese interests, and negotiations continued during the rest of 1938 and into 1939. Nippon Soda eventually dropped out of the negotiations, apparently believing that German hydrogenation processes obtained from I.G. Farben were sufficient for their needs. But by the fall of 1939, H.P.C. was on the verge of consummating an agreement with another Japanese firm, Toa Nenryo Kogyo K.K. The agreement called for a 12,000-barrel-a-day license at a cost of $2,850,000. The Japanese would obtain the use of technical developments made by Houdry for ten years and be allowed to use such developments until the expiration of the patents. The main point of disagreement was H.P.C.'s demand for an advanced royalty. In exchange for eliminating this provision, H.P.C. wished to have the Japanese take out a 30,000-barrel license.[82]

At this point, Frank Howard, president of Standard Oil Development and a rapidly worsening thorn in the side of the Houdry interests, approached the State Department on the subject of licensing processes to Japanese firms. As a result, Joseph C. Green, Chief of the State Department Division of Controls, summoned representatives of Sun Oil, E.B. Badger, M.W. Kellogg, Phillips Petroleum, Shell Oil, Inc., the Shell Union Oil Company, Socony-Vacuum, Standard Oil Development, and Universal Oil Products to a meeting in Washington.[83] Green eventually held three meetings, on November 20, December 7, and December 14, 1939. On December 19, his office issued a memorandum stating the government's position that "It is considered that the national interest suggests that for the time being there should be no further delivery to interests of certain countries of plans, plants, manufacturing rights, or technical information required for the production of high quality aviation gasoline."[84]

Immediately following receipt of this memorandum, Badger wrote to the Japanese interests breaking off all negotiations "due to circumstances which have recently arisen."[85] Frank Howard had ostensibly gone to the State Department to obtain clarification of Standard's position regarding the sale of iso-octane plants, tetraethyl lead, and hydrogenation and alkylation processes to the Japanese. Furthermore, the State Department was evolving a policy of limited economic sanctions toward Japan, which the Roosevelt administration adopted in the period 1939–1941.[86] To George Hargrove of E.B. Badger, however, "It was the consensus of opinion of the Houdry group that the reason for the meeting was more a question of competition than patriotism."[87] As wartime events later demonstrated, this would not be the last time that competition between the Houdry and the Standard groups became enmeshed in issues of American national interests.

HOUDRY ON THE EVE OF WAR

Since it had the only commercially operating catalytic plants, the Houdry group was in a very good position when war broke out. Houdry base stock became the heart of America's 100 octane aviation fuel program and brought large profits to Sun. But although Sun and Eugene Houdry remained committed to the fixed-bed process in the late thirties, Socony-Vacuum had not been as satisfied with its plants, and as early as 1936, it had begun research into a moving-bed process. In this approach, the catalyst was regenerated outside the cracking environment and then reintroduced. This eventually emerged in the early forties as the Thermofor Catalytic Cracking or "T.C.C." Process. Socony became convinced much earlier than Sun of the obsolescense of fixed-bed units, and during the war, Houdry fixed-bed licenses faced competition from T.C.C. as well as Standard's fluid process.[88]

In the final analysis, Eugene Houdry and the French investors were the major losers in the successful competition presented to Houdry fixed-bed units. Sun Oil, Socony-Vacuum, and other licensees obtained the benefit of the process in their refineries when the process reigned supreme. But the huge royalties envisaged by Houdry and Arthur E. Pew, Jr. never materialized. Enos estimates the rate of return on investment in the Houdry fixed-bed process at approximately twenty-six percent, a low figure when compared to the fifty percent originally yielded on the first Burton thermal stills and the whopping two hundred and ten percent on the Jersey Standard Tube and Tank process. He cites three reasons for the relatively less profitable Houdry development. The first was the low price level of the Depression years when the fixed-bed process first appeared. The price differential between gas oil and gasoline was much lower than when Burton introduced thermal cracking. In a more favorable economic environment, the Houdry return would probably have been higher. Second, efficient cracking processes were in general usage during the thirties. The quantitative improvement of Houdry over advanced thermal cracking made over older distillation processes in 1913. Third, the main advantage of the Houdry Process, improvement of octane quality of gasoline, is much harder to measure in economic terms. But it was exactly this last point that favored Sun. Since the Houdry Process enabled Sun Oil to continue marketing non-leaded gasoline exclusively, the investment Sun made had a very great return.[89]

Sun initially blended Houdry-processed gasoline with its other stocks to maintain a sufficiently high octane product. With the addition of more Houdry units at Marcus Hook and Toledo, however, Sun decided in 1939 to market a new gasoline, one largely produced from catalytic cracking. In November 1939, Sun announced the sale of "Nu-Blue Sunoco" gasoline with an extensive campaign in newspapers, billboards, and magazines. The effort was successful, and Sun estimated that it added as many as half a million customers.[90] Very soon, however, the necessities of war found Sun converting all of its Houdry units to the manufacture of high octane gasoline stock for the military.

NOTES

1. "Sun Oil Company Product Sales, 1937-1947, May 10, 1948," Acc. #1317, Eleutherian Mills Historical Library, File 21-A, Administrative, Box 100.

2. A good overview of current literature on innovation as well as an extensive bibliography is contained in Patrick Kelly and Melvin Kranzberg, (eds.), *Technological Innovation: A Critical Review of Current Knowledge* (San Francisco: The San Francisco Press, 1978). For an examination of process innovation in the petroleum industry, see John Lawrence Enos, *Petroleum Progress and Profits: A History of Process Innovation* (Cambridge, Mass.: The M.I.T. Press, 1962), and George Foster, "An Examination of Process Innovation in Petroleum Refining," in Harold F. Davidson, Marvin J. Cetron, and Joel D. Goldhar, (eds.), *Technology Transfer* (NATO Advanced Study Institute Series, Series E: Applied Sciences—No. 6, Leiden: Noordhoff International Publishing, 1974), pp. 165-74.

3. See Kelly and Kranzberg, *Technological Innovation*, Part I: *The Ecology of Innovation*, especially chapter 1, "The Ecology of Innovation," pp. 1-17; chapter 2, "The World Outside," pp. 18-46; and chapter 3, "The Process of Innovation: The Organizational and Individual Contexts," pp. 47-118.

4. "Sun Oil," *Fortune* 23, 12 (February 1941), p. 114; "Blue Sunoco 25th Anniversary, 1927-1952," *Our Sun* 17, 2 (Spring 1952), p. 31.

5. E.W. Webb, Ethyl Corporation, to Irénée duPont, January 7, 1930, Papers of Irénée duPont, Eleutherian Mills Historical Library, File V.C. #25.

6. E.W. Webb to Irénée duPont, october 7, 1931, Papers of Irénée duPont, EMHL, File V.C. #25, Re: Ethyl Corporation.

7. Memo, "Proposed Agreement Adopted by Board of the Ethyl Corporation," November 19, 1931; E.W. Webb to Irénée duPont, October 18, 1932; A.E. Mittnacht, Ethyl Corporation, to Irénée duPont, February 9, 1934; Memo: November 19, 1931; E.W. Webb to Alfred P. Sloan (Copy), November 19, 1931, Papers of Irénée duPont, EMHL, File V.C. #25.

8. E.W. Webb to Irénée duPont, October 6, 1931, Papers of Irénée duPont, EMHL, File V.C. #25. The relationship between the "octane race" and the development of higher compression engines by the automobile industry is discussed in Harold F. Williamson, Ralph L. Andreano, Arnold R. Daum and Gilbert C. Klose, *The American Petroleum*

Industry, volume II, *The Age of Energy, 1899-1959* (Evanston: Northwestern University Press, 1963), pp. 605-7. Data on the octane-engine compression ratio is contained in Appendix Table 2c, "Gasoline Prices and Octanes at Retail, and Engine Compression Ratios, 1901-1957," in Enos, *Petroleum Progress and Profits,* p. 289 (adapted in this study as Appendix Table 11).

9. E.W. Webb to Irénée duPont, October 9, 1931, enclosed memo of October 7, 1931, Papers of Irénée duPont, EMHL, File V.C. #25.

10. *Ibid.*

11. Williamson, et al., *Energy,* p. 606; see Appendix Table 11.

12. Memo, Blue Gasoline, 1931, Acc. #1317, EMHL, J. Howard Pew papers, Box 56; Williamson et al, *Energy,* pp. 687-9.

13. W.C. Teagle to J. Howard Pew, August 13, 1931; J. Howard Pew to D.J. Moran, Continental Oil Company, September 19, 1932; Moran to Pew, September 24, 1932; Pew to Moran, September 27, 1932; Acc. #1317, EMHL, J. Howard Pew Papers, Box 56.

14. M.H. Leister, Sun Sales Manager, to Edward L. Greene, National Better Business Bureau, November 5, 1930, Acc. #1317, EMHL, J. Howard Pew Papers, Box 56.

15. Folder, "National Better Business Bureau, 1932-1933"; M.H. Leister to Edward L. Greene, November 5, 1932, Acc. #1317, EMHL, J. Howard Pew Papers, Box 57.

16. M.H. Leister to Greene, November 5, 1932.

17. Folder, "National Better Business Bureau, 1932-1933"; J. Howard Pew to Atlantic Refining Company et al., December 29, 1932, Acc. #1317, EMHL, J. Howard Pew Papers, Box 57.

18. *Ibid.,* "National Better Business Bureau"; J. Howard Pew to Atlantic Refining et al., December 29, 1932.

19. R.C. Holmes, Texas Company, to J. Howard Pew, January 27, 1933; M.H. Leister to R.C. Holmes, March 24, 1933, Acc. #1317, EMHL, J. Howard Pew Papers, Box 57.

20. J. Howard Pew to Frank Busser, October 11, 1930, Busser to Pew, October 13, 1930, Acc. #1317, EMHL, J. Howard Pew Papers, Box 28.

21. "J. Howard Pew Testimony," *Pennsylvania Oil Investigation 1937,* Book 4, pp. 936-7, Acc. #1317, EMHL, J. Howard Pew Papers, Box 48; A.E. Mittnacht, Ethyl Corporation, to Irénée duPont, February 7, 1936, Papers of Irénée duPont, EMHL, File V.C. #25.

22. *Interesting Oil Facts and Rumors* 4, 6 (June 1, 1933), p. 5, Acc. #1317, EMHL, File 21-A, Box 100. Copies of this publication and other similar materials may be found in the folder marked "Blue Gasoline" in File 21-A, Box 100.

23. Samuel Eckert to J.N. Pew, Jr., August 14, 1933, A-c. #1317, EMHL, File 21-A, Box 100.

24. Clarence Thayer, interview held at Media, Pennsylvania on November 25, 1975; Enos, *Petroleum Progress and Profits*, pp. 145-6.

25. *Ibid., Annual Report of the Sun Oil Company*, 1931, 1933.

26. B.R. Tunison, "The Future of Industrial Alcohols," *The Journal of Industrial and Engineering Chemistry* 12, 4 (April 1, 1920), pp. 370-6; Conger Reynolds, "The Alcohol-Gasoline Proposal," paper presented at the Twentieth Annual Meeting of the American Petroleum Institute, Chicago, Ill., November 14, 1939, pp. 1-2, in Acc. #1317, EMHL, J. Howard Pew Papers, Box 52; William A. Scheller, "Tests on Unleaded Gasoline Containing 10% Ethanol-Nebraska GASOHOL," Paper Presented at the International Symposium on Alcohol Fuel Technology, Methanol and Ethanol, Wolfsburg, Federal Republic of Germany, November 21-23, 1977, p. 1; Reynold Millard Wik, "Henry Ford's Science and Technology for Rural America," *Technology and Culture* 3, 3 (Summer 1962), pp. 248-9.

27. Tunison, "Future of Industrial Alcohols," p. 374; see also Ralph C. Hawley, "The Forests of the United States as a Source of Liquid Fuel Supply," *The Journal of Industrial and Engineering Chemistry* 13, 1 (November 1, 1921), pp. 1059-60.

28. Reynolds, "Alcohol-Gasoline Proposal," pp. 1-5; "Alcogas," *The Business Week* (Feb-

ruary 8, 1933), p. 9; "Farm-Brewed Fuel," *The Business Week* (March 15, 1933), pp. 14-6; "Iowa Alcohol-Gasoline Proposal Tabled Temporarily as Idea Grips West," *National Petroleum News* 25, 9 (March 1, 1933), pp. 11-2. For discussion of the role of chemists and chemical engineers in the farm chemurgy movement, see Carroll W. Pursell, Jr., "The Farm Chemurgic Council and the United States Department of Agriculture, 1935-1939," *Isis* 60, 3 (Fall 1969), pp. 307-17, and Christy Borth, *Pioneers of Plenty* (New York: Bobbs-Merrill, 1939).

29. Gustav Egloff, "Alcohol-Gasoline Motor Fuels," address Delivered at the Semi-Annual Meeting of the National Petroleum Association, Cleveland, Ohio, April 21, 1933, pp. 1-6; "Analysis of Technical Aspects of Alcohol Gasoline Blends," report prepared by American Petroleum Institute Special Technical Committee, April 10, 1933, pp. 1-5; "Analysis of the Economic Aspects of Alcohol-Gasoline Blends," report prepared by American Petroleum Institute Special Economics Committee, April 10, 1933, pp. 1-4; L.S. Bachrach, "Facts and Arguments Presented Against Alcohol-Gasoline Motor Fuel Legislation at Hearings Before Iowa State Senate, February 21, 1933," pp. 1-5, all in Acc. #1317, EMHL, J. Howard Pew Papers, Box 52; Oscar C. Bridgeman, "Utilization of Ethanol-Gasoline Blends as Motor Fuels," *Industrial and Engineering Chemistry* 28, 9 (September 1936), pp. 1102-12; Gustav Egloff and J.C. Morrell, "Alcohol-Gasoline as Motor Fuel," *Industrial and Engineering Chemistry* 28, 9 (September 1936), pp. 1080-8.

30. *Alcohol Gasoline Blends*, American Petroleum Institute Industries Committee Pamphlet (May 1, 1932), pp. 1-16; Paul E. Hadlick, Executive Secretary, American Petroleum Industries Committee, to J. Howard Pew, May 1, 1933; "Memorandum Re Alcohol-Gasoline Blends," American Petroleum Industries Committee, April 15, 1933, pp. 1-27; R.P. Anderson, A.P.I., to J. Howard Pew, April 10, 1933; W.M. Irish, Atlantic Refining Co., to J. Howard Pew, April 20, 1933; telegram, J.N. Pew, Jr. to H.D. Collier, Standard Oil Company of California, April 26, 1933; telegram, Lewis Mars, Industrial Alcohol Institute, to J.R. Pew (*sic*), April 26, 1933; Representative Everett M. Dirkson to Lowell Thomas, May 8, 1933; Thomas to Dirkson, May 18, 1933; telegram, J. Howard Pew to G.P. Crosby, Sun Oil Company, April 25, 1933; G.D. Jones, Cleveland Tractor Company, to J.H. Pew, April 26, 1933, all in Acc. #1317, EMHL, J. Howard Pew Papers, Box 52. Sun's broadcast on the Lowell Thomas program was apparently part of a coordinated A.P.I. effort. On April 26, 1933, Walter Teagle, president of Standard Oil (New Jersey), wrote to J. Howard Pew enclosing a transcript of "what Hard, our announcer, said at the opening of our radio hour last evening." The arguments presented in this text are essentially the same as those stated by Thomas in the Sun broadcast (see J. Howard Pew Papers, Box 52).

31. Inter-office Memo, R.B. Davidson to A.E. Pew, Jr., April 26, 1933; Interoffice Memo, H.F. Angstadt to A.E. Pew, Jr., May 1, 1933, Acc. #1317, EMHL, J. Howard Pew Papers, Box 52.

32. J. Howard Pew to W. King White, Cleveland Tractor Company, April 28, 1933, Acc. #1317, EMHL, J. Howard Pew Papers, Box 52.

33. "Monsieur Houdry's Invention," *Fortune* 19, 2 (February 1939), p. 132; D.R. Lamont, Socony, to C.S. Teitsworth, Socony, July 25, 1951, re: Sun-Socony-Houdry relationship (copy), Acc. #1317, EMHL, File 21-B, Box 31.

34. "Monsieur Houdry," *Fortune*, pp. 127-30; Enos, *Petroleum Progress and Profits*, pp. 131-2.

35. "Monsieur Houdry," *Fortune*, p. 130.

36. *Ibid.*, p. 130; Williamson et al., *Energy*, pp. 614-5; "German Combine Working to Assure Independent Motor Fuel Supply," *National Petroleum News* 18, 39 (September 29, 1926), pp. 63-4.

37. "Monsieur Houdry," *Fortune*, p. 130.

38. *Ibid.*, p. 132; Thayer interview; Williamson et al., *Energy*, p. 615; Enos, *Petroleum Progress and Profits*, pp. 135-6.

39. "Monsieur Houdry," *Fortune,* p. 132; Larson, Knowlton, and Porter, *New Horizons,* pp. 154, 166–7.
40. "Monsieur Houdry," *Fortune,* p. 132; Lamont to Teitsworth, July 25, 1951, p. 1; Thayer interview.
41. Thayer interview; de Chazeau and Kahn, *Integration and Competition,* p. 297; "Monsieur Houdry," *Fortune,* p. 132.
42. Thayer interview; R.B. Davidson to H.F. Angstadt, April 11, 1933; A.E. Pew, Jr. to Clarence Thayer, April 13, 1933; John G. Pew to A.E. Pew, Jr., May 27, 1933; Acc. #1317, EMHL, Papers of Arthur E. Pew, Jr., Box 18.
43. Thayer interview; Enos, *Petroleum Progress and Profits,* pp. 147–8.
44. "Sun Oil," *Fortune* 23, 12 (February 1941), p. 112
45. A.E. Pew, Jr. to John G. Pew, June 5, 1933; T.B. Prickett to A.E. Pew, Jr., June 7, 1933; A.E. Pew, Jr. to T.B. Prickitt, June 8, 1933; Memo, "Houdry Process," A.E. Pew, Jr., Acc. #1317, EMHL A.E. Pew, Jr. Papers, Box 18; Board Minutes, Sun Oil Company, Office of the Assistant Secretary, Sun Oil Company, St. Davids, Pennsylvania, June 30, 1933, Minute Book #5, p. 126.
46. Frank S. Busser to J. Howard Pew, June 20, 1933, Acc. #1317, EMHL, J. Howard Pew Papers, Box 40.
47. Eugene Houdry to A.E. Pew, Jr., June 29, 1933; Telegram, A.E. Pew, Jr. to J. Howard Pew, July 25, 1933, Acc. #1317, EMHL, A.E. Pew, Jr. Papers, Box 18; Busser to J. Howard Pew, June 29, 1933, Acc. #1317, EMHL, J. Howard Pew Papers, Box 40.
48. Telegram, A.E. Pew, Jr. to J. Howard Pew, July 25, 1933, Acc. #1317, EMHL, A.E. Pew, Jr. Papers, Box 18.
49. Eugene Houdry to J. Howard Pew, September 28, 1933, Acc. #1317, EMHL, J. Howard Pew Papers, Box 40.
50. Memo: "H.P.C.," April 26, 1934, Acc. #1317, EMHL, A.E. Pew, Jr. Papers, Box 1; Lamont to Teitsworth, July 25, 1951, pp. 2–3; Board Minutes, Sun Oil Company, February 25, 1934, Minute Book #5, p. 146; *Annual Report of the Houdry Process Corporation,* 1934.
51. Enos, *Petroleum Progress,* p. 148; Lamont to Teitsworth, July 25, 1951, pp. 2–3; Thayer interview; R. Anderson "Andy" Pew, interview held at St. Davids, Pennsylvania, on October 8, 1975.
52. Thayer interview; Enos, *Petroleum Progress,* p. 148–50.
53. *Ibid.*
54. Williamson et al., *Energy,* p. 617; Enos, *Petroleum Progress,* p. 150; Thayer interview; see J.E. Evans and Raymond C. Lassiat, "Combustion Gas-Turbine in the Houdry Process," *Petroleum Refiner* (November 1945); Adolf Meyer, "The Combustion Gas Turbine: Its History, Development, and Prospects," *Institute of Mechanical Engineering Proceedings* 141 (1939), pp. 197, 202–4.
55. Lamont to Teitsworth, July 25, 1951, p. 3; Enos, *Petroleum Progress,* p. 150; Thayer interview.
56. F.S. Reitzel to J. Howard Pew, January 10, 1935, Acc. #1317, EMHL, A.E. Pew, Jr. Papers, Box 6; *Annual Report of the Houdry Process Corporation,* 1935; "Monsieur Houdry," *Fortune,* p. 134.
57. Williamson et al., *Energy* p. 646; Roy C. Cook, *Control of the Petroleum Industry by Major Oil Companies,* Temporary National Economic Committee Monograph No. 30 (Washington, D.C.: U.S. Government Printing Office, 1941), pp. 75, 80.
58. Enos, *Petroleum Progress,* p. 151; Lamont to Teitsworth, July 25, 1951, p. 4.
59. "Monsieur Houdry," *Fortune,* pp. 132, 134.
60. *Annual Report, H.P.C.,* 1935; A.E. Pew, Jr. to Eugene Houdry, December 23, 1925, Acc. #1317, EMHL, A.E. Pew, Jr. Papers, Box 20; Cook, *Control of the Petroleum Industry,* p. 75.
61. Enos, *Petroleum Progress,* pp. 151–2.

62. *Ibid.*, p. 156; Thayer interview. Although Socony later claimed credit for the molten salt heat-transfer system, Clarence Thayer argued that the idea originated with a Sun engineer, L.W.T. Cummings. Socony did choose the proper salt and developed the flow system using the hot molten salt to generate steam for the steam purge cycle (see Lamont to Teitsworth, July 25, 1951, p. 4).

63. Board Minutes, Sun Oil Company, October 17, 1937, Minute Book #6, p. 219; "The First Fifty Years," *Our Sun* 16, 4 (Autumn 1951), p. 7; Enos, *Petroleum Progress*, pp. 151–2; Lamont to Teitsworth, July 25, 1951, pp. 4–5.

64. Board Minutes, Sun Oil Company, June 29, 1937, Minute Book #6, p. 97. Sun did not obtain refineries in the southwest and a truly national marketing structure until after the 1969 Sunray-DX merger.

65. Eugene Houdry, Wilbur F. Burt, A.E. Pew, Jr., and W.A. Peters, Jr., "Catalytic Processing of Petroleum Hydrocarbons by the Houdry Process," paper presented before the group session on refining, Nineteenth Annual Meeting of the American Petroleum Institute (Chicago, November 18, 1938), Acc. #1317, EMHL, A.E. Pew, Jr. Papers, Box 2.

66. Williamson et al. *Energy*, p. 621.

67. Larson, Knowlton, and Popple, *New Horizons*, p. 167; "Memorandum on H.P.C. Discussions," December 17, 1937, Acc. #1317, EMHL, A.E. Pew, Jr. Papers, Box 17; "Monsieur Houdry," *Fortune*, P. 137; J.A. Brown, Socony-Vacuum, to W.S. Farish, Standard (New Jersey) (Copy), March 16, 1938, Acc. #1317, EMHL, A.E. Pew, Jr. Papers, Box 18; Thayer interview.

68. This account of Standard's position is based on two slightly different versions of a memo sent by Frank Howard, Standard Oil Development Company, to Arthur E. Pew, Jr., on December 17, 1937. Both accounts are located in Acc. #1317, EMHL, one in the A.E. Pew, Jr. Papers, Box 17, and the other in the J. Howard Pew Papers, Box 42.

69. Thayer interview; Larson et al., *New Horizons*, p. 167.

70. Eugene Houdry to A.E. Pew, Jr., February 21, 1938, A.E. Pew, Jr. Papers, Box 18.

71. Thayer interview; J.A. Brown to W. S. Farish (Copy), March 16, 1938, Acc. #1317, EMHL, A.E. Pew, Jr. Papers, Box 18; Farish to Brown, April 8, 1938; Brown to Farish, April 12, 1938; Brown to J. Howard Pew, April 27, 1928, Acc. #1317, EMHL, J. Howard Pew Papers, Box 42; Williamson et al., *Energy*, p. 646.

72. W.S. Farish to J.A. Brown, April 8, 1938, Acc. #1317, EMHL, J. Howard Pew Papers, Box 42.

73. Farish to Brown, April 8, 1938; Brown to J. Howard Pew, April 27, 1938; Thayer interview.

74. J.A. Brown to J. Howard Pew, May 4, 1938, Acc. #1317, EMHL, J. Howard Pew Papers, Box 42.

75. Larson et al., *New Horizons*, pp. 167–8; Enos, *Petroleum Progress*, p. 196.

76. Enos, *Petroleum Progress*, p. 196.

77. See Enos, *Petroleum Progress*, Chapter 5, "The TCC and Houdri-flow Processes," pp. 163–186 and Chapter 6, "The Fluid Catalytic Process," pp. 187–224.

78. Folder, "E.B. Badger and Sons," Acc. #1317, EMHL, A.E. Pew, Jr. Papers, Box 1; Agreement, E.B. Badger and Sons, June 26, 1939, A. E. Pew Jr. Papers, Box 20; Thayer interview; Telegram, R.H. Patchin, Standard Oil of California, to A.E. Pew, Jr., February 10, 1939; Acc. #1317, EMHL, A.E. Pew, Jr. Papers, Box 16.

79. R.W. Hanna, Standard of California, to A.E. Pew, Jr., January 31, 1939, Acc. #1317, EMHL, A.E. Pew, Jr. Papers, Box 16.

80. Hanna to A.E. Pew, Jr., February 27, 1939; Pew to Hanna, March 1, 1939, Acc. #1317, EMHL, A.E. Pew, Jr. Papers, Box 16; Enos, *Petroleum Progress*, p. 151; Memo: "TCC and HPC Units in Operation and Building," July 14, 1943, Acc. #1317, EMHL, A.E. Pew, Jr. Papers, Box 23.

81. Memo, "Japanese Negotiations," August 18, 1938; Eugene J. Houdry to E.E. Butterworth, Petroleum Administration for War, October 12, 1943, Acc. #1317, EMHL, A.E. Pew, Jr. Papers, Box 14.

82. Houdry to Butterworth, October 12, 1943; George Hargrove, E.B. Badger and Sons, to A.E. Pew, Jr., October 18, 1939, Acc. #1317, EMHL, A.E. Pew, Jr. Papers, Box 14.

83. Memo, Hargrove to A.E. Pew, Jr., November 21, 1939; Memo, Hargrove to A.E. Pew, Jr., October 8, 1939; Joseph C. Green, United States Department of State, to Sun Oil Company, December 19, 1939, Acc. #1317, EMHL, A.E. Pew, Jr. Papers, Box 14.

84. Green to Sun Oil Company, December 29, 1939.

85. Eugene J. Houdry to Butterworth, October 12, 1943.

86. Green to Sun Oil Company, December 19, 1939. On the subject of Roosevelt's policy toward Japan prior to Pearl Harbor, the earliest revisionist attack is in Charles Beard, *President Roosevelt and the Coming of War, 1941* (New Haven: Yale University Press, 1948); for a defense of the Roosevelt policies, see Basil Rauch, *Roosevelt: From Munich to Pearl Harbor* (New York: 1950); the best statement from the newer economic revisionists is Lloyd C. Gardner, *Economic Aspects of New Deal Diplomacy* (Madison: University of Wisconsin Press, 1967), while a more recent realist view is given in Paul W. Schroeder, *The Axis Alliance and Japanese-American Relations, 1941* (Ithaca: Cornell University Press, 1958).

87. George Hargrove to A.E. Pew, Jr., October 21, 1939, Acc. #1317, EMHL, A.E. Pew, Jr. Papers, Box 14.

88. Lamont to Teitsworth, July 25, 1951, pp. 6–7; H. Heller to A.E. Pew, Jr., September 13, 1940, Acc. #1317, EMHL, A.E. Pew, Jr., Papers, Box 26; Thayer interview; Enos, *Petroleum Progress*, pp. 164–6.

89. Enos, *Petroleum Progress*, p. 154. George Foster suggests that the main reason that the Houdry process did not receive wider initial acceptance was that it did not promise a higher yield of gasoline along with higher quality. In fact, it did indeed provide increased yield, but this was not readily apparent at the time. Firms using tetraethyl lead in their gasolines did not have to look for increased qualitative gains from their cracking. This was, of course, just the opposite for Sun. Thus, there was no perceived need to innovate by the several refiners who had turned a cold shoulder to Eugene Houdry in the early 1930s (see Foster, "An Examination of Process Innovation," p. 171).

90. Board Minutes, Sun Oil Company, October 17, 1939, Minute Book #6, p. 219; *Annual Report of the Sun Oil Company*, 1939.

CHAPTER VIII

Sun Oil, the Petroleum Industry, and The New Deal, 1930–1940

The Great Depression of the 1930s was a vital watershed in the evolution of business-government relations in the United States. The New Deal economic programs of Franklin D. Roosevelt coalesced forces of business-government cooperation that had been developing since World War I and established patterns of regulatory activity that have continued into the present. Interpretations of the larger meaning of these events have reflected the changing times. Whereas some still view the government's intervention in the economy as the beginning of dangerous socialist experimentation, others interpret the New Deal as the triumph of a "mixed economy" better equipped to deal with the complexities of modern industrial society. More recently, a younger generation of historians have critically interpreted the Roosevelt program as a movement with only one goal—the preservation of capitalist America.[1] Thus, what J. Howard and J.N. Pew, Jr. perceived as creeping socialism in the 1930s, some contemporary writers have judged a victory of conservative capitalism. Although there remains disagreement about the goals as well as the long-term implications of the New Deal, most analysts agree that it brought government into the economy as never before.

In the petroleum industry, the New Deal established the same basic system of production controls to achieve economic stabilization that governs the domestic industry today. Industry and government reached a consensus for solving the problems of uneven crude production and its attendant economic instability, but they did not achieve this agreement easily. Policy struggles at both the state and federal levels continued to reflect the particular concerns of different elements in the oil business and the often conflicting philosophies of government officials. Many oil executives never came to accept the idea of government regulation and were highly critical of the New Deal, while others embraced the stabilization programs that emanated from Washington or the producing states.

Most, like the Pews, were highly selective concerning which aspects of government regulation they would accept. However, Sun's managers stood out as a result of their vociferous attacks on those aspects of New Deal programs perceived as alien to their philosophy. More than any other oil company, Sun became identified as an anti-Roosevelt, anti-New Deal firm.

This perception was largely accurate. Sun Oil was owned by the Pews, members of the family held the key executive positions in the firm, and Sun's business practices reflected their beliefs and values. The family and the firm were intertwined in ways more typical of the nineteenth than the middle of the twentieth century. Thus, in order to fully assess the role that Sun played in the major policy issues of the New Deal era, one has to consider the underlying philosophy that motivated its managers.

PETROLEUM AND THE GREAT DEPRESSION

The 1920s ended with an increased degree of cooperation between oil producers and state governments to meet and need for combating economic as well as physical waste. The Seminole boom of 1927 had again focused attention on overproduction as the main economic problem facing the industry. Disagreements on precisely what form regulation should take, however, prevented the emergence of a broad consensus in the industry. The failure of President Hoover's attempt to bring relief at the Colorado Springs Conference in 1929 had highlighted the mistrust independent producers felt for the large integrated companies and the general aversion to federal regulation that most industry leaders shared. Many of them, including J. Edgar Pew, had come to accept conservation regulations at the state level and the development of voluntary unitization plans among oil drillers. However, they were still unwilling to open the door for federal regulation. Hoover's and Mark Requa's plan for an interstate compact to restrict oil production collapsed in the face of disagreements over federal participation.[2]

In 1931, history again repeated itself, and the vast East Texas field brought in record amounts of crude oil. The new pool, first drilled in the fall of 1930, had proven acreage more than ten times that of Oklahoma's Seminole field by the summer of 1931. Its discovery could not have come at a worse time for the industry. The Depression had hit almost all sectors of the economy hard, and price levels declined steadily throughout 1930 and 1931. Now disaster threatened independent producers and major companies alike as the price of crude dropped dramatically. As the unrestricted flow accelerated, East Texas crude that sold for three dollars a barrel in 1919 fell to ten cents a barrel in the summer of 1931.[3]

Affiliates of the major companies rapidly purchased a total of between forty and fifty percent of the East Texas lease rights (Sun obtained approximately a five percent holding in the field).[4] As in the previous decade, the larger firms were anxious to implement some kind of production control program. The vast assortment of smaller independent operators, who owned and operated the rest of the field, viewed curtailment plans as discriminatory. They naturally wanted a good price for their oil but felt that they could not afford to give up their livelihood while controlled production gradually brought prices upward. Moreover, history had demonstrated that glut was only temporary, and these small producers believed that they should obtain oil while they could at any price available in the marketplace.

Unitization had become the key approach to conservation and stabilization ever since Henry Doherty had proposed it in the early 1920s. But voluntary unitization (operating each pool on a cooperative regulated basis according to the relative shares or leases held) proved totally inadequate in stemming the flood of East Texas crude. More extreme measures were required. "Prorationing" now became the remedy favored by many producers. This involved limiting or reducing output from flush fields under the authority of a state agency.

A mechanism for prorationing rampant East Texas production existed in the form of the Texas Railroad Commission. Since 1919, that body had issued and enforced conservation regulations throughout the state, and had achieved a moderate degree of success in establishing such practices as the wider spacing of wells in oil fields. Prior to the East Texas strike, the Commission had begun to develop a definition of conservation that encompassed the economics of market demand as well as physical waste. In August 1930, it issued its first state-wide prorationing order, limiting daily production to 750,000 barrels a day, 50,000 barrels less than the previous year.[5] The Railroad Commission announced its first East Texas prorationing order in April 1931, but it ran into strong opposition. There was physical resistance including several threats of violence, as well as the institution of hundreds of court challenges to the order.[6]

The Commission couched its orders in terms of eliminating waste and did not explicitly refer to market demand. In order to expand the Commission's power, Governor Ross Sterling summoned a special session of the Texas legislature in July 1931 to consider a new conservation law that would address the need to bring production into line with demand. While the lawmakers debated this proposal, a bombshell burst on July 28 in the form of a Federal District Court ruling that the Railroad Commission's first East Texas order was invalid. The court's action had the effect of nullifying the prorationing order and chaos again ruled in

August as unrestrained production resumed. A divided legislature then passed a weak conservation law on August 12, 1931 that empowered the Railroad Commission only to prohibit "physical waste."[7]

The East Texas debacle did not occur in isolation. The action of the Federal District Court in Texas in striking down the prorationing efforts of the Railroad Commission had served to complicate a similar controversy that had been brewing in Oklahoma. Since the early 1920s, the Oklahoma Corporation Commission had gradually assumed more power in regulating the production of oil in that state under the banner of conservation. A new law enacted in 1929 had given the Commission power to directly regulate production, and the Oklahoma Supreme Court had upheld the statute. However, on August 3, 1931, the Federal District Court in Oklahoma ruled that the Corporation Commission did not have the authority to directly prorate production. There seems little doubt that the court's action in Texas the previous week had some influence on the Oklahoma decision. The next day Oklahoma's colorful governor, "Alfalfa Bill" Murray, declared martial law in his state and ordered the closing of all wells that had been affected by the decision. The presence of troops in the Oklahoma oil fields was one of the more dramatic events that highlighted the failure of the economic system in 1931. The East Texas glut had exacerbated an already difficult situation that had existed since the Seminole field had created chaos in Oklahoma in the 1920s. The entire southwest was in a panic.[8]

Murray's bold action had an immediate effect in Oklahoma. On August 5, two federal judges announced that they now viewed the Oklahoma law valid, although their formal decision was not released until October. However, Murray kept the National Guard in the oil fields to police the Corporation Commission's prorationing orders through the spring of 1932. In March of that year, the United States Supreme Court upheld the Oklahoma Conservation Act and tensions diminished. The Texas situation, however, remained more complicated.[9]

After the Texas court's ruling had negated the Railroad Commission's orders, crude prices again dropped to ten cents a barrel and there was talk of vigilante action to dynamite wells and pipelines. To the hard-pressed Texas producers, government seemed unable to act, and it appeared that law and order was on the verge of breaking down completely. In the face of this renewed crisis, Governor Ross Sterling followed the lead of Governor Murray and declared marital law in the East Texas oil fields. On August 17, 1931, four thousand National Guard troops under the command of General Jacob Wolters went into the fields to restore order and close down all producing wells.[10]

While the troops were in place, the Texas Commission began hearings on a prorationing plan that would be consistent with the weak conservation law that the legislature had passed in early August. A new Commission order could take one of two forms. The larger integrated companies wanted restrictions established on an acreage rather than a per-well basis. This would more effectively limit the number of wells drilled and benefit the long-term stabilization goals of the larger lease holders. The smaller independents, however, favored restrictions on a well basis so that they could continue to drill and establish a basis for later exploitation of the pool. Since these men had relatively small lease-holdings, it was imperative for them to drill quickly or lose the oil under their property. Most involved parties did agree on a goal of restricting total East Texas production to under 400,000 barrels a day.[11] The problem was how to arrive at this target.

On September 2, 1931, the Railroad Commission issued an order restricting East Texas production to 400,000 barrels and Governor Sterling committed the National Guard troops to its enforcement. The limitation restricted each well to a maximum of 225 barrels per day initially, and then to 185 barrels. The Commission envisioned lowering this to 100 or 75 barrels in order to maintain the target of 400,000 barrels per day. Governor Sterling was sympathetic to arguments for prorationing on an acreage basis but feared the strong opposition of the independent producers.[12] Sterling was himself a former president of the Humble Oil Company, one of the largest producers in the field and a part of Jersey Standard's operation since 1919, while General Wolters was on leave from his regular position as an attorney for the Texas Company.[13] Thus, even though the per-well order encouraged further drilling, Sterling felt that this plan was better than nothing.

J. EDGAR PEW AND PRORATIONING

J. Edgar Pew played a significant role in this debate as both an executive in a major East Texas operating company and as Chairman of the American Petroleum Institute's Committee on Conservation. Pew had championed prorationing in the wake of the disastrous glut of 1930–1931, and specifically favored the plan in New Mexico, Texas, and Oklahoma. However, in October 1931, he tried to get Sterling and the Railroad Commission to adopt an acreage plan rather than the less restrictive well program then operating in East Texas.[14]

In addition to writing directly to Chairman C.V. Terrell of the Railroad Commission, Pew worked through a lobbyist, Judge Ben H. Powell, in Austin. On October 7, 1931, Pew sent a circular letter to eighteen companies representing both large and smaller integrated firms, and some independent producers. In this letter, J. Edgar summoned a conference on October 21 to discuss a compromise acreage plan outlined in an accompanying memorandum. Pew's proposal would allow small operators to produce oil equivalent to the amount allowed for a twenty-acre leaseholding under the acreage plan. One barrel would be deducted from their quota for every acre of the lease under twenty. There was some objection to having the conference at all, and some specific criticism of Pew's compromise acreage plan, but the meeting nevertheless was held in Dallas on the 21st.[15]

Pew's efforts led to a meeting of the Texas Oil and Gas Conservation Association, an organization of independent East Texas producers, on October 27. Pew's initiative had indicated that the majors were willing to accommodate the independents in an acreage proposal, and eighty independent operators showed up at the conclave. After a full and heated discussion, the group voted unanimously to recommend the compromise acreage plan to state authorities. Edward T. Moore, an independent who chaired the meeting, wrote J. Edgar Pew that "This was the first meeting of the East Texas producers that has resulted in any unanimous approval of any plan for operating that field."[16] Those at the meeting appointed a committee of seven individuals to present the plan to Texas authorities. Chaired by Roeser, the committee included two independent operators, H.L. Hunt and J.P. Pearson; Colonel J. Lewis Thompson of the Royalty and Land Owners of East Texas, another association of independent operators; Judge Beeman Strong of the Yount-Lee Oil Company; T.J. Donoghue of the Texas Company; and J. Edgar Pew.[17]

J. Edgar immediately cabled leaders of the major companies represented in the East Texas field, urging their cooperation with the compromise acreage plan. On October 30, the Committee of Seven forwarded its proposal to Governor Sterling in Austin, seeking his support. On November 2, Pew sent a more detailed memo to the twenty largest lease holders in the East Texas field outlining the purpose of the program. Pew also sought to obtain the backing of the American Petroleum Institute board of directors at its Chicago meeting in early November.[18]

At this point, J. Edgar stepped into the background and let Charles Roeser, an independent, take center stage in the movement to get the plan adopted in Texas. However, Pew offered his continued service to Roeser and expressed the view that:

> We do not know how long East Texas will be under martial law, but while it is under martial law is the best time for us to put into practice, and convince the producers of its soundness, our acreage recovery basis program. We can put this over under martial law when it might be very difficult otherwise.[19]

Pew's efforts were to no avail, however, as the plan ran into further opposition from some smaller independent interests.

The independents claimed that the majors had increased their strategic position by drilling intensively on their large holdings and could now safely fall back on their acreage allotments. The independents, on the other hand, would have to be content with the low production levels to be allowed from their leases. Moreover, the still considerable independent opposition to the acreage plan influenced Governor Sterling against it. At the same time that the plan was being submitted, the Governor's more popular well-basis program was under constant legal attack by some producers who opposed any form of restriction.[20]

J. Edgar used his influence to oppose some of these court challenges, one in particular lodged by Clint Murchinson's Brock Lee Oil Company. This was an antitrust suit charging Governor Sterling and the large Humble and Texas Companies with violating Texas law in the martial law restrictions. These court challenges continued to hinder the prorationing efforts and specifically blocked attempts to push Governor Sterling into adopting an acreage plan. To further complicate matters, in May 1932, the Federal District Court for East Texas issued an injunction against the use of troops, an order which Governor Sterling and General Wolters ignored. Now a jurisdictional dispute between state and federal government had been added to the already bitter disagreements between majors and independents, prorationists and anti-prorationists.[21]

This new action of the Federal District Court in Texas was similar to a challenge that had been raised to the imposition of martial law in Oklahoma by Governor Murray. Murray also had ignored the courts and continued to regulate production by the force of the National Guard in that state. On May 16, 1932, however, the U.S. Supreme Court voted to uphold the Oklahoma prorationing act, arguing that it was proper to eliminate physical waste by limiting production to reasonable market demand. This decision had no immediate results in Texas, however, since the Texas law was written to specify physical waste alone. Confusion reigned in Texas when, on October 23, 1932, the Federal District Court for East Texas again enjoined the Railroad Commission from enforcing its prorationing orders.[22]

In November, Governor Sterling again called a special session of the legislature to draft a new conservation law that would enable the Commission to resume its regulatory functions. This new law, the Market Demand Act, was passed with little opposition. Economic waste as well as physical waste had now fully become a target of the state of Texas. The way was now clear for a comprehensive plan of prorationing on an individual well allotment basis.[23]

During the great turmoil in the Texas and Oklahoma fields in 1931–1932, J. Edgar Pew addressed the A.P.I. Division of Production at this Dallas meeting in November 1931. In his speech, "The New Conception of Oil Production," Pew condemned the law of capture, praised Henry Doherty's earlier foresight, and strongly supported programs of unitized oil field development. Most revealing of the change he had undergone was his concluding statement:

> It is no longer of any use to inveigh against governmental interference with business, or to denounce as revolutionists any who disagree with us. The country, the world, looks upon us as trustees for a vital resource. Our administration of that trust has not been satisfactory. If we are to gain our trusteeship, if we are to keep our hold on this industry, we must win back a public confidence that we have lost.[24]

The realities of the East Texas debacle had moved J. Edgar Pew further toward supporting business-government cooperation, but his Philadelphia cousins were not so enthusiastic.

A NEW DEAL FOR OIL: THE N.R.A.

The chaos in East Texas and Oklahoma had demonstrated several things. Most importantly, it showed that a significant number of oil men wanted some sort of government action to bring order to the industry, but the Hoover administration seemed unwilling or unable to act. Hoover himself had become bitter after the failure of the Colorado Springs Conference, but he was also opposed on philosophical grounds to extending federal control. A large segment of the oil industry was therefore ready for a political change in 1932. State action in Texas and Oklahoma had been successful in raising the price of crude. East Texas oil was selling for eighty-five cents a barrel in July of 1934, a significant recovery from the ten cents it commanded in August of 1931, but significant problems still remained.[25] One of the major difficulties that state regulators faced was that of "hot oil," oil produced in excess of mandated prorationing and either sold locally or shipped interstate. It was this latter problem, the interstate shipment of hot oil, that called for some form of federal participation if any curtailment program would ultimately remain successful.

Much of American industry was waiting to see which way the wind would blow in the spring of 1933. Similarly, Roosevelt and his economic advisors felt their way slowly before committing themselves to a definite oil policy. On March 8, soon after his inauguration, F.D.R. instructed new Interior Secretary Harold Ickes to call a meeting of the governors of the oil producing states later in the month. A.P.I. leaders and other industry groups were also invited to the meeting in Washington. On the same day, the A.P.I. board of directors appointed a group to negotiate production limitation agreements with the Governors of Oklahoma, Texas, and Kansas. C.B. Ames of the Texas Company, W.F. Teagle of

Jersey Standard, Harry Sinclair of Sinclair Consolidated Oil, and J. Edgar Pew were all members of this group. Their hope was to effect production curtailment through the enforcement of existing state prorationing legislation.[26]

Representatives of thirteen producing states met with Ickes on March 26, and the new administration quickly learned of the diverse views within the industry. Governor Alfred Landon of Kansas spoke for the governors in urging strong federal action and a take-over if necessary to do something about reducing and stabilizing production. J. Edward Jones, an independent operator, opposed federal intervention but urged that the government investigate the large integrated companies with the aim of introducing legislation to divorce pipeline ownership from the majors.[27] Production curtailment programs that the A.P.I. proposed met opposition from members of both groups.

Roosevelt received advice from various quarters. Mark Requa urged him to adopt the type of federal controls that had proven successful in World War I, and Governors Landon of Kansas and "Alfalfa Bill" Murray of Oklahoma continued to urge federal intervention. Throughout April 1933, action continued at the state level in an effort to stabilize the industry. Kansas Senator Arthur Capper and Oklahoma Congressman E.W. Marland introduced similar "hot oil" legislation prohibiting the interstate shipment of oil produced over and above quotas.[28]

In May, the President decided to opt for more direct federal intervention. He requested that Congress set aside the Capper and Marland bills so that petroleum could be brought into the general industries bill, the National Industrial Recovery Act. The N.I.R.A. became law on June 16, 1933. Section 9(c), an amendment added by Senator Tom Connally of Texas, had particular relevance for the petroleum industry.[29] The Connally amendment, the basis for later "hot oil" legislation, prohibited the interstate shipment of oil produced in excess of state quotas.

Under the National Industrial Recovery Act, business representatives were to draw up codes of fair competition for each industry. Hoping to get a jump on things and possibly forestall federal initiatives if they did not act, the A.P.I. immediately called a meeting of oil industry leaders for June 17 in Chicago to draw up a preliminary petroleum code. J. Edgar Pew served on the draft committee for the production code, and J. Howard Pew on the refining and marketing sections. The precedent for the new set of standards and regulations was "The National Code for Practices for Refined Petroleum," established in 1929. This earlier code was one of several that had been promulgated by the Hoover administration and supervised by the Federal Trade Commission. The American Petroleum Institute had drafted the document and the F.T.C. had approved it. Sun Oil had accepted the code and J. Howard Pew served on the committee of enforcement and propagation for the Eastern region. This precursor of the N.R.A. was an outgrowth of the trade association movement in the 1920s and the philosophy of business voluntarism espoused by President Hoover.[30]

The tentative N.R.A. code envisioned extensive federal control, including drilling permits, production quotas, and minimum prices, all set at the federal level. The price-fixing component was the most controversial, and both J. Howard and J. Edgar Pew voted against any price-fixing provision in Chicago and later at an A.P.I. meeting in New York on June 29. N.R.A. administrator General Hugh Johnson sent the proposed code back to the A.P.I. in July because of the price-fixing dispute and a lack of labor provisions. The A.P.I. resubmitted a revised version in mid-July, but two opposing camps had emerged, one in favor of direct price-fixing and the other diametrically opposed.[31]

THE FIGHT AGAINST PRICE-FIXING

The Pews supported the A.P.I. draft of a proposed code for the sake of unity and cooperation, but they had made their position clear on price-fixing. They indicated their unhappiness with the establishment of minimum crude oil prices that were tied to the price of gasoline, and stated that Sun Oil would not cooperate with any later attempts to apply minimum prices on refined products. Sun's managers charged that price-fixing was inefficient, could possibly cause inflation, and was opposed to the best interests of the consumer. They also raised the philosophical argument that the elimination of price competition hindered business incentive.[32] No doubt they felt strongly about the merits of all of these arguments, but they were particularly sensitive to Sun's own marketing situation. The establishment of minimum prices for certain octane grades of gasoline might seriously endanger its Blue Sunoco sales, since Sun sold only one brand of unleaded fuel at a price less than most "premium" grades, but higher than that of many regular and third grade products.

Joining Sun in opposition were representatives of the largest major integrated firms, including Jersey Standard, Indiana Standard, Shell, Texas, and Gulf.[33] Here again, fundamental differences between integrated firms and the independent producers surfaced. The relative economic position of the large firms was good in 1933. They had made record profits in the 1920s and were able to absorb huge operating losses in the early years of the depression without any serious damage to their overall strength. More importantly, the integrated companies could offset losses in the producing arm of their business with profits from their refining, marketing, and pipeline activities. Cheap crude oil was in many respects a boon to refiners, since the demand for gasoline continued to increase steadily during the Depression years and profits were made from these sales. Although the average retail price of gasoline had declined from 21.4 cents per gallon in 1929 to 17.0–17.8 cents per gallon during 1931–1933, one can argue that this price decline aided in keeping the demand high for gasoline. Individual real income per capita dropped by over twenty-five percent between 1929 and 1933, but gasoline sales, although declining some, did not fare as badly during the same period. Americans appeared reluctant to give up the use of their automobiles, even in the wake of declining income. Moreover, gasoline sales increased again in 1934 and continued to grow at a steady rate for the rest of the decade.[34]

The independent petroleum producer needed immediate relief. These smaller operators argued that the integrated companies were content to achieve gradual price increases to be effected by the curtailment of production. In the meantime, they could fill their storage tanks with very cheap oil produced on their own properties or purchased in the depressed market.[35] By the same line of reasoning, most of the majors opposed formal fixed prices that would raise their costs and possibly diminish the market for their refined products.

Major oil company executives did not unanimously oppose price-fixing. James A. Moffett, vice-president of Jersey Standard, a Democrat, and friend of President Roosevelt, supported the establishment of minimum prices for crude and refined products. His position was directly opposite that held by Jersey board chairman Walter Teagle and president W.S. Farish. After a heated argument, the two top executives forced Moffett's resignation in late July of 1933, thus ending the long relationship that he and his father had had with the Jersey company. Siding with Moffett on the price-fixing issue were K.R. Kingsbury of Standard of California, Amos Beatty of Phillips Petroleum, and Harry Sinclair of Consolidated Oil.[36]

One of the arguments raised by representatives of larger companies who favored price-fixing was "cost recovery." They made particular reference to the labor provisions of the N.R.A. code, which one spokesman argued "might add $300,000,000 a year to the pay roll." The argument was that minimum prices were needed to guarantee profitable operations to the oil companies.[37] E.B. Reeser of the Barnsdall Oil Company of Tulsa argued that "A number of us feel that the same men who put a minimum price on a bushel of wheat can with equal propriety put a minimum price on crude oil."[38] Harry Sinclair, another strong advocate of price minimums, maintained that the entire recovery program in oil would fail without the enforcement of minimum prices from the wellhead to the consumer.[39]

Although most of the independents favored price-fixing, their ranks also were not united. Wirt Franklin, president of the Independent Petroleum Association, an organization representing western crude oil producers, was one of the leading advocates of a minimum price for crude. In addition to his plea for price relief for the independent producer, he also invoked the "cost recovery" argument alleging that money was needed to pay labor increased wages under the N.R.A. labor codes.[40] All "independents" were not necessarily producers, however, and all refiners were not large integrated concerns. The small and medium-sized refiners were very wary of price provisions that would significantly raise the purchase cost of crude oil in an uncertain environment for the marketing of refined products in 1933. These refiners opposed the price-fixing clause.[41]

Thus the oil industry was split into strong camps. It was not just a struggle of independents against majors, producers against refiners, or Easterners against Westerners. Size did play a role in that the largest integrated companies were in the forefront of the "anti" group and the smallest producers were in favor of the pricing plan. But aside from short-term arguments, there is a great deal of evidence to support the view that business philosophy played a large part in the debate. The Pews, Farish, Teagle, and other opponents of the price-fixing clause were strongly opposed to direct government price-fixing. They supported production regulation and the indirect effect this would have on raising prices, but drew the line on formal controls.

The full delegation of oil industry representatives became deadlocked on the pricing issue in early August, and N.R.A. Administrator Hugh Johnson appointed a committee of twenty-four individuals to remain in Washington to iron out a settlement. Seven members of this group were selected to represent the anti-price-fixing forces. Among these were C.B. Ames, board chairman of the Texas Company; W.S. Farish of Jersey Standard; E.G. Seubert, president of Standard of Indiana; W.G. Skelly, president of Skelly Oil; C.E. Arnott, president of Socony-Vacuum; W.S.S. Rogers, president of the Texas Company; and J.N. Pew, Jr. This committee was empowered to go over all provisions of the code, paying particular attention to the pricing provisions.[42]

In public statements before the code hearings in Washington that July, J.N. Pew, Jr. had made clear Sun's stand against the practice. He argued that crude producers should be given "police protection" so that their oil would not be stolen from them, but drew the line at minimum prices. "What is price-fixing," he argued, "but a subsidy under another guise? A subsidy paid by the consuming public, which is already bled white with taxes. No industry—that is subsidized, by Government or otherwise, can long prosper. It puts a premium on over-production and over-expansion, and leads but to one end -industrial suicide."[43]

While Joe Pew led the fight within the N.R.A. councils in Washington,

Sun carried its message to the public. On his evening broadcast of August 14, 1933, Lowell Thomas described the sharp differences between the price-fixing and anti-price-fixing groups and read excerpts from J.N. Pew's statement at the code hearings. He concluded with the statement that "My sponsors, the Sun Oil Company, have wished their view understood by all of their friends, and I have been glad to pass it on to you."[44] This was not the last time the Pews made use of this highly popular program to get their views across to the public.

Within the administration, Interior Secretary Ickes supported price-fixing while the N.R.A.'s Hugh Johnson opposed it. Roosevelt himself, as he did on so many issues, remained ambivalent. The President, however, increasingly came under the influence of James Moffett on oil matters, and Moffett favored price controls. This dispute became entangled with a bureaucratic power struggle going on in Washington. Ickes wanted to become the administrator of the petroleum code from within the Interior Department, while Johnson assumed that he, as N.R.A. chief, would have jurisdiction.[45] Johnson presented a draft of a code without price regulations to the President on August 17, and Ickes soon after offered one with rigid production controls and price-fixing. Roosevelt then instructed the two men to effect a compromise with the industry, an end result Johnson later referred to as "hermaphroditic."[46]

The compromise gave the N.R.A. power to recommend rather than to set production quotas, and gave to President Roosevelt the discretionary power to grant a temporary price minimum for a period of ninety days. The first part was a concession to the Sun Oil-Jersey Standard group, the latter to the Wirt Franklin-Moffett people. The President also had the power to appoint a Planning and Coordination Committee to assist the petroleum administrator in the implementation of the codes.[47] J.N. Pew, Jr. was still in Washington leading the fight against price-fixing when the final draft of the revised code reached Roosevelt's desk late in the evening of August 19, 1933. He later traced his break with the New Deal and Roosevelt to that issue. Assured that Johnson's version of the petroleum code contained no price-fixing provision, Pew was understandably upset when he learned of the compromise code. He commented that

> I went to Washington in 1933 to help draft the NRA petroleum code. We were opposed to price-fixing. We got up our code and sent it to the White House. It had no price-fixing in it. But it came back with a price-fixing clause in the President's own handwriting. His own handwriting. I blew up. I would have no part of it. It obviously was a trap baited with their own greed. It was illegaL.[48]

The handwriting had been on the wall prior to the evening of the 19th, for at noon on that day, J. Howard Pew, William Farish, Edward Seubert of Indiana Standard, and William Skelly of the Skelly Oil Company called on Ickes at the Interior Department. As Ickes related the meeting, "They were a sober lot of men. They said they were conservatives who didn't believe in price regulation but that they wanted to cooperate with the government."[49] Ickes felt that the oil men wanted to establish a friendly relationship, and the interview was pleasant on both sides. Presumably, the Pew-Farish group were already working on the next line of defense, the make-up of the important Planning and Coordination Committee of the Petroleum Code in the National Recovery Administration.

On August 19, William Farish, J. Howard Pew et al. (a total of fifteen men representing fifteen companies) sent a formal protest to Hugh Johnson objecting to the price-fixing clause and asking for a fair representation on the Planning and Coordination (P & C) Committee. On August 20, Farish and Pew released a statement to the press with their group's nominations for the committee.[50] On the same day, J.N. Pew, Jr. sent a formal letter to Hugh Johnson in which he protested "vigorously" against any clause in the petroleum code that provided in any way for the price-fixing of refined products. Sun Oil asked for a hearing with J.N. Pew, Jr. as its designated representative.[51]

Of the fifteen members appointed to the P & C Committee, twelve would come from industry and three would represent the government. By August 21, N.R.A. administrator Johnson had received several nominations for the twelve industry positions. The A.P.I. group supporting price-fixing nominated nineteen men and the Pew-Farish opposition faction put forward fifteen names. This latter group had approved the code with reservations on the price-fixing clause. A third group of independent oil producers opposed to any form of production control, led by Jack Blalock of Marshall, Texas, nominated twelve men.[52]

The Pews banked on General Johnson's having a major say in the makeup of the P & C Committee and hoped that the anti-price-fixing group would obtain a significant share of its membership. J. Howard went to Washington to meet with Teagle, Ickes, and Moffett in hopes of furthering the cause of his group, but came away only partially satisfied. Although he found Ickes still hostile to the anti-price-fixing position, "Jimmy" Moffett was very friendly and insured that all would get a fair deal. By mutual agreement, Pew did not debate price-fixing with Moffett and only discussed the overall purposes of the N.R.A. plan.[53]

On August 30, Roosevelt announced that Harold Ickes, not Hugh Johnson, would become administrator of the petroleum code. Furthermore, his nominees to the P & C Committee heavily favored the price-fixing element of the industry. Moffett, M.L. Benedum, and Donald Richberg represented the government, and the twelve industry members (among them W.R. Kingsbury, Axtell Byles, Amos Beaty, and

Wirt Franklin) all supported price-fixing.[54] Ickes implied that he would set the wholesale price of gasoline at the refinery at six cents a gallon. Under the formula established by the code, this would make the price of Mid-Continent crude $1.11 a barrel (the price of crude was to be 18.5 times the wholesale refinery price per gallon of gasoline). This was the compromise that Hugh Johnson had agreed to in Section 6 of the Petroleum Code. In fact, this minimum price never went into effect, but Ickes used the threat of price-fixing as a club over the majors to get them to conform to the other provisions of the code.[55]

Ickes's first concern was to limit the production of crude, and the P & C Committee made recommendations for monthly quotas for each of the oil producing states. Ickes also named a Petroleum Administrative Board (P.A.B.) to assist the P & C Committee. This Board consisted of a staff of technical experts who would work full time in Washington to implement the Petroleum Code. Ickes also announced that he was limiting oil imports to an amount equal to imports in the last half of 1932. This had also been a controversial issue during the code hearings; the independent producers urged quotas or tariffs on imported crude to support their declining prices, while the major firms with cheap supplies of foreign oil opposed such measures as discriminatory.[56] Sun had little opposition to the quota system since it had minimal foreign investment. The firm's Venezuelan venture had been a failure in the 1920s and Sun felt that it had sufficient domestic production. Moreover, management was very wary of the political problems involved. J.N. Pew, Jr. wrote to a friend in November 1932 that "I am very pessimistic as to the value of foreign interests of any kind, particularly in the oil industry. I believe the time is coming when the nationals of every country will control the petroleum industry of their country."[57] Pew was, of course, very prophetic in the long run, but for years, this Sun policy limited the company's share of the profitable business in foreign oil exploitation.

In addition to promulgating quotas on domestic production and imports, the Petroleum Administrator had the power to regulate the marketing of refined products and, when deemed necessary, the prices for those products. Provision 7(a) of the National Recovery Act also came under Ickes's jurisdiction, and he had the job of maintaining labor peace within the industry. Sun Oil and most of the other companies where the unions had not yet made inroads compiled with the N.R.A. by adopting employee representation schemes and company-dominated unions.[58]

The pricing issue was not yet dead; Ickes and the anti-price-fixing companies continued to struggle over the problem throughout the fall of 1933. Sun consulted its attorney, Senator George Wharton Pepper, particularly on what options it had if the firm decided to cease cooperating with the N.R.A. code after the agreed ninety-day period.[59] Sun was determined to fight any attempt to place price controls on gasoline. It was clear that Sun was becoming politically opposed to the Roosevelt Administration and to Harold Ickes in particular. The Pews felt that Ickes wanted to become czar of a federally dominated petroleum industry. At the same time, J. Howard Pew wanted to insure that Sun complied with the day-to-day regulations of the petroleum code in the event that the company was on a collision course with the administration. The firm hired Robert G. Dunlop, a young accountant from the University of Pennsylvania's Wharton School of Finance, to oversee the firm's operations within the N.R.A. codes.[60]

Sun Oil management, although excluded from membership on the P & C Committee, did participate in various subcommittees under the petroleum code. J. Howard Pew served as chairman of the transportation committee in N.R.A. Region No. 1 (Northeast and Middle Atlantic States) and as a member of the main labor subcommittee. J.N. Pew, Jr. was a member of the transportation committee in Region No. 3, and Walter C. Pew a member of the labor committee in Region No. 1. Arthur E. Pew, Jr. served on the main refining subcommittee, the refining committees in Regions No. 1 and 3, and as a member of the allocating agency in both regions. Sun Ship president John G. Pew served on the regional board of the National Labor Board under the direction of Senator Robert F. Wagner. The precursor of the National Labor Relations Board, this group attempted to settle disputes arising from Section 7(a) of the N.R.A.[61]

Sun maintained an outward stance of cooperation with the N.R.A., but the Pews remained skeptical about the recovery administration. In particular, they were aware of Ickes's potential power and nervous that he might yet invoke the price-fixing provisions of the oil code. The firm's public stance on price-fixing depended on the particular emphasis Ickes put on the subject. By the first week of October 1933, it appeared that Ickes had retreated due to fears that the courts would not sustain N.R.A. price-fixing. J. Howard was also pleased as a result of an N.R.A. cost inquiry that showed Sun's figures to be lower than those of most other companies. He felt that higher prices passed by the code authority would put the N.R.A. in a bad light.[62] For a time, it seemed that things were peaceful for Sun on the N.R.A. front.

On October 16, however, Ickes announced that he would soon issue a price-fixing order with the President's approval. This immediately threw the Sun camp into a dither, and the "anti" group began to gather information to prepare their case for November hearings to be held on the Ickes proposal. They retained noted petroleum economist Joseph E. Pogue to prepare a brief on the subject, and Sun had its Washington lobbyist gather information on the chronological development of the petroleum code and the public utterances of various public officials concerning it.[63] The Ickes plan would raise the wholesale refinery price of gasolines with octane ratings of 65–70 two cents higher than for those with ratings below 60, and one cent higher than for those with octanes between 60 and 64.9. The 18.5:1 ratio established in the compromise code would raise crude prices to a minimum $1.11 per barrel. In effect, this fixed price differential would force Sun to sell its Blue Sunoco at the same price as other premium grades and would force the company out of its unique marketing strategy.[64]

Sun submitted a formal brief protesting Ickes's order on November 14, 1933. The firm argued on legal and constitutional grounds that the August 19 code gave no authority to establish prices on refined products. The Pews were in a difficult position since, as their lawyer pointed out, they had given approval to that code even though they publicly objected to the price-fixing clause.[65] Caught in a squeeze play by the *fait accompli* of the Ickes plan, Sun wrote to P & C Chairman Wirt Franklin on December 7, formally agreeing to the plan, but objecting to the "unfortunate . . . restrictive and unprogressive step."[66]

Fortunately for the Sun Oil Company, the chameleon-like Ickes had by then decided to drop his insistence on the price-fixing order. Sun's and the other majors' opposition had exerted some influence, but the views of some independent producers who now opposed the plan were more decisive. Ickes opted for voluntary agreement on price levels and the adoption of a new scheme whereby the majors would put up $10,000,000 for a pool to purchase excess or "distress gasoline" to further stabilize the price. The contributions to the pool were determined by the total gallonage sales of each company. Sun's share was $255,000, and Jersey Standard contributed the largest amount of $1,124,000.[67]

The anti-price-fixing group still did not know exactly where it stood, and it appeared that Ickes might yet impose prices on refined products in January or February 1934. Aware that it was probably futile, J. Howard Pew decided to continue speaking out on the subject. In January, Sun submitted another brief to the P.A.B. protesting price-fixing, and circulated an article signed by J. Howard Pew attacking the principle of price-fixing. Sun Public Relations Director Judson C. Welliver had put the piece together, basing much of it on an article published by Harwood F. Merrill in the November 15 issue of *Forbes.*[68] Sun's article referred to price-fixing as a "fascinating plaything," and cited historical examples and contemporary economic opinion to stress the incffectual or even dangerous results of the tactic. J. Howard Pew later stressed this theme in his continued attacks against the New Deal.

Despite the many assaults on the N.R.A.'s stabilization program, it did achieve significant results in the area of production restriction. Federal enforcement of "hot oil" prohibition proved vital to successful prorationing in the several states. Although Ickes incurred the wrath of part of the industry by his flirtation with price-fixing, he moved more and more toward advocating greater federal control in 1934. He worked for a new petroleum bill granting him greater powers and gave his support to a rapidly expanding movement to divorce pipelines from the integrated oil companies. In the late spring of 1934, Ickes received help in the form of the Thomas-Disney Bill, a statute designed to increase the Oil Administrator's power to enforce production quotas in the states.[69] To Sun, the bill represented the first move to make the oil industry a public utility.

The Pews's opposition to formal price-fixing appears philosophically inconsistent with their willingness to accept prorationing and hot oil legislation. Any control of production at the wellhead was a form of indirect price-fixing aimed at raising oil prices. But even here an important distinction existed between the views of J. Edgar Pew, a proponent of prorationing, and his Philadelphia cousins, who thought and acted more like refiners. Although accepting the reality of regulated production, J.N. Pew, Jr. commented to a friend in 1932 that "I am afraid that proration is with us for at least as long as you and I are going to live and there is not the slightest hope in my mind of ever getting back to the days when free and open production will be the fasbion."[70] But, as J. Howard Pew later wrote in 1934, "All Petroleum needs is production of crude controlled to prevent theft of oil by one man from another and to have excessive taxes reduced. Industry can regulate and stabilize itself."[71] J.N., Jr. and J. Howard could justify production quotas as acts of conservation, but the step toward formal pricing by the federal governent was a quantum jump for them.

The fear of federal encroachment that had motivated the Pews's opposition to the Federal Oil Conservation Board and compulsory unitization in the 1920s also partially explains their resistance to the apparent attempts by Harold Ickes to centralize controls over the petroleum industry in 1933–1934. To Sun's managers, there were substantive differences between regulation of overproduction at the wellhead in the name of conservation and the price-fixing of petroleum and all of its products. In the particular case of price-fixing, however, there were important economic reasons for Sun's opposition. Sun had carved out an important niche for itself by the successful marketing of its single-grade, unleaded Blue Sunoco gasoline, and the Pews were not about to surrender their entire market strategy without a fight. On balance though, Sun's fight against certain provisions of the N.R.A. was really a mixture of its economic self-interest and the Pews's conservative business philosophy.

THE DEMISE OF THE N.R.A.

Ickes was unsuccessful in obtaining the extended power he sought in the spring of 1934, but he acted independently in October of that year to increase the N.R.A.'s control over hot oil with the creation of a system of Federal Tender Boards for the East Texas field. These boards issued tenders (permits) to allow producers to ship oil out of the state. As an extension of Section 9(c), the system functioned well. However, Ernest Thompson, chairman of the Texas Railroad Commission, Senator Tom Connally, and Representative Sam Rayburn were all Texans able to exert enough influence to thwart Ickes's goal to move the locus of power to Washington.[72]

The N.R.A. encounted more serious obstacles in 1934. In September, a federal district judge ruled that the N.R.A. had no authority to dictate where wells should be drilled. This decision, the Eason Case, seriously reduced the power of the petroleum administration.[73] However, the real death knell for the oil provisions occurred with the Panama Refining decision in January 1935. In this instance, the United States Supreme Court upheld the decision of the Federal District Court in Texas, which had ruled Section 9(c) of the National Recovery Act an unconstitutional delegation of authority. Because of a rather strange technicality, the Panama decision did not invalidate the entire N.R.A. When the petroleum code was revised in September 1933, Section 9(c) was inadvertently left out of the document. Thus the N.R.A. was spared the final *coup de grace* until the Schechter decision in May 1935. But with 9(c) no longer in force, destructive rampant production again appeared on the industry's horizon.[74]

Within days of the Panama decision, Senator Tom Connally introduced legislation that passed both houses within a month. The resultant law, the Connally Hot Oil Act, re-established 9(c) and met the court's criticism by specifically spelling out the authority of the federal government. Congress enacted the law for a series of two-year periods, and in 1942 extended it indefinitely. The Connally Act still functions to enforce state quotas of oil production in interstate commerce.[75]

Thus, the heart of the N.R.A. oil stabilization program was preserved. Even the Pews, who had fought Ickes on the issue of price-fixing and greater federal control of the industry, did not oppose the "police protection" that state production quotas and the prohibition of hot oil provided. But Sun's managers were glad to see the N.R.A. come to an end. Howard wrote to H.H. Anderson, former chairman of the Labor Sub-Committee under the Petroleum Code, in February 1936 that "No one realizes more than I do that the hand of Government on business is the cold, clammy, hand of death and it must eventually destroy everything that it touches."[76]

Following the *Schechter* decision, Sun's vice-president Samuel Eckert wired all divisions of the Sun Oil company to cease attending N.R.A. code meetings. In his circular, he expressed the view that Sun management hoped to adopt an A.P.I.-sponsor code.[77] In particular, Sun management urged continuation of the wages and hours provision of the petroleum code. J. Howard Pew, as a member of the N.R.A. labor subcommittee, had consistently maintained that these were the most important provisions of the code. He believed that only by guaranteeing decent wages could economic recovery be achieved, a view consistent with his feeling that the Depression had been largely caused by a maldistribution of wages in the 1920s.[78] Pew even incorporated this view into his anti-price-fixing statement:

> . . . the whole basic purpose of NRA has been betrayed because of the concessions that have been made to selfish interests which demanded increases of prices sufficient to offset, and in most cases more than offset the increases in wages. Had the codes been restricted to increasing wages and eliminating child labor, their purpose could have been achieved. If a few years ago when prosperity abounded, more of its profits had gone to wages, there would have been more buying power, more consumption and less speculation. Capital would still have had ample reward, and would have continued available to finance every need of legitimate enterprise.[79]

However, J. Howard's opposition to the "cost recovery" argument put forth by E.B. Reeser, Harry Sinclair, and other price-fixers did not extend to section 7(a) of the N.R.A. He was glad to see this pro-union provision go by the boards.

THE COLE COMMITTEE

Even before the death of the N.R.A., a significant element of the industry began to favor a production stabilization program maintaining authority in the states rather than the federal government. This position was enunciated strongly during the hearings held in Washington in 1934 on the Thomas-Disney Bill to extend federal power in the petroleum industry. Growing out of these hearings, the House of Representatives established a special investigative subcommittee to examine the petroleum industry in detail. Under the chairmanship of Congressman William P. Cole of Maryland, this committee conducted the most extensive investigation of the industry held up to that time. It incorporated its findings in a detailed report published in January 1935.[80]

The vast testimony and information gathered by the hearings provides insight into internal industry debates and coincidentally offers a look into the sometimes differing perceptions of the issues held by Sun Oil management. On September 19, 1934, J. Howard Pew testified before the Cole Committee to oppose federal legislation in the petroleum industry. In his prepared statement, Howard emphasized his opposition to increased government regulation, his aversion to price-fixing, and his concern with rising state gasoline taxes. He supported an interstate compact to regulate drilling in the producing states and "insure maximum yield" from each oil field.[81] The threat of massive federal intervention pushed Pew to support the same program that he had opposed when Hoover and Mark L. Requa proposed it in 1929. At that time, he opposed the creation of a federal Interstate Compact Commission as an unconstitutional incursion into state sovereignty. By 1934, the compact idea seemed much the lesser of two evils, since the states would set all production quotas and the federal government would only be participating in the state administration of the compact.

In both his testimony and in response to specific questions from the committee, however, J. Howard gave enlightening insights into his personal views of the conservation issue and the regulation of oil production. On the question of excessive supply of petroleum, Pew stated that "What is good for the consumer is good for the carrier. . . . It follows that production in this industry cannot be excessive except from the point of view of a producer anxious to limit supply and keep up prices. My company had no such desire."[82] This statement alienated independent producers and confirmed the fact that J. Howard Pew still thought as an eastern refiner. Oil at cheap prices was good for him and enabled Sun to maintain its low refining costs. He maintained that unlimited supply of oil was not necessarily injurious to commerce and that "no system of regulation is sound which sacrifices public good to private gain."[83] Even though these statements were intended to deflate the Thomas-Disney Bill's approach to extend federal control, they also reflected Howard's philosophical aversion to artificial regulation of the marketplace at any level.

On the specific issue of conservation of resources, Pew stated that "Regarding future supplies of oil, frankly I may say I feel but small concern. Nature has been bountiful in the supply of this resource, and we have barely scratched the surface. I am not in favor of preserving supplies of petroleum for the use of generations yet unborn."[84] He supported state attempts to eliminate the "law of capture" on the grounds that each individual operator deserved the right to "police protection" of the oil that was underground in his "warehouse," and argued that the interstate compact was the most palatable way to prevent the "theft" of oil from one leaseholder by another.[85] Congressman Samuel Pettingill of Indiana, a member of the Cole Committee, later commented that Pew's testimony was a voice "heard crying in the wilderness of a 'planned society.' An opponent of the 'economy of scarcity,' Pew believed that the future of the nation lay not in increasing prices, but in reducing costs."[86]

J. Edgar Pew testified several weeks later, also opposing the Thomas-Disney Bill and opting for the interstate compact to continue production controls. In the course of his testimony, however, Representative Charles A. Wolverton of New Jersey confronted Edgar with Howard's previous statements made in opposition to production regulation. In particular, Wolverton questioned J. Edgar on Howard's position that production controls raised prices for the consumer. Pew answered by saying "that is right, temporarily, for the present moment, but over any period of time if we are able to increase our production in any field beyond the point where we will realize the greatest ultimate recovery, the public will pay many times for it."[87] When Wolverton pursued J. Edgar Pew on this point and quoted further from J. Howard's testimony that regulation brought higher prices, Sun's vice-president retreated behind the admonition that regulation was justified "Only in the interest of conservation—proper conservation—that is all, Mr. Wolverton."[88]

Although both J. Howard and J. Edgar had ostensibly supported the same thing, defeat of the Thomas-Disney Bill and passage of an interstate compact, there were important differences in their testimony. Their divergent opinions reflected the outlooks each had developed as a result of his own experience. As his Cole testimony indicated, J. Howard Pew believed that if left alone, the free market would ultimately determine future reserves and fair prices for oil and its products.

The Cole Committee also investigated a number of other issues, including the popular independent producers' cry for pipeline divorcement legislation. In his prize-winning book, *Petroleum Pipelines and Public Policy, 1906–1959,* Arthur M. Johnson praises J. Edgar Pew's testimony on pipelines as the best summary of the integrated companies' case against divorcement legislation.[89] Pew emphasized that the pipelines were originally built to obtain oil and not to be common carriers. As such, early pipeline developments in the southwest were highly competitive and required great amounts of investment capital. Moreover, he argued that independent producers obtained from the integrated pipeline owners a dependable purchaser of their oil, readily available pipeline connections, and better pipelines and prices as the majors competed among themselves to keep their pipelines full. Pew argued that none of these advantages would accrue from independently owned pipelines, and that the specialized character of such lines would not attract the necessary investors to make them a success.[90]

THE INTERSTATE COMPACT

The Cole Committee submitted its findings to the Congress in January 1935. Its conclusions rejected the extension of federal regulatory power in the oil industry. Rather, it came out strongly for the formation of an Interstate Oil Compact among the producing states of the southwest.[91] The support given by the Cole Committee had a large impact on Congress's adoption of the Compact in 1935, following the rejection of the

N.R.A. in the courts. The Compact, still in effect today, became the second of the two important tiers of modern production regulation along with the Connally Hot Oil Act.

Although J. Edgar Pew had also been lukewarm on the Interstate Compact in 1929, he had gradually adopted a more liberal attitude toward the concept. In the spring of 1932, he had served on an A.P.I. Interstate Compact Committee to investigate a compact proposal that was still circulating within the Hoover administration.[92] But no one did more to change the Pews' and other oil men's views on the Interstate Compact than did Harold Ickes. His repeated threats of further federal control made the interstate compact appear a much more attractive alternative. In November 1934, Ickes had addressed the Dallas meeting of the A.P.I. and told the assembled oil men that he favored legislation making oil a public utility subjected to stringent federal controls. Following the alarming speech, Governor E.W. Marland of Oklahoma invited the governors of the oil producing states to his home to discuss the compact idea.[93]

J. Edgar Pew had been an associate of Marland's for several years and had worked with him on both the A.P.I. Committee of Eleven and the Committee of Seven, which had proposed the first gas conservation legislation in 1927. He wrote to Marland giving his support to the compact plan, stating "It seems to me that it is the only alternative if we are to avoid any semblance of Federal control." A main obstacle was Governor Alfred Landon of Kansas, who still advocated federal control and supported the Thomas-Disney legislation. Governor James Allred of Texas was more ambivalent, and Pew felt that he could be brought around.[94]

After the Cole Committee report became public in January, representatives of the governments of Oklahoma, Texas, California, New Mexico, Arkansas, Colorado, Illinois, Michigan, and Kansas met in Dallas on February 15 to formulate a compact agreement. J. Edgar's son Jno. G. "Jack" Pew represented Sun Oil in Dallas and was able to put Sun's view across to "some of the prominent parties" at the conference, presumably Governor James Allred of Texas. Allred was opposed to explicitly adopting a plan to stabilize prices and wanted the goals of the compact limited to the prevention of physical waste. The wording of the February agreement reflected a compromise with Allred's position.[95]

While Harold Ickes continued to lobby for increased federal control of the industry, President Roosevelt characteristically remained "above the battle." In August 1935, a joint resolution of the Congress sponsored by Congressman Cole ratified the compact. Congress renewed it at two-year intervals until 1943 and has done so at four-year intervals ever since. Roosevelt remained sphinx-like for a time, but in November 1935, he came out in favor of both the Compact and the Connally Act. This was clearly a defeat for Ickes within the administration. An Interstate Compact Commission composed of representatives of the oil producing states would coordinate state conservation legislation with Congressional approval.[96]

The Pews had struggled against those aspects of New Deal stabilization that they found to be most obnoxious, the price-fixing of crude and refined products and the centralization of regulatory control in Washington. J. Edgar Pew, enthusiastically, and J. Howard and J.N., Jr., somewhat more reluctantly, went along with the program of state production quotas and the prohibition of hot oil. Although the New Deal had offered little that was truly new in the area of public petroleum policy, Roosevelt was able to accomplish more than his predecessors because of the crisis atmosphere of the depression. Herbert Hoover remained bitter toward the oil industry for a long time because it had rejected the Interstate Compact in 1929 and supported it in 1935.[97] What he failed to appreciate was that East Texas ten-cents-a-barrel oil had accomplished more toward bringing cooperation than all of the cajoling and logic of Mark L. Requa a decade earlier. Harold Ickes's threats of a federal take-over of the industry further cemented the union of independents and majors alike toward stabilization of the industry. Like it or not, conservation had become a comprehensive program in the oil fields and oil men would continue to raise its banner to accomplish economic stabilization as well as combat physical waste.

THE TAXATION ISSUE

Prior to Roosevelt's election in 1932, the Pews had become involved with other issues affecting the oil industry. While J. Edgar Pew was becoming deeply entangled in the prorationing movement in East Texas, J. Howard Pew became a vocal leader in the fight against increased gasoline taxes at the state level and against the imposition of a gasoline excise tax by the federal government. Following the crash in 1929, state governments had been under tremendous financial pressure to provide funds for relief programs. One source of revenue was increased excise taxes on gasoline, a form of tax levy that had emerged in the 1920s to finance the road-building campaign associated with the automobile revolution.[98]

J. Howard Pew first publicly took a stance on this issue by organizing a fight against Pennsylvania Governor Gifford Pinchot's attempt to impose a two-cents-per-gallon increase on the gasoline tax. Pew testified in Harrisburg, organized a writing campaign, and distributed literature through Sun filling stations attacking the proposal.[99] Pew felt that these taxes discriminated against the oil companies and that it was unconstitutional to apply them to general revenue rather than to highway construction.

Although they were unsuccessful in this fight in Pennsylvania, both J. Howard and J.N. Pew, Jr. continued to speak out against gasoline taxes in the Sun Oil marketing area. One phase of their campaign was an attack against tax evaders: refiners and marketers who pocketed tax monies and did not forward them to the state governments. Sun's Judson Welliver coined a new title, "gas bootlegger," to describe these individuals, and the A.P.I. launched a major campaign against "bootlegging" in 1932.[100] Most of the alleged evaders were small operators whom the government did not supervise as closely as it did the major companies. In a company publication, Sun asked its employees to "Take a hand in the battle of Marketeers vs. Racketeers."[101]

In the course of their fight against gasoline "bootleggers," Sun's leaders also became allied with the trucking industry in its economic warfare with the railroad interests. Sun was an active member of the Highway Users Conference, a group comprising representatives of the automobile, trucking, and petroleum industries, and opposed the railroad's attempts to pass what the Pews saw as discriminatory taxation against the truckers. The railroads had supported legislation to restrict truck operation and have trucks pay higher license fees, all attempts to help the railroads gain back some of the business they had lost to the trucking industry.[102] Convinced that the United States Chamber of Commerce had joined with the railroads in urging higher gasoline taxes, Sun Oil withdrew from the Chamber in 1932. In response to Sun's withdrawal, *The Nation's Business,* the organ of the Chamber of Commerce, offered Pew space to state his case.[103]

In October 1932, the magazine published an article by J. Howard Pew which it entitled "A Squawk from the Petroleum Goose." Pew argued that the petroleum industry was being taxed unfairly by both state and federal government. Not only was the industry paying a disproportionate share of taxes, he maintained, but this was seriously hurting business.[104] In an editorial in *National Petroleum News* entitled "On Behalf of Howard Pew, of Philadelphia," the influential Warren C. Platt attacked "The Chamber" for the derogatory title used in *The Nation's Business* article, and for its editorial note that labeled Pew's comments as "frankly a prejudiced article." Platt suggested that if the U.S. Chamber of Commerce did not think of the oil industry as part of American business, individual companies could save a lot of money by withdrawing membership. This was of course just what Howard Pew wanted. He was delighted with the impact of the article. In addition to having Lowell Thomas announce its publication over the air, Sun distributed 200,000 copies through its gasoline stations.[105]

Pew's *Nation's Business* piece also criticized the failure of the Congress to pass a general manufacturers' sales tax rather than the federal gasoline tax in 1932. A significant segment of the business community, especially the firms associated with the National Association of Manufacturers, supported the establishment of a general sales tax rather than an increase in "nuisance" or general excise taxes to balance the federal budget. The Hoover administration failed to pass the general manufacturer's tax despite the lobbying campaigns of Pew and others in the fall of 1932.[106]

The Pews continued to attack gasoline taxation and the role that the railroad industry seemingly had in the campaign throughout 1933 and 1934. J. Howard addressed meetings of the Highway Users' Conference and the Resource Section of the United States Chamber of Commerce in 1933, and J.N. Pew, Jr. attacked the federal gasoline tax "outrage" with a radio address on Philadelphia station WLIT.[107] The Chamber of Commerce had attempted to mediate the railroad-trucking dispute with the establishment of a "Committee on Competing Forms of Transportation," but made little headway in settling the issue. Sun Oil decided to continue its break with the Chamber in view of "their pro-railroad position."[108]

Sun Oil's executives carried the fight against the use of this tax revenue for general purposes into the New Deal era. Sun also used the influence of the A.P.I. to urge Harold Ickes to study the tax problem in 1934. On the state level, the Pews continued the fight against gasoline tax increases; J. Howard again testified in Harrisburg in 1935 in oposition to raising the Pennsylvania tax.[109]

The tax issue brought Sun Oil into the forefront of the industry's political issues during the early years of the Depression. Although their opposition reflected the economic interests of their firm, the Pews viewed the arbitrary raising of gasoline excise taxes as an unfair attack by government on one industry. As the New Deal embarked on experiments with increased government regulation, the Pews found many more struggles on their hands.

ANTITRUST

The Roosevelt administration's ambivalence toward the problem of monopoly in the 1930s is readily apparent in its relations with the oil industry. Ellis Hawley has explained much of the New Deal's changing approaches to this question as an internal struggle among three competing groups of New Dealers: the business planners (represented in the N.R.A. cooperative approach), a more radical group who advocated comprehensive national planning (as in the T.V.A.), and the "antitrusters."[110] The comprehensive regulation of the National Recovery Administration required a suspension of the antitrust laws so that industrial cooperation could achieve recovery. A flurry of antitrust prosecutions following the demise of the N.R.A. in 1935, however, signaled an important change in the administration's relations with the business community. Among the reasons suggested for this change in policy have been the failure of the N.R.A. to achieve significant recovery, Roosevelt's growing disenchantment with business, and the rising influence of antitrusters like Felix Frankfurter and Thomas Corcoran within the New Deal inner circle.[111]

The first evidence of this change in attitude as it related to the petroleum industry was the "Madison Case," a major antitrust investigation initiated on August 1, 1935 by Attorney General Robert H. Jackson. In July 1936, a Madison, Wisconsin grand jury indicted twenty-three oil companies, including Sun, and three industry trade publications for allegedly fixing prices. The charges arose from the complaints of independent gasoline jobbers who claimed that the major companies had joined together to fix jobber margins. The problem grew out of the gasoline stabilization program mentioned previously that began under the N.R.A. in December 1933. The large companies had agreed to buy some of the surplus or "distress" gasoline in given marketing areas in an attempt to stabilize prices. Now, after the demise of the N.R.A., they were continuing the program initiated by Harold Ickes and up to then deemed legal.[112]

Ickes had encouraged the industry to continue the pricing system, but he was now reluctant to testify on behalf of the large oil firms for fear of embarrassing the Roosevelt administration. Unsure of what the former petroleum czar might say if subpoenaed, the defendants did not force him to testify in Madison. In September 1938, the court found sixteen companies guilty of a misdemeanor in violation of the Sherman Act. The convicted companies lost an appeal to the U.S. Supreme Court in 1940, and they each ultimately paid a fine of $5,000.[113]

Sun Oil had been included in the list of twenty-three firms initially charged, but during the course of the trial, the government dropped it and a few others from the indictment.[114] The Pews' consistent stand against price-fixing had no doubt played a role in the decision to take their firm out of the indictment. Sun Oil was becoming more and more a symbol of independence and competitive business tactics in the industry. This was an image that the American Petroleum Institute was quick to exploit when it could, and an opportunity soon arose.

On the heels of the Madison case, Senator Guy Gillette of Iowa introduced a bill to divorce marketing operations from the integrated oil companies. There had long been complaints from independent marketers that the majors used their size for unfair competitive advantage. The large firms could cut prices and offset marketing losses with large profits made in production and transportation. Senator Gillette introduced the bill in April 1938 and arranged for hearings later that month. The A.P.I. chose J. Howard Pew to testify on behalf of the integrated firms as a summary witness.[115] His outspoken stance against price-fixing, the N.R.A., and monopoly, plus Sun's image as an independent, competitive company made Pew an ideal spokesman.

In Howard's preliminary statement, he attempted to deflate the basic case for marketing divorcement by citing Sun Oil's record in operating a profitable marketing operation:

> At the outset I want to deny, on behalf of our company, that we lose in one department of the business, but make it up in profits of another. Ours is one of the smaller, independent so-called integrated companies; but it is not true that we use the profits from transportation and production to compensate for losses on marketing. Rather, we earn profits in every one of the industry's branches.[116]

In fact, Sun had been experiencing difficulties in its production department and its profit margin there was very slight. Although the Gillette Bill did not pass, it was indicative of increasing antitrust activity directed against the oil industry.

At about the same time, Senator William E. Borah again sought approval of a pipeline divorcement bill that he had first introduced in 1934. This bill resulted from complaints that the major integrated firms used excess profits from their pipelines to offset losses in other sectors of their business. J. Howard Pew had met with Borah in the spring of 1934 to argue against the pipeline bill and apparently succeeded in convincing him to refrain from pressing for legislation at that time. Pew and Borah had come to an understanding as a result of their support of the antitrust laws and their common opposition to price-fixing. In fact, prior to their spring meeting, Howard had supplied Borah with copies of both his personal argument and Sun's legal brief against price-fixing filed with the P.A.B. Pew later enlisted Borah's support in his fight against granting further power to the P.A.B. and P & C Committee on the grounds that it would entail further suspension of the antitrust laws under the N.R.A.[117] But by 1938–1939, Borah, the old Progressive, was once again on the trail of the monopolists.

THE TEMPORARY NATIONAL ECONOMIC COMMITTEE

The high point of antitrust activity in the New Deal came with the creation of the Temporary National Economic Committee in June 1938.

Headed by Senator Joseph C. O'Mahoney of Wyoming, the Committee contained representatives from the Justice Department, the Securities and Exchange Commission, the Federal Trade Commission, and the Departments of Commerce, Labor, and the Treasury. The T.N.E.C. hearings began in December 1938 and continued until April 1941. Spokesmen for the oil industry testified from September 25 to October 30, 1939, and in the course of the hearings, they aired virtually every aspect of the business.[118]

J. Howard Pew also played an important role in these hearings. By agreement with the American Petroleum Institute, six men acted as industry spokesmen. J. Howard opened the testimony with a broad general view of the industry, E.L. DeGolyer presented the case for production; Fayette B. Dow, transportation; Dr. Robert E. Wilson, refining; Sidney Swensrud, marketing; and H.H. Anderson, employment. William S. Farish of Jersey Standard then concluded with a general summary. Several experts, independent oil men, and other interested parties also aired their views.[119] Pew prepared his testimony in the spring and summer of 1939 and sent the draft of his opening statement to other A.P.I. executives and academic consultants for criticism and suggestions. Through five full sessions, the Sun president presented an unwavering resistance to the monopoly charges directed by committee investigators against the top twenty integrated companies.[120]

A key figure in the hearings was Roy C. Cook, a member of Thurman Arnold's staff in the Antitrust Division of the Justice Department. While a member of the economics department at George Washington University, Cook had prepared a study of concentration in the petroleum industry that was the basis for much of the oil investigation. In January 1941, the Commission published Cook's study as Monograph No. 39, *Control of the Petroleum Industry By Major Oil Companies.*[121] This was one of forty-three monographs which the T.N.E.C. published to trace the problem of growing consolidation in American industry.

Cook's study highlighted a phenomonon readily apparent to most observers of the petroleum industry. Twenty large integrated companies had gained the lion's share of the production, transportation, refining, and marketing of petroleum and its products in the thirty years since the dissolution of the old Standard Oil Trust. The "independents" of 1939 were the non-integrated producers, jobbers, marketers, and refiners. These were the groups most critical of the "major" interests. Clearly, a pattern of oligipolistic competition had emerged following the period of Standard Oil dominance.[122]

The A.P.I. representatives protested Cook's conclusions, maintaining that the industry had remained efficient and competitive despite the encroachment of government. They defended prorationing as a legitimate conservation device and argued that vertical integration was the competitive tool used to destroy monopoly rather than a vehicle for stifling competition. Because of this strong protest, the T.N.E.C. agreed to a unique arrangement. The O'Mahoney Committee published another monograph (No. 39–A), entitled *Review and Criticism on Behalf of Standard Oil Company (New Jersey) and Sun Oil Company of Monograph No. 39 with Rejoinder by Monograph Author.*[123] Although staff members of the Jersey Standard and Sun Oil staffs were largely responsible for compiling the study, William Farish's and J. Howard Pew's names appeared on the monograph as authors. The study attacked Cook's work point by point, reiterating several of the major issues that the A.P.I. experts had made in the hearings. Cook's rejoinder recapitulated his previous arguments.[124]

Thirty years previously, Sun's founder, J.N. Pew, Sr., had opposed the Standard Oil interests and rejoiced at the dissolution of the trust. In a speech given in January 1936 celebrating the death of the N.R.A., his son, J. Howard Pew, had stated: "In 1890 Congress passed the Sherman Anti-Trust Act to protect competition and prevent monopoly. I regard that law as one of the most beneficient enacted in our time. . . . The National Recovery Act . . . undertook to suspend the anti-trust law in a wide range of activities and interests; and the country had been grateful for the interposition by which the Supreme Court served preserved the integrity of the anti-trust legislation."[125] Now the son stood side by side with the head of the nation's largest petroleum company in defense against the antitrust attacks of the T.N.E.C.

In 1939, Jersey Standard was of course not the monopolistic power that the old Rockefeller trust had been, and Sun was by this time a fully integrated "major" firm. Both Farish and Pew represented an industry whose structural changes reflected the enormous growth of petroleum in the age of the internal combustion engine. J. Howard Pew argued that the industry was not monopolistic. Not only were there numerous "independent" producers, refiners, and marketers, but the top twenty firms were intensely competitive. Certainly Sun Oil had demonstrated this by its own business practices. But Pew was also in agreement that large integrated units represented the most efficient way to run the industry. The question in 1939 remained much the same as it is today—how competitive is the oligopoly of large companies in the oil industry?

The first phase of New Deal oil policy had not attempted to break up the industry, but rather to impose a system of comprehensive regulation supported by many industry leaders. The N.R.A., the Connolly Act, and the Interstate Compact were the important achievements of New Deal oil policy. The T.N.E.C., like the Madison case before it, represented a departure from this approach, but neither succeeded in denting the oligopolistic structure of the industry by the eve of World War II. The legacy of the Depression was in the area of business planning through cooperative regulation, not antitrust.[126]

THE PEWS: BUSINESS AND POLITICAL PHILOSOPHY

The Pews had gradually learned to accept regulation of crude oil production in the face of massive overproduction in the 1930s. They thus acquiesced in what proved to be the most important and long-lasting programs in New Deal oil policy. But at the same time, they staunchly opposed other New Deal ideas, including direct price-fixing, the increased centralization of industry control in Washington, and the divestiture of marketing and pipeline operations from the major integrated companies. Pew opposition to those New Deal programs seen as destructive to the free enterprise economy was so strong that Sun Oil gained a reputation as an anti-New Deal firm and "Sun money" was identified with political organizations that the Pew family supported.

The New Deal experience polarized Americans of different backgrounds and socio-economic classes. There were, and still are, few Americans with a neutral view of Franklin D. Roosevelt and his policies. But more than any other single group, American businessmen turned quickly against the New Deal. They feared the encroachment of government regulation into their private affairs and saw New Deal experimentation with "collectivism" as a dangerous socialist menace to American capitalism. Today, however, some scholars argue that more businessmen would have supported the President's program "if they had understood Roosevelt's purposes."[127] These purposes, they point out, consisted of the preservation of capitalism, democracy, and the prevailing economic class structure; in short, the prevention of radical change and revolution. In other words, American businessmen should have been able to see past the empty rhetoric of Roosevelt's "economic royalist" charges and perceive that the New Deal was really on their side.

One can, of course, argue the above interpretation with the advantages of hindsight. The New Deal does not seem very radical by today's standards. Yet the Pew family consistently opposed most of Roosevelt's program in the 1930s and after. The constant theme running through their public utterances and writings was opposition to governmental involvement in the economy. They maintained that the loss of economic freedom was the beginning of the loss of personal individual liberty. To them, N.R.A. regulations, the Tennessee Valley Authority, and various other New Deal alphabet agencies were not designed to preserve capitalism, but represented a drift toward socialism and the planned economy. In his speeches, J. Howard Pew used historical examples to demonstrate that planned economies, price-fixing, and collectivism had failed from the Egyptian Pharaohs and the Emperor Diocletian to the present.[128] Pew argued that government and business did not mix "any more than you can mix pure water with contaminated water and get anything but contaminated water."[129]

In a new post-Depression era of professional management, government cooperation with business, and the "mixed economy," the Pews appeared to be relics of the past. Owner-managers of the nineteenth-century mold, they espoused a belief in *laissez faire* and economic freedom that exhibited itself in their business activities. Although their business and political philosophy had evolved over many years, the Depression experience crystallized these attitudes.

POLITICAL INVOLVEMENT

In order to make known their strong anti-New Deal views, the Pews went beyond their activities in petroleum industry circles and increasingly became involved in the political arena. Although all of the Pews supported the Republican party in the 1930s, some family members took more active political roles than others. J. Howard Pew, as president of the Sun Oil Company, generally maintained a somewhat lower profile than his brother Joe, but he nevertheless spoke out on many controversial issues. J.N. Pew, Jr. never admitted voting for F.D.R. in 1932, but he openly acknowledged that he had been willing to cooperate with the new administration at first and see what it had to offer. When he later became identified as one of the leading "Roosevelt haters" in the country, some journalists suggested that he had initially supported Roosevelt.[130] The extent of his support remains doubtful, but it is apparent that Roosevelt's charm captured J.N. more than it did his brother J. Howard. Although disillusioned somewhat with Hoover, Howard had written in March of 1932 that "the only man who can win the election for the Republican party this year is Calvin Coolidge."[131]

J.N. Pew, Jr.'s correspondence in the fall of 1932 indicates that he supported the Republican party, but was very much aware of a likely Democratic landslide. He commented on the possibility of a Roosevelt win and expressed a view that many businessmen possessed at the time:

> The country is frightened of Democratic, wild unsound ideas, and above all of a lack of intelligence in the party. I certainly know that as far as our company is concerned we will pull in our horns and conserve our resources until we find out which way the wind blows. This, I think, expresses the general feeling throughout the east.[132]

In a letter written right after the election, however, J.N. noted that "if the Democrats had to get in it, it is just as well that it was a landslide."[133]

At first willing to give the New Deal a chance, J.N. broke completely with Roosevelt over the price-fixing dispute in the N.R.A. petroleum code. He became convinced that the New Deal had to be defeated, but after a trip to the Republican National Headquarters in Washington, he was dismayed. The R.N.C. was in disarray, short of funds, and a spirit of defeatism hung in the air.[134] At that time, Joe Pew made the decision to become politically active.

The Pew family became large contributors to the Republican Party, first in Pennsylvania, and later nationally. Between 1934 and 1940, it was estimated that they had contributed $2,000,000 to the Party.[135] When subpoenaed by the Black Senate Committee investigating campaign contributions in October 1936, J. Howard Pew defended his extensive gifts to the Republican Party and anti-Roosevelt organizations by stating: "I thought if contributors to the Roosevelt campaign fund could properly give huge sums to get the country into trouble, I might properly contribute relatively small sums to get it out again."[136] The "relatively small sum" J. Howard personally contributed in 1936 was $61,050. This was all the more impressive when one considers that his brother and two sisters contributed approximately equal amounts and other members of the family individually gave more.

Among J. Howard Pew's list of contributions in 1936 was a $5,000 check to the American Liberty League, a supposedly nonpartisan group of disaffected conservative businessmen formed in August 1934. This group, ostensibly formed as an educational organization, became one of the most active organizations working for the defeat of Roosevelt and the New Deal in the election of 1936. Howard served on the National Executive Committee of the League and members of his family contributed large sums to the organization. So vociferous was the League in its denunciation of F.D.R. and so representative of the economic elite was its membership, that the Republican National Committee asked it not to endorse Landon for President in 1936 on the grounds that League support would cost votes.[137]

Although the League became fragmented after the Roosevelt victory and never regained its peak membership of 1936, J. Howard Pew was still contributing money to and supporting the League's efforts as late as 1940.[138] This was at a time when even the League's early supporters, the duPonts and John Raskob, had largely given up on the organization.

The family's growing interest in politics and their support of the Liberty League had a common thread. The Pews increasingly identified New Deal Programs with the "isms" most feared in their day—collectivism, fascism, socialism, and communism. Moreover, they saw the Democratic landslide victories of the 1930s as a serious attack on the two-party system. J.N., Jr. in particular frequently commented on the political debacle that he had witnessed in France during the 1930s because of its multi-party government.[139] In the United States, the system seemed threatened by the Roosevelt dictatorship. J.N. became a powerful Republican figure, and national publications labeled him the "political boss of Pennsylvania" in 1940.[140] He had achieved this prominence by becoming a major bank-roller of the party during the difficult years of the 1930s. As he somewhat dramatically stated in 1940, "The Republican Party stands today where the continental Army stood at Valley Forge, and if Haym Solomon and Robert Morris could empty their purses to keep that army alive, so can we."[141]

The Pews also used the media to put forth their political philosophy as well as support programs beneficial to Sun Oil. In most cases, these two goals coincided. In 1936, J.N. Pew, Jr. acquired control of *The Farm Journal* and later obtained *The Farmer's Wife,* both magazines aimed at the rural audience. As owner of *The Farm Journal,* J.N. greatly influenced its editorial policy.[142] He wrote to a friend shortly after the acquisition that "generally, I think it will be a fine organ for putting over sound Americanism to the farmer and it also will be useful in our gasoline tax fight."[143] The Pews also obtained a financial interest in the Chilton Class Journal Company, publishers of industry trade magazines like *Iron Age, Hardwood Age, Dry Goods Economist,* and various automobile trade journals. Already geared to a conservative businessmen's audience, these publications continued an editorial policy consistent with the Pews's belief in limited government.[144]

Another vehicle for the Pew family and Sun Oil was the Lowell Thomas "Three Star Extra" radio news and commentary program that was broadcast five nights a week over the NBC network. The program had begun under Sun Oil sponsorship on June 13, 1932, and remained one of four national news and commentary shows on the airways during the Depression.[145] On several occasions, Thomas aired program segments that reflected the particular viewpoint of his sponsor. In addition to his presentation of Sun's opposition to price-fixing in the N.R.A. that was previously mentioned, Thomas criticized railroads for running trucks and buses off the highway (part of the highway lobby's struggle with the railroads); attacked the use of alcohol as a motor fuel (an agricultural recovery program opposed by the oil industry); and devoted a large segment of one program to praise of General Juan Vincente Gomez (the Venezuelan dictator, with whom Sun still had dealings).[146] When listeners wrote in protesting the "propaganda" in Thomas's broadcasts, either Howard, J.N., or Thomas himself replied that the broadcaster was his own man and that Sun Oil had no input into program content.[147]

In fairness to Lowell Thomas, the relationship between sponsor and performer was assumed to be much closer in the 1930s than it is today. People would be shocked to hear an announcer take an editorial position favoring one of his sponsors on a network's evening news, but it was not so exceptional in 1933. Moreover, Lowell Thomas usually indicated when he was stating his sponsor's position, and since his show was a news and comment program, he presumably genuinely advocated many of the positions that he broadcast. In a recent study of radio broadcasting during the 1930s, David Culbert rates Lowell Thomas's New Deal coverage "bland" when compared with his chief competitor, Boake Carter of CBS. In contrast with the broadcasts of Carter, Edwin C. Hill, and H.V. Kaltenborn, Culbert concludes that Thomas's program was almost devoid of political comment.[148] Lowell Thomas, however, was very popular during the 1930s, and when Sun chose to issue a statement over the air, it was assured of a wide audience. At least a portion of this audience was in tune with the Pews, for it was later reported that "Mr. Hoover (Herbert) makes no bones about the fact that he is an enthusiastic listener of '3-Star Extra.' He carries a portable he said, for the express purpose of catching the broadcast when he is located where there is no radio."[149]

CONCLUSION

It is impossible to divorce the role of the Sun Oil Company in the 1930s from the highly personal views of its owner-managers. The Pews' vision of "sound Americanism" was diametrically opposed to their perception of the basic thrust of the New Deal. They believed in limited government, the free operation of the market, and individual freedom. Economic planning, price-fixing, and federal regulation did injury to these fundamental values. The paradox was that while J. Howard and J.N. Pew, Jr., were launching their attacks on Roosevelt and the New Deal, they supported the petroleum industry's efforts to achieve economic stabilization through government regulation of crude oil production.[150] It is significant that J. Edgar Pew remained silent on many of the political issues that his cousins raised. He had supported industry-government cooperation in the oil fields for quite some time and did not actively campaign against the idea of government in business. However, J. Edgar did draw a line between state and federal regulation, a distinction that many oil men made at that time. Prorationing by the Texas Railroad Commission was acceptable, but not when it was directed by Harold Ickes in Washington.[151]

J. Howard Pew and J.N., Jr. did not find their views to be inconsistent. They came to accept prorationing and hot oil legislation as attempts to preserve the natural law of supply and demand. Each owner had the right to "police protection" of his underground property, and government regulation could be a legitimate way to obtain this protection. In making the distinction between "regulation" and "control," they argued that certain regulations may be justified or even necessary to preserve natural laws, but control was wrong.[152] In the final analysis, as the British petroleum economist P.H. Frankel remarked after interviewing J. Howard, it boiled down to a question of terminology. Regulation was what one approved of and control that method of cooperation with which one found fault.[153]

NOTES

1. See Jerold Auerbach, "New Deal, Old Deal, or Raw Deal: Some Thoughts on New Left Historiography," *The Journal of Southern History* 35 (February 1969, pp. 18–30; Paul Conklin, *The New Deal* (New York: Thomas Y. Crowell, 1975); Barton J. Bernstein, "The Conservative Achievements of Liberal Reform," in Bernstein (ed.), *Towards a New Past* (New York: Random House, 1967); and Howard Zinn, *New Deal Thought* (New York: Bobbs-Merrill, 1966), especially Zinn's introduction.

2. The best discussion of these events is contained in Gerald D. Nash, *United States Oil Policy, 1890–1964* (Pittsburgh: University of Pittsburgh Press, 1968), especially Chapters 4 and 5; see also Harold F. Williamson, Ralph L. Andreano, Arnold R. Daum, and Gilbert C. Klose, *The American Petroleum Industry,* Volume II, *The Age of Energy, 1899–1959* (Evanston: Northwestern University Press, 1963), pp. 336–7.

3. Norman Emanuel Nordhauser, "The Quest for Stability: Domestic Oil Policy, 1919–1935" (Ph.D. Dissertation, Stanford University, 1970), p. 98; Carl Coke Rister, *Oil: Titan of the Southwest* (Norman: University of Oklahoma Press, 1949), pp. 306–11; Nash, *Oil Policy,* p. 115; Jno. G. "Jack" Pew, interview held at Dallas, Texas, November 18, 1975.

4. Nordhauser, "Quest," p. 98; "In the Oil Fields With Sun," *Our Sun,* 50th Anniversary Issue, 3, 1 (1936), p. 34.

5. Nash, *Oil Policy,* pp. 113–4.

6. American Petroleum Institute Press Release, May 30, 1931, Accession #1317, Eleutherian Mills Historical Library (EMHL), Greenville, Wilmington, Delaware, File 21-A, Box 57; Nash, *Oil Policy,* p. 115.

7. Lawrence E. Smith, " 'Poor Boy' Operators in East Texas, Promises Falling, Are on Way Out," *National Petroleum News* 23, 26 (July 1, 1931), pp. 25–7; Paul Wagner, "Operators Split on Plan for Curtailment of East Texas at Commission Hearing," *National Petroleum News* 23, 26 (July 1, 1931), pp. 19–20; "Short Sightedness of a Few Individuals Responsible for 10-Cent Crude," *National Petroleum News* 23, 28 (July 15, 1931), pp. 19–20; Wagner, Paul, "Court Decision Voiding Texas Proration Challenges Special Session," *National Petroleum News* 23, 30 (July 29, 1931), pp. 35–6; "No Provision for Estimating Crude Demand Will Stand in Final Texas Bill," *National Petroleum News* 23, 31 (August 5, 1931), pp. 19–20, 22; "Sterling Says He Will Veto Market Demand Provision," *National Petroleum News* 23, 31 (August 5, 1931), p. 23; Nash, *Oil Policy,* pp. 115–6.

8. Nash, *Oil Policy,* pp. 123–4; "Murrayism: A Dramatization of Popular Unrest," *National Petroleum News* 23, 33 (August, 1931), pp. 19–20, 22.

9. Nash, *Oil Policy,* pp. 124–5.

10. Lawrence E. Smith, "Scramble for Crude as East Texas Output is Shut Off Under Martial Law," *National Petroleum News* 23, 33 (August 19, 1931), pp. 19–20, 22; Smith, "Troops' Occupation of East Texas Fields Marked by Order and Efficiency," *National Petroleum News* 23, 34 (August 26, 1931), pp. 32–3; Nash, *Oil Policy,* pp. 117–8; Nordhauser, "Quest," pp. 113–21; "Jack" Pew interview.

11. Nash, *Oil Policy,* p. 118; Jno. G. Pew to J. Edgar Pew, August 8, 1939; Ben H. Powell to J. Edgar Pew, October 1, 1931, Acc. #1317, EMHL, File 71, Box 6; "Texas Commission Meeting to Frame Oil Regulations," *National Petroleum News* 23, 34 (August 26, 1931), p. 34.

12. Nash, *Oil Policy,* p. 118; Ben H. Powell to J. Edgar Pew, October 1, 1931, October 6, 1931; M.B. Sweeney to J. Edgar Pew, October 15, 1931, Acc. #1317, EMHL, File 71, Box 6.

13. Nash, *Oil Policy,* p. 114; Smith, "Troop's Occupation of East Texas Fields," *National Petroleum News,* p. 31.

14. J. Edgar Pew to Ralph C. Holmes, the Texas Company, June 14, 1930; J. Edgar Pew to T.L. Foster, July 8, 1931, Acc. #1317, EMHL, File 71, Box 6; "Acreage Proration Plan Proposed to Check East Texas Drilling," *National Petroleum News* 23, 43 (October 28, 1931), p. 43.

15. Ben H. Powell to J. Edgar Pew, October 1, 1931; R.O. Garrio to J. Edgar Pew, October 8, 1931; F.R. Coates to J. Edgar Pew, October 23, 1931; J. Edgar Pew to Coates, October 24, 1931, Acc. #1317, EMHL, File 71, Box 6; "Acreage Proration Plan Proposed," *National Petroleum News,* p. 43.

16. Moore to J. Edgar Pew, October 27, 1931, Acc. #1317, EMHL, File 71, Box 6.

17. *Ibid.;* "Acreage Proration Plan for East Texas Submitted to Sterling," *National Petroleum News* 23, 44 (November 4, 1931), p. 23.

18. Night letter, J. Edgar Pew to Harry F. Sinclair et al., October 27, 1931; Roeser, Pew et al., to Governor Ross T. Sterling, October 30, 1931; J. Edgar Pew to Sinclair, Donoghue et al., November 2, 1931, Acc. #1317, EMHL, File 71, Box 6; "Proration Control, Its Need, Legality, Technique Explained to A.P.I.," *National Petroleum News* 23, 45 (November 11, 1931), pp. 23-5.

19. J. Edgar Pew to C.F. Roeser, November 14, 1931, Acc. #1317, EMHL, File 71, Box 6.

20. W. Mumphford, East Texas Refining Company, to J. Edgar Pew, Roeser et al., October 28, 1931; telegram, Ben H. Powell to J. Edgar Pew, December 1, 1931, Acc. #1317, EMHL, File 71, Box 6; telegram, Ben H. Powell to J. Edgar Pew, November 20, 1931, File 71, Box 15; "East Texas Acreage Plan Delayed by Numerous Protests," *National Petroleum News* 23, 45 (November 11, 1931), p. 26.

21. Telegram, Powell to J.E. Pew, November 20, 1931; T.L. Foster to J. Edgar Pew, November 23, 1931; J. Edgar Pew to Nathan Adams, November 23, 1931, Acc. #1317, EMHL, File 71, Box 15; Nash, *Oil Policy,* p. 118.

22. Nash, *Oil Policy,* pp. 118-9.

23. *Ibid.,* pp. 119-20; Williamson et al., *Energy,* pp. 542-3.

24. J. Edgar Pew, "The New Conception of Oil Production," speech delivered at the meeting of the A.P.I. Division of Production, Dallas, Texas, 1931, Folder #1, J. Edgar Pew, Library File, Sun Oil Company Library, Marcus Hook, Pennsylvania, p. 11. J. Edgar's speech is also reported in "Production Control, Its Need, Legality, Technique," *National Petroleum News,* pp. 23-5.

25. Nash, *Oil Policy,* p. 118.

26. *Ibid.,* p. 130.

27. Nordhauser, "Quest," pp. 142-5; Nash, *Oil Policy,* p. 131; J. Edward Jones, *And So—They Indicted Me!* (New York: J. Edward Jones Publishing Corporation, 1938), pp. 55-62.

28. Nash, *Oil Policy,* pp. 130-1.

29. Ibid., pp. 133-4; Williamson et al., *Energy,* pp. 548-9.

30. Nash, *Oil Policy,* p. 125; "Oil Industry Will Wash its Linen on the Capitol's Doorstep," *National Petroleum News* 25, 27 (July 5, 1933), pp. 11-2; "Code of Ethics for the

Petroleum Industry Now in Effect," *Our Sun* 6, 6 (September/October 1929), pp. 5-9; Nash, *Oil Policy,* p. 136.

31. Board Minutes, Sun Oil Company, June 30, 1033, Minute Book #5, p. 127, Office of the Assistant Secretary, Sun Oil Company, St. Davids, Pennsylvania; Nash, *Oil Policy,* p. 136; Roger B. Stafford, "Oil Industry Leaders Split on Price Regulation Under N.R.A.," *National Petroleum News* 25, 32 (August 9, 1933), pp. 7-8; 10-2.

32. Memo, "Price Fixing," May 1933, Acc. #1317, EMHL, J. Howard Pew Papers, Box 69.

33. Nash, *Oil Policy,* p. 138.

34. Williamson et al., *Energy,* pp. 655-6.

35. *Interesting Oil Facts and Rumors* 4, 6 (June 1, 1933), Acc. #1317, EMHL, File 21-A, Box 100.

36. Nash, *Oil Policy,* pp. 137-8; "J.A. Moffett Resigning as Standard Official, Mentioned as Probable Oil Administrator," *Oil and Gas Journal* 32, 11 (August 3, 1933), p. 9; Bennet H. Wall, and George Gibb, *Teagle of Jersey Standard* (New Orleans: Tulane University Press, 1974), p. 283-5.

37. "Code Fails Without Price Fixing Say Cost Recovery Advocates," *National Petroleum News* 25, 32 (August 9, 1933), pp. 14-5.

38. *Ibid.,* p. 15.

39. "General Johnson at Work on Revised Oil Code Following Stormy Hearings at Washington," *Oil and Gas Journal* 32, 12 (August 10, 1933), p. 32.

40. "Code Fails Without Price Fixing," *National Petroleum News,* p. 15; Nash, *Oil Policy,* p. 138.

41. "Price Sections of Code Continue Recent Refinery Losses," *National Petroleum News* 25, 34 (August 23, 1933), pp. 17-8.

42. Stanford, "Oil Industry Leaders Split." *National Petroleum News,* pp. 7-8, 10-2; "General Johnson at Work on Revised Oil Code," *Oil and Gas Journal,* pp. 9-10, 32.

43. Statement on price-fixing (no date), Acc. #1317, EMHL, J. Howard Pew Papers, Box 69; *Oildom* 161, 4 (July 28, 1933), p. 783.

44. Text of Lowell Thomas broadcast, August 14, 1933, Acc. #1317, EMHL, J. Howard Pew Papers, Box 69.

45. Nash, *Oil Policy,* pp. 138-9; "Price-Fixing Section Brings Most Debate in the Approved Oil Code," *National Petroleum News* 25, 34 (August 23, 1933), pp. 11-2, 14-6.

46. Hugh S. Johnson, *The Blue Eagle From Egg to Earth* (New York: Doubleday, 1935), p. 240; Nash, *Oil Policy,* p. 139.

47. Nordhauser, "Quest," pp. 176-9; Nash, *Oil Policy,* p. 140.

48. Potomachus, "Pew of Pennsylvania," *The New Republic* 110, 19 (May 8, 1944), p. 626.

49. Harold L. Ickes, *The Secret Diary of Harold L. Ickes,* Volume I, *The First Thousand Days, 1933-1936* (New York: Simon and Schuster, 1953), p. 81.

50. Farish, Pew et al. to Hugh S. Johnson, August 19, 1933, Acc. #1317, EMHL, J. Howard Pew Papers, Box 71; J.C. Welliver to J. Howard Pew, August 20, 1933; Press Release, Sun Oil Company, Standard Oil Company (New Jersey) et al., August 20 1933, J. Howard Pew Papers, Box 65.

51. J.N. Pew, Jr. to Hugh S. Johnson, August 20, 1933, Acc. #1317, EMHL, J. Howard Pew Papers, Box 71.

52. Judson Welliver to J. Howard Pew, August 20, 1933, Acc. #1317, EMHL, J. Howard Pew Papers, Box 65; clipping, "Oil Groups Name Choices for Board," J. Howard Pew Papers, Box 66.

53. J.N. Pew, Jr. to J. Howard Pew, August 20, 1933; J. Howard Pew to J.N. Pew, Jr., August 29, 1933, Acc. #1317, EMHL, J. Howard Pew Papers, Box 66.

54. "Roosevelt Names Ickes to Administer the Provisions of the Oil Code," *National*

Petroleum News 25, 35 (August 30, 1933), pp. 3-4, 6; "Advocates of Price Fixing Control Code's Planning Committee," *National Petroleum News* 25, 36 (September 6, 1933), p. 12; "General Johnson Discusses Oil Code," *National Petroleum Association Newsletter* (August 19, 1933), in Acc. #1317, EMHL, J. Howard Pew Papers, Box 65; J.C. Royale to J. Howard Pew, September 2, 1933, J. Howard Pew Papers, Box 68.

55. "Oil Code is Effective as Price Fixing Postponed," *National Petroleum News* 25, 36 (September 6, 1933), pp. 7-8; "Halt in Spread of Price Wars Causes Ickes to Postpone Price Fixing," *National Petroleum News* 25, 38 (September 20, 1933), pp. 11-2, 14-5; Johnson, *The Blue Eagle,* p. 224.

56. Nash, *Oil Policy,* p. 140.

57. J.N. Pew, Jr. to Clarence Bamberger, November 21, 1932, Acc. #1317, EMHL, File 21-A, Box 277.

58. Nash, *Oil Policy,* pp. 140-1.

59. J. Howard Pew to George Wharton Pepper, September 25, 1933; Pepper to J. Howard Pew, September 25, 1933, Acc. #1317, EMHL, J. Howard Pew Papers, Box 69.

60. Robert G. Dunlop, interview held at St. Davids, Pennsylvania, October 31, 1975. Dunlop distinguished himself here and in other posts with the firm and later succeeded J. Howard Pew as Sun's president in 1947.

61. "NRA Participation," Acc. #1317, EMHL, J. Howard Pew Papers, Box 66; Senator Robert F. Wagner to John G. Pew (no date, presumably 1933), J. Howard Pew Papers, Box 68.

62. W.S. Farish to J. Howard Pew, September 28, 1933; J. Howard Pew to George Wharton Pepper, October 6, 1933, Acc. #1317, EMHL, J. Howard Pew Papers, Box 69; J.H. Pew to P & C Committee, N.R.A., October 10, 1933, J. Howard Pew Papers, Box 67.

63. Nash, *Oil Policy,* p. 141; Joseph E. Pogue, Memo on Price-Fixing Hearings, October 19, 1933; J. Edgar Pew to J. Howard Pew, November 2, 1933, Acc. #1317, EMHL, J. Howard Pew Papers, Box 69; J.C. Royale to J.C. Welliver, November 1, 1933, J. Howard Pew Papers, Box 68.

64. Department of Interior Press Release, November 2, 1933; George Wharton Pepper to J. Howard Pew, November 2, 1933, Acc. #1317, EMHL, J. Howard Pew Papers, Box 69.

65. Sun Oil Company to P & C Committee, November 14, 1933; George Wharton Pepper to J. Howard Pew, November 15, 1933, Acc. #1317, EMHL, J. Howard Pew Papers, Box 69.

66. J.N. Pew, Jr., to Franklin, P & C Committee, December 7, 1933; Franklin to Pew, December 8, 1933; Pew to Franklin, December 9, 1933, Acc. #1317, EMHL, J. Howard Pew Papers, Box 70.

67. Norman L. Meyers, P.A.B., to J. Howard Pew, November 22, 1933, Acc. #1317, EMHL, J. Howard Pew Papers, Box 69; Nash, *Oil Policy,* p. 142; "Tentative Figures for Participation in Pool Agreement, December 7, 1933," J. Howard Pew Papers, Box 70.

68. George Wharton Pepper to J. Howard Pew, December 31, 1933, Acc. #1317, EMHL, J. Howard Pew Papers, Box 70; "Draft of Brief to be Submitted by Sun Oil Company to Petroleum Committee;" Memo, Price Fixing, November 22, 1933, Acc. #1317, EMHL, J. Howard Pew Papers, Box 69; "Price-Fixing of Petroleum Products, An Analysis of the Proposal in Light of History and Economic Experience, Concluding With a Protest Against the Venture by J. Howard Pew," J. Howard Pew Papers, Box 68; Merrill, "Price Fixing Right or Wrong," *Forbes* 32, 10 (November 15, 1933), pp. 11-2, 36.

69. Nash, *Oil Policy,* p. 144; J.C. Welliver to J. Howard Pew, April 16, 1934; J.C. Royale to J.C. Welliver, April 30, 1934, Acc. #1317, EMHL, J. Howard Pew Papers, Box 68.

70. J.N. Pew, Jr. to Clarence Bamberger, November 26, 1932, Acc. #1317, EMHL, File 21-A, Box 277.

71. "Price Fixing for Petroleum Products, etc." Acc. #1317, EMHL, J. Howard Pew Papers, Box 68.
72. Royale to Welliver, April 30, 1934, Acc. #1317, EMHL, J. Howard Pew Papers, Box 68; Nash, *Oil Policy,* p. 145.
73. J. Edgar Pew to J. Howard Pew, October 1, 1934, Acc. #1317, EMHL, J. Howard Pew Papers, Box 69.
74. Nordhauser, "Quest," pp. 219–20.
75. Nash, *Oil Policy,* p. 146.
76. Pew to Anderson, February 20, 1935, Acc. #1317, EMHL, J. Howard Pew Papers, Box 67.
77. Interoffice Directives, Samuel Eckert, May 28, 1935, and June 29, 1935, Acc. #1317, EMHL, J. Howard Pew Papers, Box 69.
78. J. Howard Pew to R.L. Lund, N.A.M., June 5, 1933, Acc. #1317, EMHL, J. Howard Pew Papers, Box 76; J. Howard Pew to H.H. Anderson, September 7, 1934, J. Howard Pew Papers, Box 67.
79. "Price Fixing for Petroleum Products, etc.," Acc. #1317, EMHL, J. Howard Pew Papers, Box 68.
80. Representative William P. Cole, introduction to Samuel B. Pettengill, *Hot Oil* (New York: Economic Forum Company, 1936), pp. vii–xiii.
81. "Testimony of J. Howard Pew," September 19, 1934, U.S. Congress, House, *Petroleum Investigation: Hearings Before a Sub Committee of the Committee on Interstate and Foreign Commerce,* 73rd Congress Part 1 (Washington, D.C.: U.S. Government Printing Office, 1934), p. 378.
82. *Ibid.,* p. 377.
83. *Ibid.*
84. *Ibid.,* p. 378.
85. *Ibid.,* pp. 378–9.
86. Pettengill, *Hot Oil,* p. 166.
87. Testimony of J. Edgar Pew, November 19, 1934, *Petroleum Investigation,* Part 4, p. 2044.
88. *Ibid.*
89. *Ibid.,* pp. 2049–53; Arthur M. Johnson, *Petroleum Pipelines and Public Policy* (Cambridge: Harvard University Press, 1967).
90. Johnson, *Petroleum Pipelines,* pp. 229–30.
91. William P. Cole, introduction to Pettengill, *Hot Oil,* pp. xi–xii; Nash, *Oil Policy,* p. 145.
92. Amos L. Beaty to J. Edgar Pew, April 13, 1932, Acc. #1317, File 71, Box 15.
93. Nash, *Oil Policy,* pp. 148–9.
94. J. Edgar Pew to Hon. E.W. Marland, December 17, 1934, Acc. #1317, EMHL, File 71, Box 15. Pew continued to use his influence in support of the compact (see J. Edgar Pew to John Naylor, *Fort Worth Star Telegram,* Colonel Carl Estes, *Longview Daily News,* December 11, 1934, Acc. #1317, EMHL, File 71, Box 15).
95. "An Interstate Compact to Conserve Oil and Gas," February 1935; telegram, Jack Pew to J. Howard Pew, February 11, 1935, Acc. #1317, EMHL, File 71, Box 15.
96. E.W. Marland, "The Interstate Oil Compact," paper presented at the 16th meeting of the A.P.I., November 13, 1935, Acc. #1317, EMHL, File 71, Box 15; Nash, *Oil Policy,* pp. 150–1; Nordhauser, "Quest," p. 226.
97. Herbert Hoover, *The Memoirs of Herbert Hoover,* Volume II, *The Cabinet and The Presidency, 1920–1933* (New York: MacMillan, 1953), pp. 237–9.
98. See John C. Burnham, "The Gasoline Tax and the Automobile Revolution," *Mississippi Valley Historical Review* 58, 3 (December 1961), p. 435.

99. See folder labeled "Pennsylvania Tax, 1931," Acc. #1317, EMHL, File 21-A, Box 63; *National Petroleum Association Bulletin,* November 19, 1931, File 21-A, Box 61; J. Edgar Pew to J. Howard Pew, inter-office, December 2, 1931; S.B. Eckert to J. Howard Pew, November 23, 1931, File 21-A, Box 63.

100. J. Howard Pew to W.R. Boyd, July 2, 1932, Acc. #1317, EMHL, J. Howard Pew Papers, Box 65; J. Howard Pew to Hon. Elwood Turner, December 13, 1931, File 21-A, Box 62; J. Howard Pew to William Irish, Atlantic Refining, September 15, 1932, J. Howard Pew Papers, Box 72.

101. "You Are Asked to Take a Hand in the Battle of Marketers vs. Racketeers," *Our Sun* 9, 3 (March 1932), pp. 3-4.

102. J. Howard Pew to R.C. Holmes, March 9, 1932; J. Howard Pew to Judson C. Welliver, February 6, 1932, Acc. #1317, EMHL, File 21-A, Box 71; see folder labeled "Highway Users Conference," File 21-A, Box 67.

103. J. Howard Pew to Axtell J. Byles, October 4, 1932, Acc. #1317, EMHL, File 21-A, Box 72; J. Howard Pew to K.R. Kingsbury, March 26, 1934, J. Howard Pew Papers, Box 1.

104. J. Howard Pew, "A Squawk From the Petroleum Goose," *Nations Business* (October 1932), pp. 20-1, 56, reprint in Acc. #1317, File 21-A, Box 72.

105. Warren C. Platt, "On Behalf of Howard Pew, of Philadelphia," *National Petroleum News* 24:40 (October 5, 1932), p. 13; J. Howard Pew to Axtell J. Byles, October 4, 1932, Acc. #1317, EMHL, File 21-A, Box 72.

106. Thomas J. Spellman to H. Higeman, Jr., Sun Oil, December 19, 1932, Acc. #1317, EMHL, J. Howard Pew Papers, Box 71; Thomas Charles Longin, "The Search for Security: American Business Thought in the 1930's" (Ph.D. Dissertation, University of Nebraska, 1970), p. 114; J. Howard Pew to Ellen C. Talbot, March 29, 1932, Acc. #1317, EMHL, File 21-A, Box 71.

107. J. Howard Pew, "The Highways and the Railways," address before the Philadelphia Highway Users' Conference, April 12, 1933; J. Howard Pew, "Unjust Tax Burdens of the Oil and Automobile Industries," address before the U.S. Chamber of Commerce Natural Resources Section, Washington, D.C., May 3, 1933, Acc. #1317, EMHL, File 21-A, Box 79; clippings, *Philadelphia Evening Ledger,* January 31, 1933; *Philadelphia Record,* January 31, 1933, File 21-A, Box 74.

108. H.I. Harriman, President, U.S. Chamber of Commerce, to J. Howard Pew, January 13, 1933; J. Howard Pew to Axtell Byles, October 4, 1933; J.C. Welliver to J. Howard Pew, October 11, 1933, Acc. #1317, EMHL, File 21-A, Box 74.

109. J.C. Welliver to J. Howard Pew, Memo on N.R.A. Codes and Taxes, March 15, 1934, Acc. #1317, EMHL, J. Howard Pew Papers, Box 68; testimony of J. Howard Pew before the State Senate Finance Committee, Harrisburg, March 25, 1935, J. Howard Pew Papers, Box 54.

110. *The New Deal and the Problem of Monopoly* (Princeton: Princeton University Press, 1966), pp. 472-90.

111. William E. Leuchtenburg, *Franklin D. Roosevelt and the New Deal* (New York: Harper & Row, 1963), pp. 147-50, 259-60.

112. Williamson et al., *Energy,* pp. 695-6; Nash, *Oil Policy,* pp. 152-3.

113. *Ibid.*

114. Williamson et al., *Energy,* p. 696; "Jack" Pew interview.

115. Nash, *Oil Policy,* p. 154; J. Edgar Pew to J. Howard Pew, April 6, 1938, Acc. #1317, EMHL, J. Howard Pew Papers, Box 46.

116. "Testimony of J. Howard Pew," Gillette Bill on Petroleum Marketing Divorcement, p. 100, Acc. #1317, EMHL, J. Howard Pew Papers, Box 45.

117. George Wharton Pepper to W.E. Borah, March 15, 1934; E.C. Talbot to J. Howard Pew, May 3, 1934, Acc. #1317, EMHL, J. Howard Pew Papers, Box 68; J. Howard Pew to

Hon. W.E. Borah, January 12, 1934, Acc. #1317, EMHL, J. Howard Pew Papers, Box 69; Pepper to Borah ,March 15, 1934, J. Howard Pew Papers, Box 68.

118. Nash, *Oil Policy,* p. 154.

119. "Testimony of J. Howard Pew," U.S. Congress, Temporary National Economic Committee, *Hearings Before the Temporary National Economic Committee,* 76th Congress, Part 14, Section I (Washington, D.C.: U.S. Government Printing Office, 1940), pp. 7163-4; Dunlop interview.

120. Pew sent a draft of his prepared statement to Axtell Byles and Fayette B. Dow of the A.P.I. and Stanley F. Teele of the Harvard Business School (see folder labeled "TNEC Testimony" Acc. #1317, EMHL, J. Howard Pew Papers, Box 49); Dunlop interview. Pew received congratulations from several industry spokesmen who commented on the excellence of his performance (see W.R. Boyd to J. Howard Pew, October 14, 1939; Axtell J. Byles to J. Howard Pew, October 16, 1939, Acc. #1317, EMHL, J. Howard Pew Papers, Box 45).

121. Roy C. Cook, *Control of the Petroleum Industry by Major Oil Companies,* T.N.E.C. Monograph No. 39 (Washington, D.C.: U.S. Government Printing Office, 1941).

122. *Ibid.,* pp. 3-7.

123. T.N.E.C. Monograph No. 39-A (Washington, D.C.: U.S. Government Printing Office, 1941).

124. *Ibid.,* pp. 1-96; W.S. Farish to J. Howard Pew, June 2, 1941; Pew to Farish, June 4, 1941; Farish and Pew to Sen. O'Mahoney, T.N.E.C., June 2, 1941; Farish to Pew, April 7, 1941; Pew to Farish, April 14, 1941, Acc. #1317, EMHL, J. Howard Pew Papers, Box 45.

125. Quoted in the draft of a letter from J. Howard Pew to Assistant Attorney General Robert H. Jackson (not sent), December 31, 1937, Acc. #1317, EMHL, J. Howard Pew Papers, Box 5.

126. Nash, *Oil Policy,* pp. 155-6.

127. Conkin, *The New Deal,* p. 75.

128. "J. Howard Pew is Champion of Man's Rights and Freedom," *The Oil Daily* (November 19, 1969), p. 28; J. Howard Pew, "Which Road to Take," American Liberty League Publication No. 53 (July 12, 1935); see also No. 75 (November 1, 1935) (most of the Liberty League pamphlets are in a scrapbook at the Eleutherian Mills Historical Library); Howard, J. Pew, *Governmental Planning and Control as Applied to Business and Industry* (Princeton: Guild of Brackett Lectures, 1938).

129. Quoted in George Wolfskill, *The Revolt of the Conservatives* (Cambridge, Mass.: Houghton Mifflin, 1962), p. 129.

130. "Mr. Pew at Valley Forge," *Time* 35, 19 (May 6, 1940), p. 17 (in a confidential letter to *Time,* Pew claimed that he did not vote for F.D.R. in 1932, p. 18); Potomachus, "Pew of Pennsylvania," *The New Republic* 110, 19 (May 8, 1944), p. 626.

131. J. Howard Pew to Ellen C. Talbot, March 29, 1932, Acc. #1317, EMHL, File 21-A, Box 71.

132. J.N. Pew, Jr., to Clarence Bamberger, October 11, 1932, Acc. #1317, EMHL, File 21-A, Box 277.

133. J.N. Pew, Jr., to Clarence Bamberger, November 26, 1932, Acc. #1317, EMHL, File 21-A, Box 277.

134. "Mr. Pew at Valley Forge," *Time* 35, 19 (May 6, 1940), p. 17; Potomachus, "Pew," p. 626. In a letter to Peter C. Hess on October 20, 1932, J.N. regretted his contribution to the 1932 Republican National campaign (see Acc. #1317, EMHL, File 21-A, Box 277).

135. Marquis W. Childs, "Pennsylvania's Boss Pew a G.O.P. Power Out of Grim Hate of New Deal," *St. Louis Post-Dispatch* (May 26, 1940), p. C-1; "Pew at Valley Forge," *Time,* p. 17.

136. J. Howard Pew statement, October 16, 1936, Acc. #1317, EMHL, J. Howard Pew Papers, Box 53.

137. Wolfskill, *Revolt of the Conservatives,* p. 60; Leuchtenburg, *Franklin D. Roosevelt and the New Deal,* pp. 91–2; "Platform and Organization of the American Liberty League," Papers of Pierre S. duPont, EMHL, Longwood Manuscripts 10, File 771, Box 1.

138. William H. Stayton, American Liberty League, to Pierre S. duPont, January 25, 1940, EMHL, Longwood Manuscripts 10, File 771, Box 1.

139. Wolfskill, *Revolt of the Conservatives,* p. 109; Dunlop interview; J.N. Pew, Jr., to Raymond Moley, November 23, 1940, Acc. #1317, EMHL, File 21-B, Box 26.

140. "Pew at Valley Forge," *Time,* pp. 15–8.

141. *Ibid.,* p. 17.

142. Potomachus, "Pew," p. 626; Graham Patterson to J.N. Pew, Jr., February 9, 1937 and July 16, 1937, Acc. #1317, EMHL, File 21-B, Box 23.

143. J.N. Pew, Jr. to F.H. Bedford, Jr., January 21, 1937, Acc. #1317, EMHL, File 21-B, Box 23.

144. Potomachus, "Pew," p. 626.

145. See folder labeled "Lowell Thomas," Acc. #1317, EMHL, J. Howard Pew Papers, Box 70; David Holbrook Culbert, *News for Everyman: Radio and Foreign Affairs in Thirties America* (Westport, Conn.: Greenwood Press, 1976), pp. 15–6.

146. Lowell Thomas Broadcast, August 14, 1933, Acc. #1317, EMHL, J.H. Pew Papers, Box 69; H.L. Bowyer to J. Howard Pew, October 17, 1932 (re: Thursday, October 14 broadcast), File 21-A, Box 67; Everitt Dirkson to Lowell Thomas, May 18, 1933 (re: May 18 broadcast), J.H. Pew Papers, Box 52; Lowell Thomas to "Mr. Pew," November 30, 1932 (re: November 19 broadcast), File 21-A, Box 72.

147. J.L. Bowyer to J. Howard Pew, October 17, 1932, Acc. #1317, EMHL, File 21-A, Box 67; J.N. Pew, Jr. to Lowell Thomas, May 18, 1933, J. Howard Pew Papers, Box 52; J. Howard Pew to J. Harris Worthman, October 11, 1932, J. Howard Pew Papers, Box 72.

148. Culbert, *News for Everyman,* p. 9. Boake Carter was a consistent and outspoken critic of the New Deal until he was taken from the air in 1938 (see Culbert, Chapter 2, "Boake Carter: Columbia's Voice of Doom," pp. 34–59). Culbert quotes from an interview Thomas gave in 1969 (Oral History Interview, October 7, 1969, Herbert Hoover Library, West Branch, Iowa) where he stated that "My radio sponsors seemed uncertain as to what my politics were" (p. 9).

149. Memo of conversation between Ray Henle, Editor-in-Chief of Sunoco Three-Star-Extra, and Herbert Hoover, February 17, 1948, Acc. #1317, EMHL, File 21-B, Box 25.

150. See "Comments on Mr. Cook's Chapter on Production," Pew and Farish, *Review and Criticism of Monograph No. 39,* T.N.E.C., pp. 17–29.

151. "Jack" Pew interview.

152. J. Howard Pew, "Government in Business," *Our Sun* 2, 5 (April 1936), pp. 2–4, 26.

153. P. H. Frankel, *The Petroleum Times* (London: June 23, 1946), as quoted in "Prophet With Honor . . . Both at Home and Abroad," *Our Sun* 12, 3 (October 1946), p. 23.

CHAPTER IX

World War II: Industrial Cooperation for the National Effort

Even before Pearl Harbor, the outbreak of general European war in 1939 confronted the petroleum industry with an entirely new set of challenges. Whereas the chief problem facing the industry in the New Deal era had been the need to curtail crude oil production, unprecedented wartime demand now required dramatically increased output of petroleum products and a fully coordinated system of transportation, refining, and distribution. By 1939, almost all naval vessels and fully eighty-five percent of the world's merchant fleet burned fuel oil. The average mechanized division of World War II required fuel to operate equipment totalling 187,000 horsepower. In World War I, the Allies had "floated to victory on a sea of oil." Yet, by 1945, the United States Army Air Force alone daily consumed fourteen times the total volume of gasoline shipped to Europe for all purposes between 1914 and 1918! In a letter addressed to the industry in November 1945, the Joint Chiefs of Staff stated that "The fulfillment of this gigantic task was without question one of the great industrial accomplishments in the history of warfare."[1]

The United States oil industry was in a relatively strong position to meet this challenge at the outbreak of war. Domestic crude reserves were high, refining capacity was large and modern, research efforts had successfully developed new products and processes, and the large firms had a sound financial base. Like other large oligopolistic businesses during the Great Depression, oil had emerged from the 1930s in generally excellent health.[2] Furthermore, strategic needs for petroleum products during the war years brought the oil industry and government together in a massive program of economic mobilization.

The achievement of cooperation between business and government was both a particular response to the pressures of war and a culmination of three decades of development. Since the dissolution of the Standard Oil Trust in 1911, public oil policy had moved away from an emphasis on antitrust and increasingly focused on the related issues of conservation and economic stabilization. New policies that evolved at both the state and federal levels furthered cooperative regulation. The oil companies found that they could work together toward common goals, and the industry slowly came to accept the partnership of government in achieving effective stabilization. World War II brought together the strands of cooperation that had developed from the World War I experience, the conservation and stabilization movements of the 1920s, and the more recent period of New Deal petroleum policy. Although there were serious bottlenecks that developed in the wartime petroleum program, the story of oil in World War II is basically one of overwhelming success.[3]

The war also had important long-range effects on the Sun Oil Company. The firm made outstanding contributions to war production in many areas, but stood above the rest of the industry in two categories: aviation gasoline production and tanker construction. Sun's leadership in the adoption and development of catalytic cracking placed the firm in a strong position to supply the American military and the Allies with the high octane base-stock needed for the manufacture of 100 octane aviation fuel. The Sun Shipbuilding and Drydock Company had survived the difficult interwar period to emerge in World War II as the country's leading producer of oil tankers and one of the world's largest privately-owned shipyards. The entire Sun organization benefited from wartime growth and emerged in a much stronger competitive position than it had occupied in 1939.

Sun achieved these important economic gains as a result of very large government contracts during the war. Yet, compared with the rest of the industry, Sun's attitude toward cooperation with government agencies remained consistent with its past history. The Sun Oil story had shown that its managers had resisted government encroachment into the oil business. The Pews accepted regulation that they could rationalize with their conservative business philosophy (state prorationing as legal protection of private property and elimination of waste), but strongly opposed others such as price-fixing and increased federal control of the industry. Despite the general attitude of cooperation that permeated the wartime industry and the strong patriotism of the Pew family, Sun maintained a competitive posture in its business dealings, was one of the first firms to champion the elimination of controls once victory came in 1945, and opposed extension of industry planning into the post-war era.

THE SUSPENSION OF ANTITRUST

The importance and long-lasting achievements of New Deal oil policy were represented in the Connally Hot Oil Act and the Interstate Compact Commission. The revival of antitrust prosecutions in the late 1930s and the creation of the Temporary National Economic Committee were aberrations from the major thrust of cooperative regulation. Yet, there was one final gasp of the Brandeisians, even as war appeared on the immediate horizon. The famous "Mother Hubbard" antitrust suit lodged against the petroleum industry in October of 1940 ended the New Deal on a negative note for the oil industry.

Given its colorful name because of its all-encompassing nature, this action by the Antitrust Division of the Justice Department charged twenty-two integrated oil companies, including Sun, and the American Petroleum Institute with participating in a range of activities aimed at fixing prices and restricting competition. A second suit, also filed in October, accused the Great Lakes, Phillips, and Stanolind pipeline companies of paying rebates in effect to themselves, in violation of the Elkins Act, when they paid dividends on their pipeline operations. The government interpreted these dividends as illegal rebates on published transportation rates. Although the Justice Department prepared similar complaints against fifty-nine pipeline companies, it selected these three for prosecution. Soon, however, the threat of war intervened and aided the petroleum industry.[4]

Beginning in December 1940, an industry committee headed by Harry T. Klein of the Texas Company opened negotiations aimed at the settlement of both cases. It became clear in the course of these negotiations that the government desired good relations with the industry. Both J. Howard and J. Edgar Pew were very anxious to settle the cases, and they exerted pressure within the A.P.I. to force certain hard-line elements to make concessions to the government.[5] In particular, they urged the Phillips Petroleum Company, one of the three defendants in the Elkins case, to lower its pipeline charges. Howard confessed that "I am one of those who believe that our pipeline rates have been too high and am quite prepared, as far as our group is concerned, to make substantial reductions."[6]

These negotiations continued into the spring of 1941 and reached a climax in May. On May 28, President Roosevelt created the office of Petroleum Coordinator for National Defense and named Interior Secretary Harold Ickes to the post. Shortly thereafter, Ickes appointed Ralph K. Davies, vice-president of Standard Oil of California, as Deputy Coordinator. These were the first steps in the creation of a wartime petroleum agency. Ickes had been prodding the President to formulate a petroleum policy and urging him to cooperate with the industry. In May 1941, the Interior Secretary was particularly anxious to get industry agreement to construct a major crude pipeline for national defense. In July, both the government and the industry reached a tentative settlement of the Elkins Act cases and agreed to a further delay of the "Mother Hubbard" suit. Ultimately, in April 1942, Roosevelt suspended all antitrust actions for the duration of the war.[7] With this issue out of the way, the road was clear for the development of further cooperative activity.

Ickes began a series of meetings with industry leaders in the summer of 1941 in an effort to generate the type of business-government cooperation that Mark L. Requa had created in World War I. He appointed a number of district committees of oil men who met in August and October of 1941 to discuss national defense issues. On November 28, 1941, this group became the Petroleum Industry Council for National Defense, later renamed the Petroleum Industry War Council (P.I.W.C.). By a quirk of fate, this group met for the first time on December 8, the day the United States entered the war.[8]

Sun had followed a cautious policy regarding Ickes's movement toward a national oil policy. In its 1940 *Annual Report,* the firm emphasized the oil industry's financial independence and its ability to meet any war demands through private initiative. Impressed by the "astonishingly liberal" overtures that the government made in the antitrust negotiations of early 1941, however, Sun opted for a wait-and-see attitude.[9] The Pews were suspicious of Ickes, and the Secretary's attempts to place greater federal control over the oil industry in the thirties were fresh in their minds. Although indicating total agreement with a June 3, 1941 editorial denouncing the Petroleum Administrator as a "dictator," J. Edgar Pew indicated that the major companies could not take an openly defiant position. He stressed that a degree of cooperation with Washington might serve to stave off more drastic attempts to take over the industry.[10]

As longtime supporters of the antitrust laws, the Pews' pleasure at having the cases dropped seemingly contradicted their traditional stance. When questioned on this point. J.N. Pew, Jr. confessed that he had been forced to take a position with which he had little sympathy. He re-affirmed Sun's belief that "all reasonable provisions of the antitrust laws should be enforced," but rationalized its shift in position by taking exception to "new interpretations of old laws to upset long established customs."[11] Clearly, in his view, the Elkins and "Mother Hubbard" cases belonged in that category.

THE PETROLEUM ADMINISTRATION FOR WAR

Throughout 1941 and 1942, Ickes lobbied to extend and clarify his power in the wartime oil program. Although Roosevelt's May 28 letter had made him Petroleum Administrator, there was no specific agency officially empowered to supervise Ickes's programs. He did make requests for more efficient use of tankers, increased crude production, and cooperation in providing better pipeline connections from the western fields to the eastern coastal ports. Although Ickes had great success in obtaining voluntary cooperation, he felt that he needed more authority. At the same time, there was a significant public clamor when supplies became short in late 1942 and Congress introduced various proposals for fuel rationing.[12]

Sun remained ambivalent on the issue of cooperation with Washington. On the one hand, it realized the patriotic necessity for some regimentation and appreciated the results that Ickes was achieving, but it still feared that the Secretary would go too far. Although Ickes related a meeting with J. Howard Pew in October 1941, in which "whether he [Pew] was buttering us or not, he was kind enough to say that we had really done a remarkable job,"[13] J. Edgar Pew soon after wrote to Warren C. Platt of the *National Petroleum News*, criticizing an editorial favorable to Ickes. J. Edgar expressed his belief in the need for cooperation, but maintained that industry should retain leadership and authority, and that oil men should be wary of the build-up of too much power in Washington.[14]

In late 1941, Sun voluntarily cut back intercoastal shipments of crude to the East Coast because of the tanker shortage. The firm switched much of its haulage to railroad tank cars and installed its first tank car unloading facilities at Marcus Hook. As demand for these facilities increased, Sun built a large unloading terminal at Twin Oaks, Pennsylvania, in January and February 1942. At the beginning of 1942, the company had approximately 1,200 tank cars, and it was operating 5,000 by the end of the year.[15] By this time also, Sun had entered into important supply contracts with the government to refine aviation gasoline base-stock in its Houdry units at Marcus Hook and Toledo. Sun converted its Susquehanna product pipeline from carrying refined products westward to moving gasoline from the interior to the eastern coast, thus altering its main distribution system. In 1941, the firm also introduced tetraethyl lead into its gasoline for the first time. Although the company continued to sell unleaded fuel in some of its midwestern marketing area, it was forced to use most of its catalytic base-stock for aviation gasoline production.[16] Sun was cooperating with the Petroleum Administrator on many fronts, but its executives still had reservations.

J. Howard Pew addressed himself to this subject at the October 7, 1942 meeting of the Petroleum Industry War Council in Washington. He specifically objected to an Ickes order centralizing marketing facilities in Eastern Region No. 1, a move Pew referred to as the introduction of the "cartel system." However, his speech moved on to the more general question of the retention of significant private initiative in the wartime administration. Howard argued that the best way to win the war was to allow the industry to function under free enterprise and not burden it with over-regimentation.[17]

Ickes attempted to reassure the Pews and others fearful of the growth of government bureaucracy:

> I have talked with Mr. Pew and his brother, and if there are any two men in this country who have a sincere conviction that we have to win this war, that we must win it, and who really want to win it because it will mean a survival of the democratic principle which we will lose otherwise, it is these two men in the Sun Oil Company. . . . We are in a time, however, where we have to close ranks. If we want into the army as soldiers, privates, with muskets, we would know that at the end of the war, after victory had come, that we would be discharged, and we would be resolved into our constituent elements. The regimentation which was necessary for the time—let us accept the word even though it is opprobious to all of us—would not need to exist any longer, and we would all go back into private life.[18]

Thus stating that the wartime controls were for the duration only, Ickes assured the assemblage that "if, God willing, I am still here, I will be on your side when it comes to the untangling."[19]

The reassurances were timely, for on December 2, 1942, President Roosevelt acceded to Ickes's demands for more power by creating the Petroleum Administration for War (P.A.W.). Ickes remained as Petroleum Administrator and was empowered to issue regulations concerning the production, transportation, and distribution of petroleum. To Ickes's displeasure, however, control of scarce materials needed for the industry (such as steel) remained with the War Production Board (W.P.B.), and price regulation with the Office of Price Administration (O.P.A.).[20]

The Pews appeared convinced, for the Sun Oil *Annual Report* of 1942 praised Ickes's restraint from "interference with the established methods of the industry." The report also warned that "We have, however, noted in certain directions some tendency of late to introduce the European cartel system, and in some other regards to impose forms of regimentation both of which are destructive of initiative and result in loss of production."[21] Sun continued to cooperate with the P.A.W., fully participated in the P.I.W.C., and publicly praised Ickes for his "understanding and cooperative attitude," but noted that "we take comfort in his repeated promises that as soon as the peace comes the government will free the industry from its wartime restriction."[22]

The national press was quick to point out that "Horrible Harold Ickes," the ogre of the 1930s, was rapidly becoming the darling of the petroleum industry.[23] On December 30, 1937, Ickes had spoken on a national network radio hook-up to denounce the insidious threat of "America's Sixty Families" as part of the Roosevelt onslaught against the forces of "economic royalism."[24] The Pew family was high on the list. This was probably the low point of Ickes's relations with Sun management, since the Pews were already convinced that he was out to make the oil industry a public utility. Now, just a few years later, the press quoted a more favorable assessment of Ickes by J. Howard Pew: "It gives real pleasure to testify to the fine, patriotic, intelligent, effective work he is doing."[25]

Obviously, patriotism played a role here, both that of the Pews and Ickes. However, other factors suggested a reason for this détente. Unlike his tenure as Petroleum Administrator under the N.R.A., Ickes now gave more freedom to the private sector to reach the goals set by the P.A.W. He established broad parameters, but the industry was for the most part allowed to reach these goals through its own cooperative efforts. Despite elements of the regimentation that the Pews so opposed, the petroleum industry had a surprising amount of freedom during the war. Ickes's chief lieutenant, Ralph Davies, had come from industry, and most of his other assistants also came from the private sector. Ickes fought for higher price ceilings on petroleum to encourage production, and he supported the retention of the oil depletion tax allowance. The P.A.W. lent its support to state conservation programs, and Congress extended both the Connally Act and the Interstate Compact Commission during the war. In accomplishing the goal of raising production levels and meeting wartime demand, the P.A.W. and the P.I.W.C. were both successes.[26]

On October 27, 1943, Harold Ickes journeyed to Marcus Hook, Pennsylvania, to deliver a speech at the dedication ceremonies of the Sun Oil Company's new 100 octane aviation plant. Both the CBS and NBC networks broadcast the speech. Ickes cited the importance of the aviation gasoline program and more particularly, the achievements of Sun Oil as a leader in this program.[27] The old antagonists, Ickes and J. Howard Pew, had their pictures taken together on the platform, smiling in tandem for the national patriotic effort.

There were still enormous differences between the ardent New

Dealer, Ickes, and the reactionary Liberty Leaguer of the thirties, J. Howard Pew, but both men realized that a working arrangement between government and business was possible. Ickes had come to appreciate the advantages of allowing a degree of freedom to the private sector, and Pew now realized the benefits to be gained from cooperating with Washington. In later commenting on a *Sun Oil Annual Report,* Ickes found much to disagree with, but stated that "these personal differences of viewpoint are submerged in my keen admiration for the contributions which your company had made to the country's war program."[28]

THE "BIG INCH" PIPELINE

From the early years of the war in Europe, it had become apparent that transportation was a critical bottleneck in the petroleum industry. The government and the oil industry had to develop efficient methods to carry crude from the flush western fields to the eastern refining centers and shipping ports. There was a tank ship shortage which was increasingly exacerbated by the dangers of submarine warfare in the North Atlantic. Harold Ickes directed most tankers to the European run, and encouraged the substitution of railroad tank cars to bring crude to the East Coast. But the railroads were unable to carry the tremendous traffic efficiently, and there were not enough tank cars available. As early as December 1940, Ickes had urged the construction of major trunk pipelines from the rich Texas fields to the east coast. This project became a major goal of his in the next two years.[29]

Sun was in a unique situation on the pipeline question. Not only did the firm operate a large tanker fleet of its own, but the Sun Shipbuilding and Drydock Company would benefit from a major tanker construction program. As late as May 1941, Sun still favored tanker construction over a new trunk pipeline system.[30] But at the end of the month, J. Howard Pew, William Farish of Jersey Standard, and Texaco President W.S.S. Rogers met with Ickes in Washington to discuss a pipeline proposal. The Petroleum Administrator had made it clear that the government favored an internal pipeline system and was willing to discuss an industry plan for construction. The industry group proposed building a major line from Mexico or East Texas to the New York-Philadelphia area, but requested antitrust immunity guarantees from the government. To their amazement, the P.A.W. administrator agreed in principle. As Ickes later related, "I think that I surprised them when I said I did not believe the petroleum industry should be operated under criminal statutes."[31] He pledged his full support to the plan and put an assistant to work to develop details.

Eleven eastern oil companies, including Sun, organized War Emergency Pipelines, Inc., to construct the line. In June 1942, this new corporation made an agreement with the Defense Plant Corporation, a division of the Reconstruction Finance Corporation. The R.F.C. funded the entire project, but the agreement stipulated that the companies would not make a profit on the basis of government contracts to transport crude. Ickes lobbied hard to obtain the necessary steel allocation from the War Production Board, and the Congressional legislation needed to grant the right of way for the line. This was the genesis of the so-called "Big Inch" and the "Little Big Inch" pipelines.[32]

Sun Oil had the second largest interest in the "Big Inch" line after Jersey Standard. This was based on the percentage of input into the line and the percentage of sales in the eastern market (as of 1940). The "Big Inch" line would connect East Texas to the Atlantic coast and traverse 1,388 miles, making it the biggest and longest pipeline in the world. The 24-inch pipe would carry 300,000 barrels of crude daily to the eastern terminus. Construction began in June 1942 on the western sections, and the entire line was completed by September 1943. The "Little Big Inch," a parallel product pipeline, was begun on April 21, 1943, and completed on December 2 of that year. These enormous technological achievements not only saved the East Coast from fuel starvation, but represented an essential link in the war effort.[33]

SUN'S AVIATION FUEL PROGRAM

Sun Oil's gamble in adopting the Houdry process for the catalytic cracking of petroleum during the Depression paid off in the Second World War. Sun's original motivation, of course, was the continued manufacture of high octane antiknock gasoline without the addition of tetraethyl lead. But when war broke out, the Houdry group possessed the only large-scale facilities for catalytic cracking in the United States, and was in an excellent position to supply aviation gasoline base-stock.[34]

Sun had constructed 40,000-barrel-a-day continuous catalytic units at Marcus Hook and Toledo in 1939. These Houdry fixed-bed plants were integrated operations, taking heavy crude in at the beginning and turning high octane gasoline out at the end. In 1940, Sun constructed a new plant at Marcus Hook specifically designed for the production of aviation fuel base-stock. The company built all of these plants with private investment capital, but obtained government money with lucrative supply contracts for high octane gasoline. The 81-plus octane gasoline coming from the fixed-bed Houdry units was not of sufficiently high octane for high performance aviation gasoline, and Sun needed other additives to bring the base-stock up to 87, 91, and ultimately 100 octane fuels. Sun at first supplied the aviation base-stock alone, but in the beginning of

1941, it signed a contract with the Ethyl Corporation to use tetraethyl lead in the manufacture of its own "Stratogas" aviation fuel.[35] Thus, the exigencies of war had forced the company to deviate from its long-held policy of competition with the Ethyl group.

Before the war, it had been common practice to add three cc. of lead to each gallon of aviation gas. As demands for higher performance increased, however, the industry began to add four cc. and later six cc. of Ethyl fluid. Lead alone was not sufficient, however, and refiners added specialized petroleum hydrocarbons to increase octane. During the thirties, research by several oil companies had developed various polymerization processes for producing iso-octane and codimer, both specialized high antiknock hydrocarbons used to upgrade motor fuel octane. Jersey Standard-I.G. Farben had also developed hydrogenation processes for making toluene to be used as a blending agent for gasoline as well as a basic substance in T.N.T. production.[36] But by far the most valuable of these auxiliary refining processes was alkylation.

By using sulphuric acid, hydrofluoric acid, or aluminum chloride as catalysts, these processes could produce hydrocarbon alkylates with extremely high octane ratings, excellent boiling ranges, a high heat combustion ratio, and low sulphur content. Next to the production of the base-stock itself with catalytic cracking, alkylation plants became the second most important key to the 100 octane program. Most significantly, refiners discovered that a symbiotic relationship existed between catalytic cracking and alkylation. They could use refinery gases generated as a by-product from catalytic cracking, primarily isobutane, butane, and isobutene, to produce alkylates and other hydrocarbon additives.[37] One of the problems facing the P.A.W. in 1941, however, was the conflicting licensing arrangements in existence for these various processes.

PATENT POOLING

Jersey Standard complicated the patent situation when it announced the commercial success of its Fluid catalytic cracking process in early 1941. The Houdry group had enjoyed a monopoly on catalytic facilities since 1937, but its failure to reach a licensing agreement with Standard had forced the Jersey company into one of the most intense research projects in refining history. The group, consisting of Jersey, Standard of Indiana, Shell, Texaco, Universal Oil Products, M.W. Kellogg, Anglo-Iranian, and Standard Catalytic Company (including the I.G. Farben patents owned by Jersey Standard), had developed its own process in a few short years. The continuous Fluid process in the long run proved to be more economical, and eventually supplanted the Houdry fixed-bed process.[38]

Although the Standard group maintained that there was no basic patent in catalytic cracking and it therefore did not infringe on the Houdry process, Eugene Houdry always felt that they had stolen his ideas and used his catalyst illegally. However, the war had intervened, and the P.A.W.'s desire to encourage cooperation and patent pooling precluded any litigation. Both Sun Oil and Socony-Vacuum continued to operate fixed-bed and thermofor catalytic cracking (T.C.C.) units during the war rather than switch over to fluid processes. The T.C.C. was a moving bed process originally developed by Socony in the late thirties and licensed by the Houdry group.[39]

Although Jersey announced the Fluid process in 1941, it did not go into significant commercial production until 1942. In the meantime, the P.A.W. took steps to reach agreement on pooling patent information and establishing uniform royalty rates. Its efforts reduced the royalty on sulphuric acid alkylation from 71 cents per barrel of refined product to 21 cents, and by the end of the war to 10.5 cents. This benefited Sun, which was able to license alkylation units for both its Marcus Hook and Toledo refineries. The P.A.W. established a similar royalty on other alkylation processes and authorized patent pooling on various hydrogenation reforming processes for producing codimer and iso-octane. It also set a rate of five cents a barrel for all catalytic cracking processes including Fluid, Houdry, and T.C.C. units. Before the war, Houdry Process Corporation had insisted on selling only fully paid-up licenses, but under government direction, it complied with the industry-wide policy. Ickes obtained Justice Department approval for P.A.W. recommendations that covered the patent pooling regulations.[40]

In the event that patent difficulties should emerge during the war, Arthur F. Corwin, vice-president of Socony, Arthur E. Pew, Jr., and Eugene Houdry signed an agreement in May 1942, whereby no one of the three parties would settle a patent infringement case without consulting the other two.[41] This was a concession to Eugene Houdry, who was not pleased by the general cooperative relationships between the oil companies and the P.A.W. Houdry represented not only himself, but one-third of the investors in Houdry Process Corporation. The only way these people could receive a return on their investment was through royalty payments and catalyst sales.

To encourage the oil companies in the production of aviation gasoline, the Defense Supplies Corporation entered into an agreement with the P.A.W., the Army, and the Navy for a program of loss reimbursement. Sun and the other manufacturers of 100 octane fuel signed supply contracts with the Defense Supplies Corporation to deliver aviation fuel at an agreed price. Because of high component, freight, and handling costs, the predetermined price often caused firms to operate at a loss.

The Aviation Gasoline Reimbursement Plant of July 24, 1942, authorized the government to compensate firms that developed losses as a result of meeting P.A.W. regulations for fuel. The final statement of the D.S.C. in 1946 revealed that it paid over $126,500,000 in claims to the industry as compensation for "extraordinary expenses."[42]

In 1942, Houdry units were refining ninety percent of all high octane aviation base-stock made by the catalytic process.[43] Although the Jersey Standard Fluid group rapidly made inroads into this percentage, Sun Oil and other Houdry licensees remained an important part of the total program throughout the war. By July, 1943, there were eighteen Houdry fixed-bed units in operation, five of them owned by Sun Oil. In addition, there were twelve others under construction, one by Sun. Twenty-nine Socony-H.P.C. T.C.C. moving bed units were under construction in various parts of the country, and Sun Oil had contracted for two of them.[44]

On October 27, 1943, Sun dedicated its new Plant 15 at Marcus Hook. This unit, designed expressly for the production of aviation gasoline, consisted of a six-case, Houdry fixed-bed catalytic cracking or treating unit, a large alkylation plant, a huge gas recovery and stabilization unit, and auxiliary equipment to operate these facilities. The alkylation plant was the largest built at that time and used the hydrogen fluoride catalyst system. The plant combined iso-butane and butylenes to form iso-octane. When the plant commenced production, it enabled Sun to triple its 1942 output of aviation gas. Unlike Socony, which had decided to emphasize the T.C.C. moving bed process, Sun Oil still had a strong commitment to the fixed-bed, semi-continuous process. It believed that it could produce results as good as either T.C.C. or fluid catalytic cracking, with greater control of reactions and less loss of catalyst into the atmosphere.[45]

By the end of the war, more than 400 refineries were contributing to the 100 octane program, but the Jersey Standard Fluid catalytic group licensed the vast majority of the new catalytic plants built after 1942. There were several reasons for this, but the most important was that these units were cheaper to build and operate than either the Houdry fixed-bed or T.C.C. units. The moving catalyst system also required much less steel than the Houdry units for construction. This was a crucial element for the P.A.W., since Harold Ickes had to fight with Donald Nelson and the War Production Board for every ton of steel. The Fluid system also enabled the operation of larger capacity plants. In the economic scarcity of World War II, these arguments far outweighed J. Howard Pew's belief in the technological superiority and reliability of the Houdry units. The much higher equipment authorization and supply contracts signed with the Fluid group licensees in the last three years of the war reflect this situation.[46]

SYNTHETIC RUBBER

Not all in the Houdry group remained convinced that government support of the fluid catalytic process was related to economic considerations. Standard Oil (New Jersey) exerted a great deal of influence throughout the industry and in Washington, and it appeared that a goodly share of politics had entered into the competitive arena. The best illustration of this conflict occurred in a direct offshoot of the catalytic aviation gasoline program: government support of synthetic rubber production.

The Pearl Harbor attack and Japanese capture of the Dutch East Indies seriously threatened future Allied supplies of natural rubber. With Malaya endangered and Java and Sumatra in Japanese hands, the situation was critical in early 1942. On January 15, 1942, the P.A.W. summoned a meeting of oil company representatives in Washington to discuss the synthetic rubber question. Shortly thereafter, it developed an arrangement with the Rubber Reserve Company, an R.F.C. subsidiary, to oversee the development of a synthetic rubber program. On April 8, the P.I.W.C. created an industry committee to address itself to the serious situation. Unless the United States could find 300,000 tons of rubber in excess of its immediate military requirements, the transportation network all over the country would be threatened.[47]

There was an existing technology for producing synthetic rubber, but one that had been little used in the United States. The DuPont Corporation had been producing neoprene by a chemical process since 1933, but this specialized product was not acceptable for tires or other heavy-duty uses. The main potential existed in the production of Buna-S rubber, a general purpose product produced from butadiene. In 1940, there were two basic ways to obtain butadiene. There was a chemical process developed in Europe to produce it from grain alcohol, and I.G. Farbenindustrie and Standard of New Jersey had developed a dehydrogenation process using the petroleum hydrocarbon, butylene (butene), as a feed stock. Farben began its research with coal in the 1920s, but during the next decade, collaborative work with Standard demonstrated that the best way to produce butadiene was from dehydrogenating butylene. The butadiene was then mixed with styrene, easily obtained from coal or alcohol, to produce Buna-S. Joint developments by I.G. and Standard also led to similar processes to produce Buna-N and butyl rubber, related substances with specialized uses.[48]

Immediately after Pearl Harbor, however, two other processes for producing butadiene appeared, both using butane rather than butylene as their raw material. The first, developed by the Phillips Petroleum Company, used butane from natural gas, and the other was a Houdry process for dehydrogenating butane and butene captured as a byproduct of Houdry catalytic cracking.[49]

Throughout the summer and fall of 1942, a bitter controversy broke out within the industry among these competitive processes. The Houdry group took a public stance in support of its process and against adoption of the Jersey approach. The most controversial point was Eugene Houdry's charge that the Standard butylene process was in direct conflict with the aviation gasoline program, since butylene was the main feed stock for hydrofluoric acid alkylation plants. He also argued that his butane process was much cheaper to run. Moreover, the Houdry commercial units needed to produce butane and butene were already in existence, while the Standard group had only begun to build its fluid cracking plants.[50]

After the fall of Singapore in February 1942, the Rubber Reserve Company recommended the creation of a synthetic rubber program able to produce 805,000 tons per year. The recommendation had included 40,000 tons of DuPont neoprene, 60,000 tons of Jersey Standard butyl rubber, and 705,000 tons of Buna-S produced by the Jersey Standard butylene process.[51]

The competitive struggle over synthetic rubber production was an extension of the competition between Houdry cracking processes and the Jersey Standard group's Fluid process, since both feed stocks for butadiene (either butylene or butane) were by-product refinery gases from cracking. Much hard feeling remained in the Houdry camp as a result of the catalytic cracking dispute. Eugene Houdry felt that Jersey's actions represented a continuation of its ruthless business tactics and evidence of unpatriotic relations with I.G. Farben prior to the war.

Much of this animosity had been submerged in national appeals for patriotism and unity, but Eugene Houdry himself was adamant on the issue. His sense of extreme patriotism and loyalty to his beloved France, now under Nazi domination, largely motivated these feelings. In 1941, Houdry had raised the issue of Standard's patriotism while defending his cracking process against Jersey's charges that it was obsolete. He wrote to Arthur E. Pew, Jr. two days after Pearl Harbor that "It is still my strong feeling that Jersey's tactics follow those used in France when they were offering an aviation gasoline process which was not developed, and, in different circumstances, may have cost the lives of many French aviators."[52]

Although Jersey and I.G. Farben had severed relations following the outbreak of the war in 1939, many rumors and accusations remained surrounding the American company's cartel relationship with the German industrial giant. In November 1941, the Federal Antitrust Division lodged a suit against Jersey, charging monopolist arrangements with Farben in many patents. Within the context of the antitrust suspension urged by Ickes, Thurman Arnold, Assistant Attorney General and head of the Antitrust Division, negotiated a settlement with Jersey Standard. On March 25, 1942, Jersey signed a consent decree with the government releasing all I.G. patents, agreeing not to enter into any future contracts in restraint of trade, and paying a minimal fine of $50,000.[53]

Jersey had thought this settled things, but on the next day, March 26, Arnold publicly charged that Jersey's "cartel arrangements with Germany are the principal cause of the shortage of rubber."[54] While Arnold negotiated with Jersey, Senator Harry S. Truman's Committee on National Defense began an investigation of the defense industry. The Truman Committee rubber hearings, held from March 5 to 24, 1942, highlighted the frustration that many American rubber firms had encountered trying to manufacture synthetic rubber prior to the war. Testimony placed much blame on government, but also charged that Standard's monopoly with I.G. Farben for butadiene technology had hindered the U.S. war effort. Standard had opted not to build a rubber capacity on its own, and had refused to license others.[55]

HOUDRY TAKES THE OFFENSIVE

In the midst of Jersey Standard's troubles with the Justice Department and the Truman Committee, its competitors led a fight to have their various processes included in the Rubber Reserve Program. Under the program, designated companies would construct plants for the Defense Plant Corporation, operate them, and sell rubber to Rubber Reserve on a cost-plus, fixed-fee basis. Houdry Process Corporation, Sun, Socony, and Standard of California lobbied for the Houdry butane process, and various chemical companies and distilleries for the Carbide and Carbon alcohol process to make butadiene.[56]

In May and June, 1942, Sun Oil submitted two separate plans for construction of a 50,000 ton butadiene plant using the Houdry dehydrogenation of butane and butene. Socony had joined together with four other oil companies to form the Neches Butane Products Company, one of the groups awarded a $650,000,000 contract by the Rubber Reserve Corporation. The Neches Company had decided in January to use the Jersey butylene process, but began investigating the Houdry system in February at Socony's urging. In the meantime, the Houdry camp, including H.P.C., Sun, and Socony, prepared a campaign to convince the government and the public of the advantages of the Houdry process.[57]

By early July, Sun learned that the Rubber Reserve Corporation planned to award the firm a butadiene contract, but to ignore all other applications for Houdry units, including those of Socony and Standard of California. This necessitated that Sun should "retire to the sidelines" for the moment and let H.P.C. and Socony lead the fight. The group went ahead, following a detailed plan including newspaper advertisements, radio broadcasts from friendly commentators (Lowell Thomas and Fulton Lewis), and a Washington press conference.[58]

On July 6, 7, and 8, 1942, H.P.C. ran national newspaper advertisements telling their side of the rubber story. Their titles, "Houdry Has It—The Key to Synthetic Rubber" (picture of a key), "America Needs Rubber—Here's the Fastest, Cheapest Way to Get It From Petroleum," (pictures of tanks, planes, jeeps), and "The American People Has a Right to Know the Truth About Synthetic Rubber," capture the flavor of the campaign. Fulton Lewis's radio braodcast of July 7, and Lowell Thomas's Sunoco program of July 8 added to the well-planned blitz. H.P.C. later admitted spending $46,000 on the July advertising.[59]

The advertising campaign had some immediate results. It brought a wide public reaction and greatly irritated the officials at the Rubber Reserve Corporation. Fulton Lewis reported in his July 7 broadcast that Donald Nelson of the W.P.B. was going to assume authority over the rubber program and "go heavily" into the Houdry process. On July 8, H.P.C. received an invitation from Senator Guy Gillette, Chairman of the Agriculture and Forestry Subcommittee investigating industrial alcohol and synthetic rubber, inviting the Houdry group to present their side of the case in Washington. On July 13, Eugene Houdry, Arthur E. Pew, Jr., Clarence Thayer, and Socony's Wilbur S. Burt appeared before Gillette's Committee. They testified to their practical experience and argued that cost benefits could be obtained with the Houdry butane process. Eugene Houdry also criticized the Jersey process and emphasized the need for butylene to manufacture aviation gasoline.[60]

On July 15, 1942, Thurman Arnold summoned J. Howard Pew and Franklyn Waltman, Sun public relations director, to a meeting at the Justice Department. Arnold sought information from Pew to use against Jersey Standard in light of the recent rubber campaign. Convinced that Jersey was intent on creating a rubber monopoly as it had in petroleum, Arnold contemplated initiating further antitrust action.[61]

J. Howard Pew told Arnold that the Gillette Committee testimony contained his group's position, and that the Houdry advertising program was not an attack on Standard, but an attempt to bring their process to public attention. He professed a personal friendship with William S. Farish and urged Arnold to make a distinction between the general business practices of a company and certain patent policies. Pew stated that "Mr. Farish had rendered a great service to the country, back in the days of the N.R.A., when Mr. Farish so solidly supported the Pews and others who undertook to resist the N.R.A. restrictions on price control."[62] Howard emphasized Eugene Houdry's patriotic motives as well as the Frenchman's belief in the superiority of his process. Arnold expressed the view that Standard interests dominated the Rubber Reserve, and both he and Pew agreed that an impartial decision should be made on the technical merits of the butadiene processes. The minutes of the meeting indicate that J. Howard Pew made no substantive charges against Standard Oil, but he did ask that his name not be used publicly.[63]

During the course of this meeting, Thurman Arnold expressed his admiration for the Sun Oil Company many times. At one point he said that "The 'Pew Company' was one of the few oil companies that shunned combinations and agreements to restrain competition."[64] Arnold was a complex person, and one should not make too much of isolated quotes, but there is evidence that he had indeed gained respect for the Pews during the T.N.E.C. investigations.[65]

Arnold had perceived the Pews' role in the rubber controversy as another step in their competitive struggle with the large oil interests and had assumed that they would be willing to help him in his investigations. Sun's managers were upset with the power Jersey Standard wielded in Washington, but were ambivalent on how to counter it. J. Howard Pew wrote to J. Edgar Pew on July 16 that "almost every position down there is so completely plugged with a group who have sworn to check mate the Houdry group at every turn irrespective of merit, and some of whom are most unprincipled, and who will of course exert a powerful interest."[66] Despite J. Edgar's urgings to continue the fight, Howard had become weary of the struggle. He felt that Sun had done its duty by bringing the matter to everyone's attention, and wished to let the matter drop.[67]

On July 17, however, J. Howard wrote a strong letter to Donald Nelson criticizing the inexperience of the Rubber Reserve's Stanley T. Crossland and attacking the government's choice of Jersey Standard men for key technical positions in the rubber program. He reiterated this charge in subsequent correspondence with others.[68] In the final analysis, however, Sun's and the Houdry group's efforts were to no avail. As a result of the public clamor and the Gillette hearings, Congress passed the Rubber Supply Act of 1942 in early August, but President Roosevelt vetoed it on August 6. Roosevelt then appointed an independent commission to study the problem and report. Chaired by Bernard M. Baruch, the other two members were James B. Conant, president of Harvard, and Karl T. Compton, president of M.I.T.[69]

The Baruch committee submitted a report on September 10, 1942, which endorsed the Rubber Reserve recommendation without reservation. DuPont would make 40,000 tons of neoprene per year, Jersey Standard 132,000 tons of butyl, and there would be an annual target of 705,000 tons of Buna-S rubber. The Jersey Standard butylene process would make the lion's share of 283,000 tons per year, and the rest would be divided up among the alcohol process, the Phillips process, the

Houdry process and another Jersey process, thermal or refinery conversion. Sun's authorization was for one small 16,500-ton-per-year Houdry butadiene plant, the only concession to the Houdry group. Sun eventually received funds to construct a 15,000-ton plant at its Toledo refinery, and Standard of California built a similar 15,000-ton facility in El Segundo, California. Both firms sold these plants back to the government after the war.[70]

The Houdry group's failure to obtain a larger share of butadiene production paralleled its efforts in selling fixed-bed catalytic plants. The Jersey Standard process for the catalytic dehydrogenation of butylene was an adjunct of its fluid catalytic process. Similarly, the Houdry butane dehydrogenation process was a direct spin-off of the fixed-bed cracking system. Both Jersey processes were continuous operations, while the Houdry processes were semi-continuous, requiring the refinery flow to be diverted while catalyst cases were purged of carbon. Thus, the government's decision to finance fluid over fixed-bed cracking doomed the Houdry group's hopes to be a major butadiene producer. Although the Houdry butane process was efficient (H.P.C. reported yields of about seventy percent butadiene from feed stock), the Baruch Report had concluded that the Jersey process would be a quicker and more economical way to produce butadiene.[71] This decision was correct in terms of plant construction costs and the relative volume produced from the competing processes.

A non-economic argument that Eugene Houdry had made before the Gillette Committee, however, proved to be prophetic. As the butadiene program accelerated in 1943–1944, demand for butene gases to manufacture alkylates needed for high octane aviation fuel increased dramatically. Consequently, Jersey Standard's refinery gases went largely for this purpose rather than the manufacture of butadiene, and increasingly larger amounts of butadiene came from the alcohol dehydrogenation process developed by the Carbon and Carbide Chemical Corporation. Almost two-thirds of all butadiene produced for synthetic rubber in World War II came from alcohol rather than refinery gases.[72]

AFTERMATH

An interesting result of the rubber controversy was the effect it had on the public relations efforts of the two main protagonists, Jersey Standard and the Houdry Process Corporation. In September 1942, Jersey transferred Robert T. Haslam, formerly manager of Esso Marketers, to the parent company headquarters to head up a Standard public relations counterattack. Historians Larson, Knowlton, and Popple give Haslam credit for reconstructing Standard's image following the I.G. Farben investigations and the Arnold antitrust onslaught.[73] The following May, Sun's Franklyn Waltman proposed a major public relations and promotional program for H.P.C. At the time, Waltman was working for both Sun and H.P.C. as public relations director. Waltman pointed out that the rubber situation showed once and for all that "the mousetrap theory is as obsolete as a Model-T Ford." If Houdry wished to become better known, it would have to take its case to the public.[74]

Hard feelings still existed between the Houdry and Standard groups, and events of early 1943 provided Robert T. Haslam with ammunition to use against Sun. On January 6, 1943, Congressman Martin Dies, Chairman of the House Un-American Activities Committee, announced on the floor of the House that he would provide information that Justice Department officials, a "bevy of interlocking left-wing and pseudo-liberal organizations," and a competitor company had joined together to ruin a major industrial company. Dies was referring to people in Arnold's antitrust division, Sun Oil, and Communist groups, all conspiring against Standard Oil of New Jersey.[75]

Dies's shotgun blast was an attempt to tie the strong campaign that had developed against Jersey Standard to his growing investigation of Communist influence in the United States. There had been a barrage of protests in the press during the winter and spring of 1942 following the publication of the news of the Standard-I.G. Farben connection. The efforts culminated in a series of open letters to John D. Rockefeller, Jr., Standard's largest stockholder, from liberal journalist I.F. Stone that appeared in the newspaper *PM* in April 1942.[76] Dies's implied references to Sun Oil referred to a press release issued on January 30, 1942, by the "Union for Democratic Action," a left-wing group associated with the American League for Peace and Democracy (formerly the American League against War and Facism). In its statement, the group attacked Standard's I.G. Farben connection, with particular reference to synthetic rubber patents. At the time, Jersey chairman William Farish concluded that these charges were tied to the efforts of "two men representing the Free French Organization" who were promoting "vicious accusations and propaganda" against his company. Farish concluded that Eugene Houdry, an active supporter of French causes in this country, was behind it all, and he approached J. Howard Pew and John Brown of Standard of California to see if they could put a stop to it. Reassurances were given and the episode died down, only to be resurrected by Martin Dies in January 1943.[77]

An internal Sun investigation into the Dies charges concluded that where there was so much smoke, there must be a little fire. The blaze was Eugene Houdry. Houdry had hired Sanford Griffith of Market Analysts and an associate, Francis Henson, to do a study of Standard Oil's publicity campaign for fluid cracking back in February 1941. Henson was associated with the Union for Democratic Action, and it was these activities to which Farish had objected. Eugene Houdry later argued that his efforts were only defensive and designed to protect himself from the public criticism that Standard Oil Development's Frank Howard had given to the "obsolete" Houdry process. Houdry had met Griffith in relation to his work as president of "France Forever," a patriotic group of which he was president. Sun and Houdry maintained that Griffith had been terminated in August 1942, and that they had no connection with the Union for Democratic Action. Moreover, both companies denied that they had conspired with anyone on Thurman Arnold's staff. J. Howard Pew and John A. Brown again successfully intervened to restore amicable relations.[78]

H.P.C.—INTERNAL CONFLICT

Eugene Houdry believed that Jersey Standard had stolen his basic process, and, had it not been for the war, he would have pressed the point. Moreover, as the conflict developed, the fixed-bed catalytic process encountered a challenge from the Socony T.C.C. moving bed process. Although Socony owned one-third of Houdry Process Corporation and licensed T.C.C. units through H.P.C. during the war, T.C.C. was a Socony project which Houdry never believed to be superior to his original units. In fact, Houdry fixed-bed units did not fare well against Fluid process and T.C.C. competition during the war. With the exception of two units sold to Russia in 1945 under Lend Lease, there were no new licensees of fixed-bed plants after 1942.[79] Eugene Houdry personally disassociated himself from the T.C.C. process and concentrated his efforts on an improved fixed-bed system, the Adiabatic Cracking Process, and an improved continuous synthetic rubber process, neither of which became a commercial success.[80]

Friction developed between Houdry and Socony in 1943 regarding the licensing of T.C.C. units and Socony criticism of H.P.C. management. Socony was critical of both Houdry and Sun's lack of long-range planning and the failure of H.P.C. to provide the type of research and development that it had promised. This last point related to what had been a continuing thorn in the side of H.P.C. Eugene Houdry and his French associates had always pushed for higher dividends on their initial investment, while Sun and Socony opted for a greater retention of earnings to be put into further research.[81]

Things did not improve the following year. Socony accused Eugene Houdry of criticizing the T.C.C. process in front of prospective licensees, and Houdry disapproved completely of Socony's new synthetic bead catalyst for the T.C.C. units. Furthermore, Eugene Houdry's relations with Sun degenerated rapidly. Arthur E. Pew, Jr. charged that H.P.C.'s research efforts were totally inadequate, and that Nicholas deRachet, head of H.P.C.'s foreign department, was a "crook." J. Howard Pew had declared deRachet *persona non grata* at the Sun Oil Company. Things reached a head in August 1944, when Sun and Socony pressured Eugene Houdry to resign as president of H.P.C. He remained on as board chairman and a member of the executive committee, but his duties were to be primarily in catalytic research and not management.[82]

Arthur V. Danner, formerly with Socony, became executive vice-president of H.P.C. and succeeded Houdry as chief operating officer. He instituted better relations with Socony, but encountered deep splits within the Houdry organization. H.P.C. had a large group of people around Eugene Houdry who supported the adiabatic fixed-bed process over the T.C.C. He was unable to resolve any of these issues during the war, and another major reshuffle of H.P.C. management occurred in 1947-1948.[83]

The Houdry group had begun the war leading the industry in catalytic cracking. By the end of the conflict, Houdry Process Corporation was in disarray, and the Jersey Standard group had gained catalytic leadership with their Fluid process. Part of this is explained by natural market forces. The early leader in any technology often becomes handicapped by his initial leadership. New entires into the field invest in the latest technological adaptations, and the original innovator must reinvest massively to keep up. Houdry was also up against an awesome competitor in the Jersey group. Furthermore, the war itself was an overwhelming factor. Patent pooling arrangements under P.A.W. supervision limited Houdry profits, and huge government investment in Fluid process units rapidly made the Houdry process obsolete.

Internal problems also contributed to the Houdry group's decline. Eugene Houdry was another great inventor who lacked business acumen. He was wedded to his own ideas and resisted change. Sun management also exhibited a limited outlook and stuck to the fixed-bed process too long. In the final analysis, however, Sun Oil received much more from the Houdry units than did Houdry himself. Catalytic cracking enabled the firm to become a major producer of aviation fuel under government contract, and to return to marketing high octane, non-leaded motor gasoline in the post-war period.

SUN SHIP

The production of aviation gasoline was one of two areas in which the Sun organization excelled in World War II. The other was shipbuilding.

A creature of Sun's expansion and the need for tankers during World War I, this subsidiary emerged during the second great world conflict as the nation's largest producer of oil tankers. This work and other contracts made it the largest privately-owned shipyard in the world. The Chester yard had eight shipbuilding ways in 1940 plus two drydocks installed in the 1920s. In 1941, it increased this number to twenty, and in 1942 to twenty-eight. From Pearl Harbor through June 1946, Sun Ship constructed 229 oil tankers, 32 freighters, and 35 car floats for the Sun Oil Company, other private companies, and the United States Maritime Commission.[84]

In the beginning of the war, the yard employed 9,000 to 10,000 men, but the labor force increased as it added more shipbuilding ways. At the peak, there were more than 35,000 people employed at Sun Ship. As part of the 1942 expansion, Sun employed 8,000 blacks from the Chester area to work in the new ways. This provided Sun Ship with a needed labor force and gave the firm an opportunity to gain good publicity in the Negro press. This was welcome from a public relations standpoint, since there had been labor difficulties at the shipyard since the late 1930s. Although Sun was able to maintain its non-union policies in the oil business, it was forced to reach an accomodation with the C.I.O. at the shipyard in 1943. After a National Labor Board-administered election in Chester, Sun Ship signed a contract with Local No. 2 of the Industrial Union of Marine and Shipbuilding Workers of America (affiliated with the Congress of Industrial Organizations).[85]

In 1940, prior to Pearl Harbor, the yard had already begun to expand due to lucrative government contracts. Sun constructed three destroyer tenders and three seaplane tenders for the Navy at a cost of $78,000,000. In addition, the firm worked on contracts for eight cargo vessels, four passenger freighters, and thirty-two tankers amounting to $114,000,000. In March 1941, however, the government decided that Sun Ship's greatest contribution to the war effort would be in tanker construction, a field in which it had pioneered. J. Howard Pew and Sun Ship representatives met with Navy Secretary Frank Knox and Navy representatives to discuss the shipyard's future. As a result of this meeting, Sun Ship received contracts for $350,000,000 to construct tankers, and government assistance in financing yard expansion on a fifty-acre site in Chester.[86]

Sun Ship had constructed the world's first all-welded ship, a small tanker, in 1931, and in 1937 had launched a large ocean-going tanker of similar construction. This was the prototype for the famous T-2 tanker, a ship specifically designed by Sun Ship for the Maritime Commission. The World War II T-2 was of 17,000 deadweight tons and held 5,500,000 gallons of gasoline. In the 1930s, Sun had also pioneered in the sub-assembly technique of building large ship sections separately under cover and later assembling them.[87]

The sub-assembly method was a compromise between the traditional tailor-made construction approach and the mass production methods employed by Henry J. Kaiser on the West Coast during the war. Sun used several ship ways and assembled units of from ten to fifty tons which had previously been built in shops. Kaiser assembled much larger units, from fifty to one hundred tons, totally outdoors and on fewer ways. His method required larger cranes and auxiliary equipment to keep the ships moving through the shipyard. In May of 1943, Sun was completing six ships a month to Kaiser's three, but his mass production techniques improved rapidly in the last two years of the war.[88]

The shipyard also expanded its facilities for constructing refinery equipment, an operation it had begun in the 1920s. Much of this refinery work was in fabricating T.C.C. reactors for the Lummus Company, a refinery construction specialist working under contract for Socony and H.P.C. Because Sun Ship was doing so much of this work, the Sun Oil Board of Directors discussed a possible merger of Sun Ship with Sun Oil rather than continuing it as a wholly-owned subsidiary. Sun's attorneys had advised that the Sun Oil Certificate of Incorporation was not broad enough to include the manufacture of refinery equipment. The board did not vote for merger, but amended the Sun Oil Company Charter to include the construction of "drills, derricks, pumps, tanks, stills, cracking plants, refineries, machinery, and apparatus of every kind of description."[89] This was a largely procedural move designed to eliminate bottlenecks in negotiating construction contracts for the Sun shipyard.[90]

Although Sun experienced some delays in receiving payment for ships under government contract, the shipyard's profitable operations during the war fully justified the parent firm's decision to keep the facility in operation during the lean interwar years. During its peak operating year of 1944, the shipyard did a gross business of more than $275,000,000, which resulted in a new income after taxes of $3,301,000. This represented a significant share of Sun's consolidated net income of $13,350,000.[91] Sun Ship's major contribution to Allied Victory was also highly profitable for the Pew interests.

WARTIME FINANCING

Although the government financed several aviation gasoline plants through the Defense Plants Corporation, a division of the Reconstruction Finance Corporation, the Pews were opposed to this method of operation. In order to receive financing, a firm had to agree to federal ownership of the facility, and the company would then lease it from the government.[92] This is how Sun financed the small butadiene plant built at the Toledo refinery. In the ten-year period since 1933, when Sun first formed an association with Houdry, the firm had spent over $11,000,000 in the development of the Houdry commercial process. From 1940 through 1943, Sun spent an additional $20,000,000 constructing new and converted facilities for the aviation gasoline program. In exchange, it received supply contracts with the government to produce virtually all of the high octane gasoline that it could. In January 1942, it did negotiate a $5,500,000 loan from the Defense Supplies Corporation at two percent interest to assist construction of the Marcus Hook catalytic-alkylation plant.[93]

By the late summer and fall of 1943, however, Sun was in a difficult credit position. The government owed over $47,000,000 to Sun: $23,000,000 to Sun Oil and $24,000,000 to Sun Ship.[94] Sun management had always followed conservative financial policies, but it now argued that the situation had extended its credit "to a point where we can go no further without violating good business practices."[95] Since the outbreak of the war, Sun had increased its investment in accounts receivable and plant facilities by some $60,000,000. It realized the impossibility of obtaining payment from the government for the time being, and reached an agreement with the Defense Supplies Corporation to finance a $3,200,000 T.C.C. unit and expansion of the alkylate unit at Marcus Hook. The firm never erected the facility, however, and thus did not spend government funds. Sun maintained its record of privately owning all of its aviation fuel equipment.[96]

In 1944, the company constructed two T.C.C. cracking units and a four-unit aviation lubricating oil plant. These installations brought the total amount invested in refinery facilities related to the war effort to $31,000,000.[97] On December 16, 1944, J. Howard Pew informed Harold Ickes that Sun Oil's Marcus Hook refinery had blended its billionth gallon of aviation gasoline sent to the armed forces since Pearl Harbor. About seventy-five percent of this amount was 100 octane fuel, and the remainder was equally divided between 91 and 87 octane gasoline. Sun's production of high octane aviation fuel as a percentage of its total refinery output led the industry. In the last six months of World War II, Sun shipped more 100 octane aviation gasoline than any other refiner in the country.[98]

Sun leadership in aviation gasoline showed up in the company's profit figures for the war years. Gross operating income rose steadily from $131,500,000 in 1939 to a peak of $600,800,000 in 1944. Net income from 1939 to 1945 averaged $11,800,000, with a high figure of $16,500,000 in 1941. Except for 1942, when Sun retained only 37.5 percent of earnings, the company continued its traditional policy of high reinvestment. Sun plowed back funds into many projects, but most of this investment went into aviation fuel facilities and tanker construction.[99] In 1945, the firm was in excellent shape to launch an expanded peacetime marketing program.

DEMOBILIZATION

When the Petroleum Administration for War and the Petroleum Industry War Council commenced operations in the early years of the war, J. Howard Pew had made Sun's position clear. It was willing to cooperate for the duration of the war, but wanted the wartime agencies dismantled once peace had returned. During the course of the war, however, many industry and government leaders became impressed by the way the industry had worked together for common goals. At its October 24, 1945 meeting, the P.I.W.C. appointed a committee to "consider and recommend a program designed to preserve the cooperative industry spirit of unity brought about by our war experience."[100]

The committee sent a questionnaire to all P.I.W.C. members soliciting their opinions. J. Howard Pew's answers were brusque and to the point. He commented that "it would be a mistake to try to continue the Petroleum Industry War Council into peacetime," and that he "can visualize no activities which the industry should undertake which cannot be better handled by the American Petroleum Institute." His answer to a query about establishing an international United Nations oil group was "no."[101]

Harold Ickes remained true to the promise given to J. Howard Pew in 1942. Within one month after V-J Day, the P.A.W. had revoked almost seventy-five percent of the sixty recommendations, orders, and directives in effect at the war's end. Ickes terminated all domestic orders by October 15, 1945, and all orders affecting foreign operations by November 1. The P.A.W. staff, which had reached a peak of 1,438, began shrinking even before V-E Day. By December 31, 1945, it was down to 161 and by April 1, 1946, to 58. In the midst of demobilization, Ickes and President Truman split on Truman's appointment of a California oil man as Under-Secretary of the Navy. The old New Dealer resigned in February, and on May 8, 1946, Truman signed an order terminating the P.A.W.[102] Although the formal structure for industry regulation was ended, most oil men had learned to accept a strong, active role of government in the various activities of the oil industry.

In Sun's 1945 report to stockholders, J. Howard Pew praised the speedy termination of government controls and lauded the P.A.W. for fulfilling its promises. But he strongly opposed the continuance of price regulation by the Office of Price Administration (O.P.A.). Pew blamed inflation on the high national debt and argued that its solution lay in increased production throughout the economy. He also maintained that the O.P.A. was an obstacle to the reorganization and extension of Sun's marketing. Howard carried his fight into the public arena and became one of the most outspoken opponents of price controls in the post-war period.[103] In reply to a radio address by C.I.O. spokesman Walter Vaughn in support of O.P.A., Pew refuted monopoly and labor exploitation charges by citing high wages rates. In his speech over the ABC radio network, Pew further argued that the O.P.A. was not in the best interests of the consumer:

> All this is still the old NRA fight that aroused our opposition thirteen years ago and will keep us protesting until the hypocritical fallacies are ended for good, until the discredited Blue Eagle . . . is buried too deep for another resurrection, and until freedom, under God, is restored to the American people.[104]

Pew's attack on the O.P.A. was not unique. Barton Bernstein points out that businessmen had always mistrusted the Price Administration because of its "New Deal flavor" and the nature of its operations. By comparison, most executives representing the large oligopolistic industries were far more comfortable with the War Production Board (W.P.B.) and similar agencies like the P.A.W., at least up to the fall of 1944 when it appeared that the Allies were indeed winning the war. These production-oriented agencies were largely staffed by businessmen and fostered goals with which the private sector could more readily identify.[105] What was different about J. Howard Pew's position (and also consistent with Sun's past policies) was that the O.P.A. represented a threat to Sun's distinctive marketing strategy.

Sun had begun planning its post-war marketing as early as the spring of 1944. Samuel Eckert presented a report to the board of directors entitled "Analysis of Motor Gasoline Business of Sun Oil Company—Pre-War, Present and Post War" in May. At the same meeting, the board authorized the expenditure of $2,000,000 for the marketing department to develop outlets for the expected excess supplies of gasoline in the post-war period. Sun's aviation gasoline facilities would enable the company to return to its earlier policy of selling one grade of un-leaded gasoline. Arthur E. Pew, Jr., announced the successful production of Sun's new product, "Dynafuel," in September 1945, and the company introduced it to the public in November.[106]

Earlier, Sun had objected to N.R.A. price-fixing because of its harmful affects on Blue Sunoco marketing; it now opposed O.P.A. for similar reasons. The war had forced the firm to use lead and to submit to O.P.A. price regulations, but now Sun wanted to see prices return to a free market operation. In this environment, the Sun leaders believed that they could still sell their product more cheaply than could their competitors. Economic as well as philosophical considerations continued to motivate the Pews' opposition to price-fixing.

THE ANGLO-AMERICAN OIL AGREEMENT

One other aspect of post-war planning illustrates how the Pews reacted to business-government arrangements that ran counter to their philosophy and economic interests. This was an international agreement negotiated between the United States and Great Britain in 1944 which defined the interests of the two countries in Middle East oil, and arranged for future collaboration on pipeline construction. J. Howard Pew publicly denounced the agreement as an international cartel and successfully led a fight against it. Although the treaty was re-drafted and submitted again following the war, it ran into continued opposition and the Senate never ratified it.[107]

Early in the war, the P.A.W. had created a Foreign Operations Committee to oversee the activities of American oil companies overseas. Its operations were limited until 1943, when military representatives and government officials became concerned with shortages. Upon Harold Ickes's recommendation, President Roosevelt created the Petroleum Reserve Corporation (P.R.C.), a division of the Reconstruction Finance Corporation, in June 1943, to acquire and develop new foreign reserves. Many members of the P.I.W.C. opposed the new agency on the grounds that the federal government should not be involved in foreign oil development. The Independent Petroleum Association also opposed the program. When Harold Ickes announced a joint agreement with Gulf Oil, Aramco, and the P.R.C. to build a government-owned pipeline from Saudi Arabia to the Mediterranean, continued opposition from within the P.I.W.C., coupled with Congressional resistance, forced the Petroleum Administrator to temporarily abandon the plan.[108]

In addition to speaking out in opposition within the P.I.W.C., Sun used other means to lobby against the Middle East proposal. In 1940, J.N. Pew, Jr. had purchased a controlling interest in *The Pathfinder*, a weekly news magazine. He quickly molded it into a politically conservative organ which, along with *The Farm Journal,* was a vehicle for Pew's own brand of Republican philosophy. J.N. had written to Robert W. Howard, editor of *The Pathfinder,* about the Arabian pipeline deal and suggested that the magazine run a story in its next issue. Pew was concerned that the project was going out of Ickes's hands and into those of Secretary of State Cordell Hull. The *Pathfinder* story warned of the dangers in Middle East politics, Zionist-Arab tensions, and United States plans for a naval base in Haifa. It also stated that "Government concessions to big oil companies is also a kick in the teeth for independent oil operators back home."[109]

Although there was an economic motive for Sun's opposition in that they were a "have not" in foreign oil, the Pews had been consistently opposed to international cartels. On May 16, 1944, for example, J.N. Pew, Jr. authorized Sun's representative, John D.M. Hamilton, to negotiate with the Turkish ambassador for oil drilling concessions in Turkey. Hamilton was one of Sun's attorneys, a former National Chairman of the Republican Party, and a personal friend of J.N. Pew, Jr., the man who had furthered his political career. J.N. informed Hamilton of Sun's interest in Turkey, but stated that

> We would negotiate as a private contractor with a free and sovereign nation for the development of petroleum resources under fair and proper laws governing free enterprise. We would not be interested in any arrangement which involved the participation of the United States government in any shape or form. We would not participate in any contract which would involve, nor (at) any time in the future, any monopolistic practices or any participation in cartels or quotas either between private companies, between governments, or both.[110]

This letter may have been written to impress the Turkish ambassador, but it also indicates the strong stand Sun had taken against multinational firms in 1943-1944.

The British government, concerned with what appeared to be a growing American imperialist threat, had also opposed the P.R.C. deal with Gulf and Aramco. Industry and government officials who wished to revive the project realized that they would have to reach an accord with Great Britain. President Roosevelt announced the creation of an Anglo-American conference in March 1944, and technical discussion began with industry advisors and government officials of both countries. The two sides hammered out an agreement in the spring of 1944, and Lord Beaverbrook arrived in Washington to sign the treaty on August 8.[111]

The agreement immediately ran into a storm of protest. On August 17, 1945, J. Howard Pew directed an open letter to Senator Tom Connally, Chairman of the Senate Foreign Relation Committee, attacking the vague wording of the document. Pew also raised questions about the treaty's implications for price-fixing and oil production controls. He labeled the agreement part of the "evil cartel system," and called it the antithesis of the Interstate Compact—established to eliminate waste and not to specifically fix prices.[112]

Opponents of the treaty strongly criticized specific provisions. Article I rather vaguely stated that petroleum supplies should be obtained with "consideration to available reserves, sound engineering practices, relevant economic factors, and the interests of the producing and consuming countries," without defining those interests. Article II established an International Petroleum Commission to estimate demand, recommend policies, and provide analyses. Article IV stated that the governments would, "in accordance with their respective constitutional procedures," endeavor to carry out the recommendations of the Commission.[113]

J. Howard Pew continued his onslaught against cartels in general and the Anglo-American treaty in particular. In a statement before the P.I.W.C. on October 24, 1944, he introduced a related issue by arguing that the treaty was "a deliberate attempt to place the American petroleum industry under the bureaucratic control of the federal government."[114] Howard insisted that government ownership of the petroleum industry would grow out of the agreement and he demanded to know the real relationship between this treaty and other international agreements, including the Bretton Woods (monetary agreements), Dumbarton Oaks (United Nations), and commodity trade agreements.[115]

It appeared that Pew and the other treaty opponents had carried the day when the P.I.W.C. met on October 24, until Chairman William R. Boyd, Jr. read a letter from Harold Ickes asking the group to draw up their objections in specific language. "This," Ickes said, "[is] in the interest of the creative and helpful approach, as opposed to a purely destructive one."[116] The P.I.W.C. complied with Ickes's request and put its National Oil Policy Committee to work drafting a special revision of the treaty. On December 6, the P.I.W.C. met again in Washington, and the Policy Committee submitted its new version.[117]

Both Dean Acheson of the State Department and Harold Ickes attended the meeting to urge the P.I.W.C. membership to submit terms acceptable to the oil industry. Before the P.I.W.C. took a vote, however, J. Howard Pew rose to take exception to the entire proceeding. He argued that neither the P.A.W. nor the P.I.W.C. had the right to discuss post-war issues, and that the clearance Ickes had secured from the Justice Department was improper. He then read a statement which John D.M. Hamilton had prepared. It maintained that only the President and the Senate had the Constitutional authority to negotiate and ratify treaties, and that the public would rightly interpret these oil industry amendments as representing the influence of special interests. Any individual had a right to make known his feelings during public hearings, Pew argued, but the P.I.W.C. should not speak for the petroleum industry. Howard then asked to be registered as not voting, and five others followed his lead.[118]

The P.I.W.C. voted to submit the proposed changes to various government departments, but by this time general treaty opposition had become very strong. Senator Connally now opposed the agreement, and

President Roosevelt recalled it from the Senate on January 10, 1945. Work continued throughout the winter and spring of 1945 to draft a new version to take to the British. The Pews maintained their opposition to the "mischief in post-war planning" during this period,[119] but government and industry reached a compromise in June, and Ickes headed a delegation to London in September to negotiate with the British. The majority of the A.P.I., P.I.W.C., and Independent Petroleum Association membership supported the revised treaty which resulted from six days of talks in London. But while the treaty was in committee, opposition developed from western oil producers, and the Senate took no action after it was introduced on the floor in 1947. After months of controversy, the treaty was finally defeated.[120]

Sun Oil's opposition to the Anglo-American Treaty is interesting in light of revisionist studies of United States diplomacy in World War II. In his *Politics of War,* Gabriel Kolko argues that one of the wartime aims of the United States was the creation of a new world economic order with this country at its head. To accomplish this, America had to defeat fascism and neutralize Stalinism, but perhaps just as importantly, it had to supersede Great Britain as the economic leader of the Western, capitalist world. Kolko suggests that U.S. oil policy revealed the true aims of American policy. The British were conscious of American imperialism in Middle East oil, but were powerless to prevent it. In this interpretation, the Anglo-American treaty proposal was an extension of this attempt to gain a competitive advantage over all other industrialized nations, including Great Britain.[121]

There is no question that agencies of the United States government aided American private oil interests in the Middle East. The Arabian pipeline and the Anglo-American Treaty are evidence. Moreover, when this aid extended into the area of post-war planning, it is difficult to explain it as an example of wartime emergency. But the industry and the government had largely acted from their own experience and knowledge of the past. They had remembered Anglo-French attempts to freeze the United States out of Mesopotamia in 1920, and were determined that the United States and American companies would have a share of the pie after this war. But regardless of one's interpretation of P.I.W.C. support of the British agreements, Sun's outspoken opposition clearly demonstrates that all large oil companies did not think alike.

THE WAR EXPERIENCE

During World War II, Sun Oil cooperated with government agencies and the private sector of the petroleum industry to attain the most productive use of America's oil resources. The firm also earned a lot of money in the process. The successes of the P.A.W. and the P.I.W.C. had firmly demonstrated to the major integrated firms that government and business could work well together to achieve mutual goals. This awareness had not come easily. The roots of business-government cooperation in oil are found in similar coordinating activities in World War I, the beginnings of the unitization and regulation movement in the 1920s, and, most importantly, in the regulatory activities that emerged from the New Deal experience of the 1930s.

The flurry of antitrust prosecution that appeared on the eve of the war did not represent the mainstream of United States oil policy. It is worth noting that those antitrust cases suspended for the duration were not seriously revived in the immediate post-war era. The dual goals of conservation and economic stabilization through government regulation remained the dominant theme of United States oil policy through the 1950s and 1960s. Thus, World War II represented an important culmination and synthesis of developments that had evolved over three previous decades. But again, the Sun experience is a story of cautious conservatism and reluctance to enter fully into the new era of business-government cooperation.

The Pews were skeptical about the role of the Petroleum Administration for War, and only a personal promise from Harold Ickes that he would disband the structure at war's end enlisted their full participation. In 1945, Sun's managers were among the first to demand a return to a free market, and they opposed those oil executives who saw benefit in continuing the P.A.W. in peacetime. J. Howard Pew's attack on cartels and his opposition to the Anglo-American Middle Eastern treaty were consistent with his firm's previous stance in the industry.

Sun also went its own way in the conduct of its business during the war. The firm continued its conservative financial policies during wartime and refrained from seeking government financing for most of its expanded refinery capacity. Sun's competitive position in the industry was shown by its continued use of Houdry fixed-bed units over the fluid catalytic process, and by its agressive activities during the rubber crisis of 1942–1943. The Pews remained cynical concerning Standard Oil of New Jersey's influence in the P.A.W. and the Rubber Reserve, and perceived Jersey very much as a competitor company, even during a period of such great interindustry cooperation. The company's foresight in keeping its shipyard in business during the interwar years paid dividends when the facility became a major source of income during the war. J. Howard Pew's comparison of O.P.A. price controls with the ill-fated price-fixing schemes of the N.R.A. reflected both his opposition to government interference with the free market and his firm's own "Dynafuel" gasoline marketing plans for the post-war period.

Both during the war and after, Sun's owner-managers continued to use the rhetoric of *laissez faire* when commenting on the economy. However, their business practices and public positions on crucial policy matters during the first half of the twentieth century demonstrate that they believed a great deal of what they said.

NOTES

1. Harold F. Williamson, Ralph L. Andreano, Arnold R. Daum, and Gilbert G. Klose, *The American Petroleum Industry,* Volume II, *The Age of Energy, 1899–1959* (Evanston: Northwestern University Press, 1963), p. 748; John W. Frey, and H. Chandler Ide, *A History of the Petroleum Administration for War* (Washington, D.C.: U.S. Government Printing Office, 1946), p. 288.

2. Robert Sobel, *The Age of Giant Corporations* (Westport, Connecticut: Greenwood Press, 1972), p. 171.

3. The Frey and Ide volume (see note 1 above) chronicles the development of business-government cooperation during the war. For a more analytical treatment, see Gerald D. Nash, *United States Oil Policy, 1890–1964* (Pittsburgh: University of Pittsburgh Press, 1968), especially chapter 8, "World War II," pp. 157–79.

4. Williamson et al., *Energy,* pp. 597–8; Nash, *Oil Policy,* pp. 154–5; Arthur M. Johnson, *Petroleum Pipelines and Public Policy* (Cambridge, Mass.: Harvard University Press, 1967), pp. 286–300; Memo, Pipeline Divorcement, September 25, 1940, Accession #1317, Papers of the Sun Oil Company, Eleutherian Mills Historical Library (EMHL), Greenville, Wilmington, Delaware, File 71, Box 37; J. Edgar Pew to J. Howard Pew, February 14, 1941, February 15, 1941, Acc. #1317, EMHL, J. Howard Pew Papers, Box 43.

5. J. Howard Pew to Harry T. Klein, February 18, 1941, Acc. #1317, EMHL, J. Howard Papers, Box 43; J. Edgar Pew to Howard Pew, February 25, 1941, File 71, Box 88; J. Edgar Pew to Edwin S. Hall, Jersey Standard, February 26, 1941; J. Edgar Pew to J. Howard Pew, February 20, 1941, File 71, Box 37.

6. J. Howard Pew to Harry T. Klein, February 18, 1941, Acc. #1317, EMHL, J. Howard Pew Papers, Box 43.

7. Frey and Ide, *Petroleum Administration,* pp. 14–5; also see the correspondence between Attorney General Robert H. Jackson and Harold Ickes on the antitrust issue in Frey and Ide, Appendix 7, pp. 382–3; Nash, *Oil Policy*, pp. 159–60; Williamson et al., *Energy*, pp. 599–601; Johnson, *Petroleum Pipelines*, pp. 300–4; "Sun Oil Denies Anti-Trust Charges in Answer to 'Mother Hubbard Suit,'" Sun Oil Company News Release, July 23, 1946; "Facts Regarding Oil Anti-Trust Suit," August 5, 1946, Acc. #1317, EMHL, J. Howard Pew Papers, Box 43.

8. Frey and Ide, *Petroleum Administration,* pp. 58–62.

9. *Annual Report of the Sun Oil Company, 1940,* p. 2; J. Edgar Pew to J. Howard Pew, February 14, 1941, Acc. #1317, EMHL, J. Howard Pew Papers, Box 43.

10. Ray L. Dudley, "The President's Action," *The Oil Weekly,* Supplement (June 2, 1941); Jno. G. "Jack" Pew to J. Edgar Pew, June 4, 1941; J. Edgar Pew to "Jack" Pew, June 6, 1941, Acc. #1317, EMHL, File 71, Box 74.

11. J.N. Pew, Jr. to O.G. Saxon, Yale University Economics Department, October 4, 1941, Acc. #1317, EMHL, File 21-B, Box 3.

12. Nash, *Oil Policy*, pp. 160, 162–6; Harold L. Ickes to J. Howard Pew, June 10, 1941, National Archives Building, Department of the Interior, Office of the Secretary, Records Group 48, File 1-188, Petroleum Administrator (May 28, 1940–June 30, 1940); also see Frey and Ide, *Petroleum Administration*, Chapter 5.

13. Harold L. Ickes, *The Secret Diary of Harold L. Ickes,* Volume III, *The Lowering Clouds, 1939–1941* (New York: Simon and Schuster, 1954), p. 630.

14. J. Edgar Pew to Warren C. Platt, *National Petroleum News*, November 26, 1941, Acc. #1317, EMHL, File 71, Box 74.

15. Harold L. Ickes to J. Howard Pew, June 10, 1941, National Archives, Department of the Interior, RG 48, File 1-188; Kelsey, John L., "A Financial Study of the Sun Oil Company of New Jersey From its Inception through 1948" (M.B.A. Thesis, Wharton School, 1950), p. 41; Memo, Coordinator's District #1, October 14, 1941, Acc. #1317, EMHL, J. Howard Pew Papers, Box 83; "Outline of Sun Oil Company Traffic Activities, World War II, 1940-1946," Folder, "World War II," Library File, Sun Oil Company Library, Marcus Hook, Pennsylvania, pp. 1-5.

16. *Annual Report, Sun Oil, 1942*, pp. 1-2; Robert G. Dunlop, interview held at St. Davids, Pennsylvania, on October 31, 1975; "Outline of Sun Oil Traffic Activities," p. 4.

17. Excerpts from remarks of J. Howard Pew before P.I.W.C., October 7, 1942, Acc. #1317, EMHL, J. Howard Pew Papers, Box 83.

18. Excerpts from remarks of Harold L. Ickes, October 7, 1942, Acc. #1317, EMHL, J. Howard Pew Papers, Box 83.

19. *Ibid.*

20. Nash, *Oil Policy*, pp. 166-7; Frey and Ide, *Petroleum Administration*, pp. 39-45.

21. *Annual Report, Sun Oil, 1942*, pp. 1-2.

22. *Annual Report, Sun Oil, 1943*, p. 8.

23. "Somebody's Sweetheart Now," *Time* 40, 24 (December 14, 1942), p. 29; "Ickes and Oil Men Sing Mutual Praises," *Washington D.C. Star* (January 3, 1943), clipping in Acc. #1317, EMHL, J. Howard Pew Papers, Box 83.

24. Thomas Charles Longin, "The Search for Security: American Business Thought in the 1930's" (Ph.D. Dissertation, University of Nebraska, 1970), p. 267.

25. "Ickes and Oil Men Sing Mutual Praises," *Washington D.C. Star* (January 3, 1943), clipping in J. Howard Pew Papers, Box 83.

26. Dunlop interview; Nash, *Oil Policy*, pp. 167-70.

27. "Address by the Hon. Harold L. Ickes," October 27, 1943, Acc. #1317, EMHL, J. Howard Pew Papers, Box 83; "Sun Oil's Newest Plant for Aviation Fuel Dedicated," *Our Sun* 10, 1 (December 1943), pp. 1-7.

28. Harold L. Ickes to J. Howard Pew, March 18, 1944, Acc. #1317, EMHL, J. Howard Pew Papers, Box 83.

29. Nash, *Oil Policy*, p. 163; "Fuel for the Fighting Fronts Via 'Big Inch' and 'Little Big Inch'" (War Emergency Pipelines, Inc., 1944), Acc. #1317, EMHL, J. Howard Pew Papers, Box 80; Frey and Ide, *Petroleum Administration*, pp. 104-7.

30. J. Howard Pew to William S. Farish, May 19, 1941, Acc. #1317, EMHL, J. Howard Pew Papers, Box 80. Earlier, in speaking at the christening of the *American Sun* at the Chester shipyards, J. Howard Pew had urged the creation of a "two ocean" tanker fleet for the war. According to Pew, "it was sea power that in the end defeated Napolean" ("Tanker Sea Power Essential Says Sun Oil Co. President," *National Petroleum News* 32, 36 [August 7, 1940], p. 22).

31. Ickes, *Secret Diary*, III, p. 528; J. Howard Pew to Ickes, July 11, 1941, Acc. #1317, EMHL, J. Howard Pew Papers, Box 83.

32. "Fuel For Fighting Fronts;" Possible Pipeline Participants, July 22, 1941, Acc. #1317, EMHL, J. Howard Pew Papers, Box 80; "Pipelines Make Headlines," *Our Sun* 9, 3 (March 1943), pp. 16-7; Johnson, *Petroleum Pipelines*, pp. 322-6.

33. Possible Pipeline Participants, Acc. #1317, EMHL, J. Howard Pew Papers, Box 80; Williamson et al., *Energy*, pp. 764-6; Nash, *Oil Policy*, pp. 164-5; "Pipelines Make Headlines," *Our Sun*, p. 17.

34. Williamson et al., *Energy*, p. 789; "Story of Sun," *Our Sun*, 75th Anniversary Issue, 26, 3,4 (Summer/Autumn 1961), p. 30.

35. Board Minutes, Sun Oil Company, October 17, 1937, Minute Book #6, p. 219, Office of the Assistant Secretary, Sun Oil Company, Radnor, Pennsylvania; John Lawrence Enos, *Petroleum Progress and Profits* (Cambridge: M.I.T. Press, 1962), pp. 151–2; Dunlop interview; Board Minutes, January 22, 1941, Minute Book #7, p. 6.

36. Williamson et al., *Energy*, pp. 636–40, 787; Frey and Ide, *Petroleum Administration*, pp. 199–201; Matthew Van Winkle, *Aviation Gasoline Manufacture* (New York: McGraw-Hill, 1944), pp. 107–10.

37. Williamson et al., *Energy*, pp. 631–2, 787–8; Van Winkle, *Aviation Gasoline*, pp. 160–80.

38. Memo, Continuous Catalytic Cracking Process Licensed by the M.W. Kellogg Co. and Universal Oil Processes Co., March 3, 1941; clipping, "Jersey Unveils New Catalytic Cracking Unit," *National Petroleum News* (February 19, 1941), Acc. #1317, EMHL, Papers of Arthur E. Pew, Jr., Box 25; clipping "Axis Cracker," *Time* (February 1941), A.E. Pew, Jr. Papers, Box 1; J. Edgar Pew to J. Howard Pew, A.E. Pew, Jr., and Eugene Houdry, January 31, 1941, A.E. Pew, Jr. Papers, Box 88.

39. William S. Farish to J.A. Brown (Copy), April 18, 1938, Acc. #1317, EMHL, J. Howard Pew Papers, Box 42; Clarence H. Thayer, interview held at Media, Pennsylvania, November 25, 1975.

40. Williamson et al., *Energy*, p. 789; Frey and Ide, *Petroleum Administration*, pp. 199–202; George G. Hargrove, Badger and Co., to A.E. Pew, Jr., September 23, 1941, Acc. #1317, EMHL, A.E. Pew, Jr. Papers, Box 10; J. Howard Pew to Wright Gary, Office of the Petroleum Coordinator, May 18, 1942, J. Howard Pew Papers, Box 83.

41. Eugene J. Houdry to Socony-Vacuum Oil Company, May 18, 1942, Acc. #1317, EMHL, A.E. Pew, Jr. Papers, Box 6.

42. Frey and Ide, *Petroleum Administration*, p. 202; Ickes to J. Howard Pew, July 7, 1943, Acc. #1317, EMHL, J. Howard Pew Papers, Box 83.

43. *Annual Report, Sun Oil*, 1942, p. 1.

44. Memo, T.C.C. and H.P.C. Units as of July 14, 1943, Acc. #1317, EMHL, A.E. Pew, Jr. Papers, Box 23.

45. Board Minutes, Sun Oil Company, January 5, 1943, Minute Book #7, p. 165; *Annual Report, Sun Oil*, 1943, pp. 1–3; "Sun Oil's Newest Plant," *Our Sun*, pp. 1–2, 6–7; *Annual Report, Sun Oil*, 1944, p. 9; Thayer interview.

46. Frey and Ide, *Petroleum Administration*, pp. 205–6; Thayer interview; Dunlop interview; Enos, *Petroleum Progress*, Chapter 6, "The Fluid Catalytic Process"; *National Petroleum News*, 44 (October 30, 1946), p. 16.

47. Frey and Ide, *Petroleum Administration*, pp. 221–2; Williamson et al., *Energy*, pp. 790–1; telegram, J.A. Brown, Socony, to J. Howard Pew, March 11, 1942, Acc. #1317, EMHL, J. Howard Pew Papers, Box 83.

48. Williamson et al., *Energy*, p. 791; Henrietta M. Larson, Evelyn H. Knowlton, and Charles S. Popple, *New Horizons, 1927–1950*, Volume III, *History of Standard Oil Company (New Jersey)* (New York: Harper and Row, 1971), pp. 170–4; R.F. Dunbrook, "Historical Overview," in G.S. Whitby, C.C. Davis, and R.F. Dunbrook (eds.), *Synthetic Rubber* (New York: Wiley, 1954), pp. 37–42.

49. James B. Conant, Karl T. Compton, and Bernard M. Baruch, Chairman, "Report to the Rubber Survey Committee," September 10, 1942, pp. 1–10, Acc. #1317, EMHL, J., Howard Pew Papers, Box 79; A.E. Pew, Jr. to Stanley T. Crossland, Rubber Reserve Corporation, June 12, 1942, J. Howard Pew Papers, Box 42; Dunbrook, "Historical Overview," pp. 42–4.

50. "Testimony of Eugene Houdry," July 13, 1942, Gillette Committee Investigation of Industrial Alcohol and Synthetic Rubber, Subcommittee of the United States Senate Committee on Agriculture and Forestry, pp. 1992–2009, Acc. #1317, EMHL, J. Howard Pew Papers, Box 42.

51. Larson et al., *New Horizons*, p. 507.

52. Eugene Houdry to A.E. Pew, Jr., December 9, 1941, Acc. #1317, EMHL, A.E. Pew, Jr. Papers, Box 18.

53. Larson et al., *New Horizons*, pp. 428–32; Joseph Borkin, *The Crime and Punishment of I.G. Farben* (New York: The Free Press, 1978), pp. 88–90; Undated memo, "The Standard Oil-I.G. Farben Cartel," Acc. #1317, EMHL, J. Howard Pew Papers, Box 56.

54. Larson et al., *New Horizons*, p. 433.

55. *Ibid.*, pp. 433–5; Borkin, *I.G. Farben*, pp. 90–4.

56. Larson et al., *New Horizons*, pp. 507–9; Dunbrook, "Historical Overview," pp. 44–6.

57. Larson et al., *New Horizons*, p. 507; A.E. Pew, Jr. to Stanley T. Crossland, June 12, 1942; Testimony of W.S. Burt, Socony-Vacuum, July 13, 1942, Gillette Committee Investigation, pp. 2068–78, Acc. #1317, EMHL, J. Howard Pew Papers, Box 42.

58. Memo, "Outline of Procedure" (undated, presumably in late June or early July), Acc. #1317, EMHL, J. Howard Pew Papers, Box 42.

59. See clippings or advertisements in the *Philadelphia Inquirer* and many other national newspapers; text of Fulton Lewis Broadcast, July 7, 1942; text of Lowell Thomas Broadcast, July 8, 1942, Acc. #1317, EMHL, J. Howard Pew Papers, Box 42; testimony of Arthur E. Pew, Jr., July 13, 1942, Gillette Committee Investigation, p. 2046, J. Howard Pew Papers, Box 42; for a brief but useful discussion of the Houdry Group publicity campaign, see Thomas C. Guider, "Sun Oil Company and the Butadiene Controversy" (unpublished manuscript, Eleutherian Mills Historical Library, 1973), especially pp. 6–11.

60. Text of Fulton Lewis Broadcast, July 7, 1942; testimonies of Eugene Houdry, Clarence Thayer, A.E. Pew, Jr., and W.S. Burt, Gillette Committee Investigation, July 13, 1942; testimony of M.J. Madigan, Special Assistant to the Under-Secretary of War, July 15, 1942, Gillette Committee Investigation, pp. 2226–7, Acc. #1317, EMHL, J. Howard Pew Papers, Box 42; Thayer interview.

61. Minutes of Meeting of Mr. J. Howard Pew and Franklyn Waltman with Thurman Arnold, Assistant Attorney General of the United States, July 15, 1942, Acc. #1317, EMHL, J. Howard Pew Papers, Box 56.

62. *Ibid.*

63. *Ibid.*

64. *Ibid.*

65. In May 1941, Arnold and J.N. Pew, Jr. had held a long discussion on the subject of war preparedness and national unity. J.N. had expressed dismay at the lack of unity and suggested that both political parties should cooperate to obtain more positive action. On the basis of this conversation, Arnold had drafted a letter to F.D.R. suggesting that Joe Pew should form a Republican organization opposed to Charles Lindbergh and the America First Committee, the isolationist group then working to keep the United States out of war. In this letter, Arnold referred to the Pews as the "Henry Ford" of the oil industry, and cited their past adherence to the antitrust laws. In reply to Arnold, however, Pew suggested that both men could best work from within their own parties and requested that Arnold not send the letter to the President (Thurman Arnold to J.N. Pew, Jr., May 12, 1941; draft of suggested confidential letter from Arnold to Franklin D. Roosevelt; J.N. Pew, Jr. to Thurman Arnold, May 15, 1941, Acc. #1317, EMHL, File 21-B, Box 2).

66. J. Howard Pew to J. Edgar Pew, July 16, 1942, Acc. #1317, EMHL, J. Howard Papers, Box 42.

67. *Ibid.*

68. J. Howard Pew to Nelson, July 17, 1942; J. Howard Pew to George A. Hill, Jr., Houston Oil Company, July 21, 1942, Acc. #1317, EMHL, J. Howard Pew Papers, Box 42.

69. Larson et al., *New Horizons*, p. 508.

70. U.S. Congress, House, *The Rubber Situation, Message From the President of the United States, September 10, 1942, 77th Congress, 2nd Session, House Document No. 836* (Washington,

D.C.: U.S. Government Printing Office, 1942); "Report of the Rubber Reserve Committee," p. 39, Acc. #1317, EMHL, J. Howard Pew Papers, Box 79; "Rubber From Oil," *Our Sun* 11, 3 (July/August 1945), pp. 12–3; Williamson et al., *Energy*, p. 791; C.E. Morrell, "Manufacture of Dienes From Petroleum," in Whitby et al., *Synthetic Rubber*, p. 60.

71. Morrell, "Manufacture of Dienes," p. 64.

72. W.J. Toussaint, and J. Lee Marsh, "Manufacture of Butadiene From Alcohol," in Whitby et al., *Synthetic Rubber*, pp. 102–3.

73. Larson et al., *New Horizons*, pp. 443–55.

74. Franklyn Waltman to A.E. Pew, Jr., May 2, 1943, Acc. #1317, EMHL, A.E. Pew, Jr. Papers, Box 3.

75. David Sentner, "Dies Will Link 10 U.S. Officials to Plot to Ruin Firm," *New York Journal American* (January 7, 1943), clipping; Memo, "Sun Investigation" (undated, unsigned), Acc. #1317, EMHL, File 71, Box 59.

76. Borkin, *I.G. Farben*, p. 93.

77. Larson et al., p. 433–4; J.A. Brown to J. Howard Pew, February 25, 1942, Acc. #1317, EMHL, J. Howard Pew Papers, Box 56; U.S. Congress, House, *Extract From Hearings Before a Special Committee on Un-American Activities, House of Representatives, 75th Congress, 3rd Session on H. Res. 282* (Washington, D.C.: U.S. Government Printing Office, 1938), pp. 51, 155–69; J.A. Brown to W.S. Farish (Copy), March 11, 1932, Acc. #1317, EMHL, J. Howard Pew Papers, Box 83.

78. Memo, "Sun Investigation," Acc. #1317, EMHL, File 71, Box 59; Franklyn Waltman to Thurman Arnold, February 16, 1943, A.E. Pew, Jr. Papers, Box 3; J. Howard Pew to John A. Brown, August 27, 1943, J. Howard Pew Papers, Box 56. Brown enjoyed good relations with the New Jersey Company, and he forwarded Pew's memoranda on the situation to Ralph Gallagher of Jersey's public relations staff. Brown suggested to J. Howard Pew that both Sun and Socony should disassociate themselves from the Houdry Process Corporation Washington office and that Franklyn Waltman should cease working for H.P.C. By the fall of 1943, the incident had pretty well quieted down (J. Howard Pew to J.A. Brown, August 27, 1943; Brown to J. Howard Pew, September 11, 1943, Acc. #1317, EMHL, J. Howard Pew Papers, Box 56).

79. D.R. Lamont, Socony, to C.S. Teitsworth, Socony (re: Houdry Process Corporation), July 25, 1951, p. 6, Acc. #1317, EMHL, File 21–B, Box 31; on negotiations with the Soviet government and the United States Department of State, see A.E. Pew, Jr. Papers, Boxes 1 and 15.

80. Lamont to Teitsworth, July 25, 1951, p. 10, Acc. #1317, EMHL, File 21–B, Box 31; "The Houdry Adiabatic Cracking Process for the Production of Aviation Base Stock" (New York: Catalytic Development Corp., 1943), A.E. Pew, Jr. Papers, Box 25; "Houdry Discloses Catalytic Process," *Bradford, Pennsylvania, Era* (January 21, 1943), clipping, A.E. Pew, Jr. Papers, Box 1; *Annual Report of the Houdry Process Corporation*, 1943, 1944.

81. D.R. Lamont to W.F. Burt, Socony, July 8, 1943, Acc. #1317, EMHL, A.E. Pew, Jr. Papers, Box 6; W.F. Burt to E.J. Houdry, November 7, 1943; Houdry to Burt, November 18, 1943, A.E. Pew, Jr. Papers, Box 26; Lamont to Teitsworth, July 25, 1951, p. 11, File 21–B, Box 31; E.J. Houdry to J.D.M. Hamilton, March 1, 1948, File 21–B, Box 30.

82. Thayer interview; E.J. Houdry to A.E. Pew, Jr., August 15, 1944; E.J. Houdry to J. Howard Pew, August 16, 1944, Acc. #1317, EMHL, J. Howard Pew Papers, Box 42; J. Howard Pew to E.J. Houdry, August 21, 1944, A.E. Pew, Jr. Papers, Box 6; *Annual Report, H.P.C.*, 1944.

83. "Matters Proposed by A.V. Danner for Consideration at H.P.C. Executive Meeting," October 16, 1944; Memo, H.P.C. Executive Committee Meeting, December 11, 1944, Acc. #1317, EMHL, A.E. Pew, Jr. Papers, Box 1; B.B. Jennings, Socony, to J.N. Pew, Jr., December 26, 1947; A.V. Danner to Clarence H. Thayer, February 26, 1948, Acc. #1317, EMHL, File 21–B, Box 30. Eugene Houdry resigned as an H.P.C. employee in 1948, but

he still used his financial influence to shape policy. Danner had left to rejoin Socony in 1947, and the New York firm withdrew from all H.P.C. management at that time. After years of more controversy, H.P.C. brought suit against Socony in the early 1950s (E.J. Houdry to J.D.M. Hamilton, March 1, 1948, Acc. #1317, EMHL, File 21-B, Box 30; Lamont to Teitsworth, July 25, 1951, pp. 13-22, File 21-B, Box 31).

84. "Sun Ship's 30 Years," *Our Sun* 12, 2 (July/August 1946), pp. 4-6; "Sun Ship . . . The World's Largest Single Shipyard," *Our Sun* 9, 3 (March 1943), pp. 13-4; "Petroleum and Ships," *Our Sun* 8, 1 (August/September 1941), p. 4; U.S. Congress, House, Remarks of Hon. Leon H. Gavin (Pennsylvania) on Food and Oil Production Problems, March 30, 1943, 78th Congress, First Session, *Congressional Record*, Vol. 89, Part 10, pp. A1521-2.

85. "Sun Shipbuilding," *Fortune* 23, 12 (February 1941), p. 54; J.N. Pew, Jr. to Ira F. Lewis, *Pittsburgh Courier*, 1942 (no other date); also see article on Negro employment at Sun Ship in *The Brown American* (Spring 1942), Acc. #1317, EMHL, File 21-B, Box 2; "Its C.I.O. at Sun," *Business Week* (July 17, 1943), p. 92; " Articles of Agreement, Sun Shipbuilding and Dry Dock Company and Industrial Union of Marine and Shipbuilding Workers of America and its Local No. 2," Acc. #1317, EMHL, J. Howard Pew Papers, Box 84.

86. *Annual Report, Sun Oil*, 1940, p. 5; Memo, meeting with Admiral Lamb and Knox, March 17, 1941, memo dated March 31, Acc. #1317, EMHL, J. Howard Pew Papers, Box 5.

87. "Sun Ship's 30 Years," *Our Sun*, pp. 47-8.

88. J.N. Pew, Jr., to Franklyn Waltman, May 6, 1943, Acc. #1317, EMHL, File 21-B, Box 32. Kaiser had sent a team to investigate Sun's welding methods, and the Chester firm had provided them with a great deal of technical information. Although competition developed between the two giant yards, Sun Ship Chairman J.N. Pew, Jr. was upset when Fulton Lewis reported on his radio program a wager that Pew had supposedly made that Sun Ship would outstrip Kaiser's production totals. This was apparently not the type of publicity that the Pews wanted, since the shipyard wished to stress the patriotic rather than the competitive nature of its business.

89. "Sun Ship's 30 Years," *Our Sun*, p. 48; Wilbur F. Burt, Socony, to A.E. Pew, Jr., May 6, 1943; Acc. #1317, EMHL, A.E. Pew, Jr. Papers, Box 24; Board Minutes, Sun Oil, January 22, 1941, Minute Book #7, pp. 7-8.

90. *New York Times* (February 31, 1941), p. 43; Kelsey, "Financial Study," pp. 38-391.

91. Board Minutes, Sun Oil, September 5, 1943, Minute Book #7, pp. 218-9; *Annual Report, Sun Oil*, 1945, p. 13; *Annual Report*, 1944, p. 10. Due to complex negotiations during the war resulting from charges of excess profits, Sun reached an agreement with the government to build ships at a flat profit of $150,000 per ship (Memo, R.G. Dunlop to author, April 6, 1977).

92. *Annual Report, Sun Oil*, 1944, p. 1; Dunlop interview; Thayer interview; Frey and Ide, *Petroleum Administration*, Appendix 6, p. 367.

93. *Annual Report, Sun Oil*, 1943, pp. 1-3; Dunlop interview; Board Minutes, Sun Oil Company, January 29, 1942, Minute Book #7, p. 106.

94. J. Howard Pew to George L. Parkhurst, P.A.W. refinery Division, August 17, 1943, Acc. #1317, EMHL, J. Howard Pew Papers, Box 83; Board Minutes, Sun Oil Company, September 5, 1943, Minute Book #7, pp. 218-9.

95. J. Howard Pew to Parkhurst, J. Howard Pew Papers, Box 83.

96. R.D. Dunlop to G.L. Parkhurst, June 22, 1943; J. Howard Pew to Parkhurst, August 17, 1943, Acc. #1317, EMHL, J. Howard Pew Papers, Box 83; Frey and Ide, *Petroleum Administration*, Appendix #6, "Plancor No. 1951, Sun Oil Company," p. 369.

97. *Annual Report, Sun Oil*, 1944, p. 8; "Ten Years of Progress, 1939-1949," *Our Sun* 14, 2 (Spring 1949), p. 3.

98. "Billionth Gallon," *Our Sun* 11, 1 (February 1945), p. 15; "Employees at Toledo Host Two Sun Officers," *Our Sun* 10, 4 (August/September 1944), p. 7; "Ten Years of

Progress," *Our Sun*, p. 3; Frey and Ide, *Petroleum Administration*, Appendix 12, Table 43, p. 457.

99. Kelsey, "Financial Study," Table 14, pp. 59, 75.

100. W. Alton Jones, P.I.W.C., to J. Howard Pew, October, 1945, Acc. #1317, EMHL, J. Howard Pew Papers, Box 81.

101. *Ibid.*

102. Frey and Ide, *Petroleum Administration*, pp. 289–90, 293, 298; Nash, *Oil Policy*, pp. 182–3.

103. *Annual Report, Sun Oil*, 1945, pp. 3–5; "Why OPA Should Be Ended," pamphlet distributed by Sun Oil Company; testimony of J. Howard Pew, House Banking and Currency Committee, March 21, 1946; Memo, "A Continuance of the OPA Can Only Be Regarded As a Mechanism Designed to Persecute the Law-Abiding and Encourage All Others," Acc. #1317, EMHL, J. Howard Pew Papers, Box 5.

104. "Pew Answers Charges Made by CIO Speaker," *National Petroleum News* 38, 27 (July 3, 1946), p. 24.

105. Barton J. Bernstein, "The Removal of War Production Board Controls on Business, 1944–1946," *Business History Review* 39, 2 (Summer 1965), pp. 243–4; see also Bernstein, "Industrial Reconversion: The Protection of Oligopoly and Military Control of the Wartime Economy," *American Journal of Economics and Sociology* 26, 2 (April 1967), pp. 159–72.

106. Board Minutes, Sun Oil, May 23, 1944, Minute Book #7, p. 268; September 5, 1945, Minute Book #8, p. 54; "Blue Sunoco Dynafuel," *Our Sun* 11, 4 (November 1945), pp. 1–2.

107. Nash, *Oil Policy*, pp. 177–8.

108. Frey and Ide, *Petroleum Administration*, pp. 65, 174–5, 275–9; Nash, *Oil Policy*, pp. 170–3; Benjamin Shwadran, *The Middle East, Oil and the Great Powers* (New York: Praeger, 1955), pp. 318–22.

109. Potomachus, "Pew of Pennsylvania," *The New Republic* 110, 19 (May 8, 1944), p. 626; Robert W. Howard to J.N. Pew, Jr., February 14, 1944; carbon copy of *Pathfinder* story for February 21, 1944, issue, Acc. #1317, EMHL, File 21–B, Box 23.

110. J.N. Pew, Jr. to John D.M. Hamilton, May 16, 1944, Acc. #1317, EMHL, File 21–B, Box 29.

111. Frey and Ide, *Petroleum Administration*, pp. 279–80; Nash, *Oil Policy*, pp. 176–7.

112. "Letter of Sun Oil President Probes Meaning of Vague Phrases in Anglo-American Pact," *National Petroleum News* 36, 34 (August 23, 1944), pp. 9–10, 14; J.N. Pew, Jr. to Lee Ellmaker, *Philadelphia Daily News*, August 14, 1944, Acc. #1317, EMHL, File 21–B, Box 4; for a discussion of J. Howard Pew's views on international cartels, see Charles R. Whittlesey, *National Interest and International Cartels* (New York: Macmillan, 1946), p. 6.

113. P.H. Frankel, *Essentials of Petroleum, A Key to Oil Economics* (New York: August M. Kelley, 1969), p. 121; Frey and Ide, *Petroleum Administration*, p. 281.

114. "Pew Denounces Oil Pact as a Vicious Cartel," *National Petroleum News* 36, 43 (October 25, 1944), p. 3.

115. *Ibid.*, p. 14.

116. Frey and Ide, *Petroleum Administration*, p. 282.

117. *Ibid.*

118. "Memo to Mr. J. Howard Pew in Re: the American-British Oil Agreement," George Wharton Pepper and John D.M. Hamilton, December 2, 1944; J. Howard Pew to George Wharton Pepper and J.D.M. Hamilton, December 12, 1944; J. Howard Pew to George Wharton Pepper, December 28, 1944, Acc. #1317, EMHL, J. Howard Pew Papers, Box 80.

119. Frey and Ide, *Petroleum Administration*, pp. 282–3; Nash, *Oil Policy*, p. 177; *Annual Report, Sun Oil*, 1944, pp. 4–5; "J. Howard Pew Hits Trend Toward Collectivism," *National*

Petroleum News 37, 21 (May 23, 1945), p. 16; "J. Howard Pew Strikes at Anglo-U.S. Oil Pact as Forerunner of Cartels, Foe of Enterprise," *National Petroleum News* 37, 13 (March 28, 1945), pp. 16, 58; J. Howard Pew, Virgil Jordan, Alvin Hansen, Senator Harold H. Burton, "Post-War Planning and the Planned Economy," *Proceedings of the National Industrial Conference Board,* 1945, pp. 46-7.

120. Nash, *Oil Policy*, pp. 177-8; Frey and Ide, *Petroleum Administration,* pp. 283-7.

121. Gabriel Kolko, *The Politics of War* (New York: Random House, 1968), especially pp. 18-20.

CHAPTER X

Epilogue

In January 1945, a group gathered at Haverford College to hear a debate on "Free Economy" as the basis for America's post-war economic life. Many of them had anticipated seeing sparks fly from the two speakers, Joseph N. Pew, Jr. and U.S. Court of Appeals Judge Thurman Arnold, former head of the Federal Antitrust Division. On the contrary, as the evening proceeded, both men advocated strikingly similar positions. They opposed international cartels and the growth of monopoly, favored the decentralization of bureaucracy, and agreed that collectivism as practiced in Soviet Russia could not succeed in the United States. Their common denominator was a professed belief in individual economic freedom and the competitive ethic in American business. Toward the end of the debate, a frustrated Haverford student asked Arnold just where he and Pew differed. Arnold answered that "We differ on our support of Mr. Roosevelt, and what kind of dismemberment we should use on Pew's oil company, but in our general philosophies, I don't think Mr. Pew and I differ at all."[1]

The Pews and Thurman Arnold both held views that were largely outside the mainstream in 1945. While the petroleum companies had moved toward increased cooperation among themselves and with government in the period after World War I, Sun Oil management had maintained an extremely competitive and independent stance. The Pews opposed many proposals for governmental regulation and embraced a business philosophy based on their own *laissez faire* principles.

Similarly, Arnold's career highlighted a major departure from the main course of the Roosevelt administration. Although Richard Hofstadter later argued that Arnold's writings represented the best example of the New Deal philosophy of pragmatism, the antitrust approach he championed accomplished little change in American business.[2] The T.N.E.C. experience and Arnold's antitrust prosecutions did not alter the move toward centralization and consolidation that came with the New Deal.[3]

The suspension of antitrust prosecutions during World War II brought an end to serious efforts to break up the large firms. The oligopoly of major, fully-integrated companies that had come to dominate the industry after 1911 was strengthened by the war experience. In their massive efforts to increase petroleum production from 1941–1945, government agencies found that cooperation with the giant firms proved to be the most effective way to mobilize the economy. Thus, the most significant public policy issue enamating from the New Deal period was not antitrust, but a continued emphasis on comprehensive regulation and business-government cooperation.[4]

Both J.N. Pew, Jr. and Thurman Arnold were part of a still significant but diminishing minority in 1945. The war had brought together forces that had been developing for years. Efficiency, bigness, economic stabilization, regulation, and business-government cooperation were the hallmarks of the new era. Yet, the similarities between the ideas of the notorious trustbuster, Arnold, and the conservative businessman, Pew, represent a significant footnote to the history of American business thought. The views of proponents of antitrust law were quite compatible with those of small, independent, and conservative businessmen. Although their collective voice had lost much of its power, it would continue to be heard as a critique of growing consolidation in American industry. The irony was that the Pews, representatives of a large and powerful company, never really thought like "big businessmen," but acted more like the managers of a smaller unit in competition with much larger economic interests.

Gerald Nash argues convincingly that World War II firmly crystallized the role of government as arbiter of the American economy in general and the oil industry in particular. Although there were developments to consolidate this relationship in the ensuing two decades, little new appeared to alter the fundamental consensus that had been achieved. This consensus consisted of at least four main areas of agreement between the industry and government: (1) that the oil business would remain in private hands, but under government supervision; (2) that government would promote the industry with tax policies, research aid, or direct financial subsidy; (3) that government had a responsibility to use its regulatory power to help the oil industry solve its problems; and (4) that government would provide support for American oil firms doing business abroad. The goals of these policies met both the industry's need to achieve economic stability and the public's interest in the conservation of oil for future needs.[5]

This *modus operandi* had its roots in the attempts of "progressive" elements of the oil industry to achieve some degree of control over their changing business in the period 1900–1920. A sector of the economy that had become rationalized during the period of Standard Oil dominance in the late nineteenth century, oil once again became more competitive in the post-Spindletop era of the internal combustion engine. This new competitive structure after 1900 forced oil men, Standard Oil executives and "independents" alike, to seek new ways to restore order in their chaotic industry. Government policy, which had been exclusively tied to antitrust since the 1880s, reached a symbolic turning point with the enforced dissolution of the Standard Oil Company by the Supreme Court in 1911.

Government and business forged a new public policy for oil during periods of economic crisis. World War I mobilization efforts created a framework of government regulation, and oil industry leaders learned that they could work together with government. Although the formal regulatory agencies disappeared in peacetime, the pattern of interindustry cooperation was continued by the American Petroleum Institute throughout the 1920s. Beyond this, the economic threat resulting from overproduction in the California and Mid-Continent oil fields during the decade led many industry spokesmen to advocate programs of unitized field development and conservation measures through state or federal regulation.

These efforts culminated during the New Deal decade when crude oil at ten cents a barrel and the threat of economic chaos did more to generate industry backing of government regulation of oil production than all of the previous urgings of some business and political leaders. It was this legacy rather than antitrust that meshed with the needs of wartime demand to establish a clear consensus on business-government cooperation during World War II.

Thus the fundamental pattern of oligopolistic competition and business-government cooperation in the oil industry had been achieved by 1945, the termination point of this study. The single most important change that occurred in the post-war period was the increasing multinational character of the oil business. This, however, did not alter the basic structure of the industry. Only the major integrated firms could afford to invest heavily in foreign exploration, and these endeavors simply served to consolidate their oligopolistic position. Furthermore, in the era of petroleum abundance that lasted through the 1960s, the basic framework of business-government cooperation was not seriously modified.

Sun Oil was not in the forefront of foreign oil development in this period because the Pews were wary of too-heavy involvement abroad. Therefore, though the post-1945 chapter of petroleum industry and Sun Oil history is important, it is not of direct relevance to the present analysis. The essential features of the domestic petroleum industry today were readily apparent by the end of World War II.

The fact that Sun's owner-managers thought of their firm as an "independent" rather than a "major" is important to consider when evaluating the company's role in the petroleum industry. The environment of Standard Oil domination shaped Sun's early history. The Pews' insistence on limited growth and their views on monopoly reflected the lessons learned from competition with the industry giants. The family's long participation in company management and the close ownership of their company explains why these attitudes permeated Sun Oil management decisions for such a long time.

The founder of the firm, J.N. Pew, Sr., had learned well the lessons of competition in a monopolistic market. Rockefeller's successes had taught Pew the importance of a transportation network in maintaining control of one's own business, and he quickly discovered that vertical integration was one way to insure this independence. To J.N. Pew, as well as to most independent Standard competitors, the Sherman Antitrust Act was a victory for the small businessman. He cheered when the courts finally dissolved the Rockefeller trust in 1911, only a few months before his death. Pew had spent his career eking out a niche for himself in the natural gas and oil businesses, and it now appeared that Sun was on the threshold of achieving great success.

J.N. Pew, Sr. passed his business philosophy to his sons, J. Howard and J.N., Jr. He taught them the importance of the independence that came with financial control and preached a philosophy of *laissez faire* that they retained throughout the many years that they ran the Sun Oil Company. Their older cousins, J. Edgar Pew and John G. Pew, were graduates of the rough-and-tumble days of the Pennsylvania and Texas oil fields and had themselves worked for a time with affiliates of the Standard Oil Company (New Jersey). These men rejoined Sun in 1918 and played important roles in the firm up through the end of World War II. They were well aware of the price one had to pay to retain an independent posture, and they advised their younger cousins accordingly.

This independence was evident in the way the Pews continued to run the Sun Oil Company after the death of the founder. Sun had initially found a place for itself as a refiner and marketer of speciality products, mostly gas oil for the enrichment of manufactured gas, and industrial lubricants. Because of their ties to the gas industry through their partnership with the United Gas Improvement Company of Philadelphia (U.G.I.), the Pews moved belatedly into the gasoline market and did not open their first service station until after World War I. Sun used the large profits made during the war to construct its own shipyard in 1916, to achieve financial independence by buying out the U.G.I. interest, and to launch itself into the automobile age by marketing automotive lubricants and gasoline.

The Pews' decision to sell only a single grade of unleaded "premium" motor fuel in 1927 was further evidence of Sun's highly competitive practices and its deviation from industry norms. The Blue Sunoco marketing decision had long-term implications for Sun Oil. Sun opposed the "patent club" on thermal cracking technology, remained outside the Ethyl group in its gasoline marketing, expended millions to develop its own refining processes, and irked much of the industry with its aggressive marketing strategy. Sun's desire to produce high octane gasoline without the addition of tetraethyl lead motivated the Pews to adopt the Houdry Process for catalytic cracking in the 1930s. This step made Sun an important leader in technological innovation, but it also pointed out the continuing difficulties of competing with the largest single unit in the industry, Jersey Standard. Unable to dictate favorable licensing terms, Standard opted to develop its own process and became a competitor of the Houdry group during World War II. Sun's investments in Houdry technology did pay off handsomely during the war, however, when the firm became one of the nation's major producers of high octane aviation gasoline. Similarly, the firm's decision to build the Sun Shipyard and Drydock Company during World War I and keep it operational during the Depression also payed dividends when the facility became a major producer of tankers under government contract.

Sun Oil's history shows that its independence probably cost it profits. From a purely economic standpoint, the Pews might have done better using lead in their gasoline and not investing heavily in cracking technology. Their conscious concern for maintaining family control inhibited them from letting Sun become too large. Sun's decision to refrain from foreign operations also cost money, at least in the short run. But although the Pews may have had a limited vision, they had definite ideas of where they wanted to be in the petroleum industry. This enabled Sun Oil to be financially successful within its self-imposed size limitations and to carve out a reputation as an aggressive and independent corporation.

The competitive tactics that the Pews exhibited in the running of their business mirrored their position *vis-à-vis* the government. When many industry leaders early advocated measures to bring government into the oil industry to effect stabilization, the Pews went their own way. Sun management, led by J. Edgar Pew, opposed enforced oil field unitization in the 1920s, and Sun became identified with the *laissez faire* segment of the industry. Although Sun eventually accepted production regulation in the 1930s, it struggled against the formal price-fixing arrangements of the N.R.A. The Pews were among the country's most outspoken opponents of the Roosevelt administration's economic policies.

When war came, the Pews cooperated with their old opponent, Harold Ickes, but opposed the extension of wartime controls into the post-war period. The Pews also spoke out against cartels and led the successful fight against the Anglo-American Oil Agreement of 1944. They feared that the entering wedge of government would deprive them of the independence that they had won over the years. As a result, the Pews increasingly became identified with the extreme, conservative wing of the Republican Party. They opposed the centralization, control, and "mixed economy" concepts that emerged from the New Deal, and continuously extolled the virtues of the free enterprise system.

The Sun Oil experience demonstrates the complexity of the key issues which the petroleum industry faced in the first half of the twentieth century. Although an industry consensus did develop to work with the government for economic stabilization, Sun Oil, more than any other "major," charted an independent course. Sun was an active participant in the affairs of the American Petroleum Institute, the largest industry trade association and main spokesman for the large integrated companies. However, its philosophical position was often more in tune with the views of the independent producers, refiners, and marketers opposed to the integrated firms. But just as the oil industry was not a monolith, neither was the Sun Oil Company. Management differed on some issues, and J. Edgar Pew in particular was much more "progressive" than his eastern cousins on the issue of production regulation. In most cases, however, the Pews dealt internally with any differences and presented a united front on the important economic and political questions affecting the industry.

Sun's consistency on the issues over the years underscores the genuineness of the Pews' outspoken advocacy of *laissez faire*. More importantly, their business practices give further weight to the conclusion that they practiced what they preached. This is not to suggest that they always deviated from the mainstream. In defending the petroleum industry from attempts to dismember the large integrated companies, the Pews argued from their own history that vertical integration was as much the tool of competition as it was of monopoly. They were also selective concerning which aspects of governmental regulation they would tolerate. Opposed to price-fixing and cartels, they finally adopted prorationing, the Connally Act, and the Interstate Compact. Sun's managers rationalized these inconsistencies by maintaining that conservation was a legitimate government function and that all oil producers had the right to "police protection" of their underground reserves. The Pews' support of tariffs and quotas on imported oil also conflicted with the classical *laissez faire* doctrines of free trade. In these instances, economic self-interest triumphed over rigid orthodoxy. But in the main, the history of the Sun Oil Company demonstrates that the firm consistently opposed government intervention and only slowly moved to accept industry-sponsored regulatory programs.

Sun Oil's history does not, of course, disprove the fundamental pattern of business-government cooperation that had evolved in the petroleum industry by 1945. It does show, however, that this consensus developed over many years in response to particular problems in the industry, and that it was accompanied by much debate. History has shown that industry-sponsored legislation and government regulation have served to aid both the private welfare of the oil companies and the wider public interest in varying degrees. When it appeared that the public interest was not being served, Congress and representatives of regulatory agencies intervened with mixed success.

In recent years, changes have occurred in the Sun Oil Company. Today the firm is much larger, more decentralized, and more heavily involved in international operations than it was when the Pews actively managed it. The company has very recently looked to its own history and changed its corporate name from the Sun Oil Company back to the Sun Company. Thus, the original name of J.N. Pew, Sr.'s New Jersey corporation, changed in 1922 to effect better product identification, has returned to reflect Sun's current emphasis on product diversification.

Although the Pew family retains a large financial interest in the Sun Company, it has a very low profile in the operation of the firm. A group of highly skilled professional managers now determine company policies. J. Howard Pew's death in 1971 brought an important chapter of the company's history to an end. It terminated an era which saw a modern industrial corporation molded by the particular economic, philosophical, and political predilections of the family who controlled it. In an age of consolidation and efficiency and the growth of the large impersonal corporation, Sun Oil remained a unique institution, personally shaped by the entrepreneurial figures who owned it.

NOTES

1. "Pew, Arnold in Economy Talk at Haverford," *Chester Times* (January 18, 1945), clipping in Folder, "J.N. Pew, Jr.," Library File, Sun Oil Company Library, Marcus Hook, Pennsylvania.

2. Richard Hofstadter, *The Age of Reform* (New York: Alfred A. Knopf, 1955), pp. 318–24; see Thurman W. Arnold, *The Symbols of Government* (New Haven: Yale University Press, 1935) and *The Folklore of Capitalism* (New Haven: Yale University Press, 1937).

3. Howard Zinn (ed.), *New Deal Thought* (New York: Bobbs-Merrill, 1966), introduction, pp. xviii–xxii.

4. See Gerald D. Nash, *United States Oil Policy, 1890–1964* (Pittsburgh: University of Pittsburgh Press, 1968), Chapter 7, "A New Deal for Oil"; Ellis W. Hawley, *The New Deal and the Problem of Monopoly* (Princeton: Princeton University Press, 1966).

5. Nash, *Oil Policy*, pp. 179, 249–50.

Appendix

Table 1: Summary of Standard Oil's Position in the American Petroleum Industry, 1880-1911

Percent of control over crude oil supplies				
Fields	1880	1899	1906	1911
Appalachian	92	88	72	78
Lima-Indiana		85	95	90
Gulf Coast			10	10
Mid-Continent			45	44
Illinois			100	83
California			29	29

Percent of control over refinery capacity				
	1880	1899	1906	1911
Share of rated daily crude capacity	90-95	82	70	64

Percent of major products sold			
	1880	1899	1906-1911
Kerosene	90-95	85	75
Lubes		40	55
Waxes		50	67
Fuel Oil		85	31
Gasoline		85	66

Source: Harold F. Williamson, Ralph L. Andreano, Arnold R. Daum, and Gilbert C. Klose, The American Petroleum Industry, Volume II, The Age of Energy, 1899-1959, p. 7 (Adapted from Williamson and Andreano, "Competitive Structure of the Petroleum Industry," Oil's First Century, p. 75).

Table 2: Refining Capacity of 30 Largest Oil Companies, January 1, 1920

Companies	Percent of total industrial capacity
1. Standard New Jersey*	9.5
2. Standard California*	7.6
3. Standard Indiana*	4.7
4. The Texas Company	4.6
5. Gulf Oil	4.6
6. Union Oil of California	4.2
7. Atlantic Refining*	3.6
8. Sinclair Consolidated Oil	3.3
9. Royal Dutch-Shell	3.1
10. Magnolia Petroleum	3.1
11. Midwest Refining	3.0
12. Associated Oil	2.8
13. Standard New York*	1.7
14. Cities Service	1.7
15. Ohio Cities Gas	1.4
16. General Petroleum	1.3
17. Pierce Oil*	1.3
18. Tidewater Oil	1.3
19. Cosden	1.3
20. Vacuum Oil*	1.1
21. Sun Oil	.9
22. Standard Kansas*	.8
23. Indian Refining	.7
24. American Oilfields	.7
25. Mid-Continent Gasoline	.6
26. Pan-American Petroleum and Transport	.6
27. White Eagle Oil and Refining	.6

(Cont.)

Table 2–Continued

Companies	Percent of total industrial capacity
28. Continental Refining	.6
29. Standard Ohio*	.6
30. Solar Refining*	.4
Total	71.6

*Members of original Standard Oil combination.

Source: Williamson et al., Energy, p. 165 (adapted from Mclean and Haigh, Growth of Integrated Oil Companies, p. 528).

Table 3: Percent of Total Domestic Daily Crude Oil Refining Capacity Owned by Major Oil Companies, 1931

	Company	Percent of total capacity
1.	New Jersey Standard	13.5
2.	Socony-Vacuum	7.4
3.	Indiana Standard	7.2
4.	California Standard	6.5
5.	Shell-Union	5.9
6.	Texaco	5.6
7.	Gulf Oil	5.5
8.	Consolidated Oil	4.3
9.	Tidewater	3.8
10.	Cities Service	3.0
11.	Union Oil of California	2.7
12.	Pure Oil	1.9
13.	Atlantic Refining	1.8
14.	Continental	1.6
15.	Sun Oil	1.5
16.	Phillip's Petroleum	1.0
17.	Mid-Continent	1.0
18.	Ohio Standard	.9
19.	Skelly	.6
20.	Ohio Oil	.1
	Ten Major oil companies	62.7
	Twenty major oil companies	75.8

Source: Williamson et al., Energy, p. 646 (adapted from TNEC Hearings, Part 14-A, p. 7801).

Table 4: Percent of Total Domestic Daily Crude Oil Refining Capacity Owned by Major Oil Companies, 1938

Companies	Percent of total capacity
1. New Jersey Standard	9.9
2. Socony-Vacuum	8.1
3. Texaco	7.5
4. Indiana Standard	6.2
5. California Standard	5.8
6. Shell-Union	5.8
7. Gulf Oil	5.0
8. Consolidated	4.8
9. Tidewater	3.0
10. Cities Service	2.5
11. Atlantic Refining	2.5
12. Richfield	2.4
13. Pure Oil	2.2
14. Union Oil of California	2.2
15. Ohio Oil	2.0
16. Sun Oil	2.0
17. Phillip's Petroleum	1.5
18. Ohio Standard	1.3
19. Continental	1.0
20. Mid-Continent	.9
Ten major oil companies	58.6
Twenty major oil companies	76.6

Source: Williamson et al., Energy, pp. 646-47 (adapted from TNEC Hearings, part 14-A, p. 7801).

Table 5: Total Assets of 20 Major Oil Companies, 1924-1938
(In Millions of Dollars)

Name of Company	1924	1929	1932	1935	1938
1. Standard Oil Co. (New Jersey)	$1,244.9	1,767.4	1,888.0	1,894.9	2,044.6
2. Socony-Vacuum Oil Co., Inc.	406.2	708.4	1,000.5	784.9	919.1
3. Standard Oil Co. (Indiana)	361.5	697.0	693.2	693.5	724.7
4. The Texas Corporation	288.3	609.9	513.8	473.8	605.4
5. Standard Oil Co. of California	352.8	604.7	578.0	575.8	601.1
6. Gulf Oil Corp.	252.0	430.8	435.9	430.2	546.9
7. Shell Union Oil Corp.	257.0	486.5	393.0	357.6	397.5
8. Consolidated Oil Corp.	346.2	400.6	368.0	328.2	357.1
9. Empire Gas & Fuel Co.	301.4	327.1	405.2	398.9	337.1
10. Phillips Petroleum Co.	78.7	145.4	178.4	174.5	226.7
11. Tide Water Associated Oil Co.	211.4	251.4	192.0	182.8	202.8
12. The Atlantic Refining Co.	131.0	166.2	156.6	163.0	199.1
13. The Pure Oil Co.	181.6	195.5	144.6	157.2	180.4
14. Union Oil Co. of California	184.2	211.2	197.7	151.7	166.0
15. Sun Oil Co.	51.5	85.3	96.7	107.1	139.1
16. The Ohio Oil Co.	97.7	110.7	177.3	139.7	138.7
17. Continental Oil Co.	93.9	198.0	87.5	91.7	125.1
18. Standard Oil Co. (Ohio)	42.9	48.7	60.4	56.9	70.5
19. Mid-Continent Petroleum Corp.	79.7	85.9	73.2	60.6	63.7
20. Skelly Oil Co.	39.9	62.8	45.2	46.1	62.0
Total	$5,002.5	7,593.6	7,685.0	7,269.2	8,107.5

Source: TNEC Hearings, Part 14-A, p. 7842 (adapted from annual reports to stockholders and Moody's Manual of Industrials).

Table 6: Crude Oil Runs to Stills in Domestic Refineries by Major Oil Companies, by years, 1929-1938

[Thousands of 42-gallon barrels]

Name of Company	1938	1935	1932	1929
United States total	1,165,015	965,790	819,997	987,708
19 Companies total	933,185	796,122	677,169	[2]686,860
Atlantic Refining Co.	34,521	30,203	25,532	26,221
Cities Service Co.	33,417	28,873	24,422	14,686
Consolidated Oil Corporation	64,616	53,787	36,263	35,729
Continental Oil Co.	13,805	12,281	10,196	([3])
Gulf Oil Corporation of Pennsylvania	76,086	61,681	53,442	67,978
Ohio Oil Co.	5,772	5,440	5,058	2,366
Phillips Petroleum Company	15,812	13,179	10,560	2,888
Pure Oil Co.	25,040	26,258	23,128	17,726
Shell Union Oil Corporation	82,835	76,329	59,504	75,476
Skelly Oil Co.	7,374	6,034	4,227	4,620
Socony-Vacuum Oil Company	96,115	82,490	68,376	75,473
Standard Oil Co. of California[1]	49,532	44,737	42,762	([3])
Standard Oil Co. (Indiana)	86,992	66,057	47,236	68,428
Standard Oil Co. (New Jersey)	135,756	116,266	117,089	146,655
Standard Oil Co. (Ohio)	15,739	12,711	12,348	9,823
Sun Oil Co.	25,780	22,856	18,555	12,890
Texas Corporation	94,715	76,132	59,395	57,323
Tide Water Associated Oil Company	44,107	38,991	35,582	48,854
Union Oil Co. of California	25,171	21,817	23,495	31,725

[1]Moody's Manuals of Investment
[2]17 Companies
[3]Not available

Source: Roy C. Cook, Control of the Petroleum Industry by Major Oil Companies, TNEC Monograph No. 30 (Washington, D. C.: U. S. Government Printing Office, 1941), p. 74 (adapted from TNEC questionnaire and Moody's Manual of Industrials).

Table 7: Number of Domestic Producing Oil Wells Owned or Operated by Major Oil Companies, by years, 1929-1938

Name of Company	1938	1935	1932	1929
Total	95,034	78,226	76,957	76,813
Atlantic Refining Co.	2,029	1,154	600	427
Cities Service Co.	7,013	6,120	6,014	5,067
Consolidated Oil Corp.	9,434	8,525	8,270	2,278
Continental Oil Corp.	2,850	2,276	2,478	2,511
Gulf Oil Corp. of Pa.[1]	6,871	5,238	4,672	4,814
Ohio Oil Co.	11,247	11,060	15,341	17,085
Phillips Petroleum Co.	4,614	3,162	3,386	2,873
Pure Oil Co.	4,969	4,595	4,811	4,997
Shell Union Oil Corp.	5,356	3,671	3,236	3,526
Skelly Oil Co.	1,802	1,412	1,113	1,300
Socony-Vacuum Oil Co.	8,497	7,225	6,849	7,585
Standard Oil Co. (Ind.)	3,818	2,147	1,136	829
Standard Oil Co. (N.J.)	10,181	8,262	7,172	7,478
Standard Oil Co. (Ohio)[1]	354	--	--	--
Sun Oil Co.[1]	1,983	1,441	986	819
Texas Corporation	8,853	7,265	7,062	7,948
Tide Water Assoc. Oil Co.	3,891	3,584	2,878	6,335
Union Oil Co. of California	1,272	1,089	953	941

[1]Segregation of oil and gas wells not made. Gas wells usually represent a small part of total wells.

Source: TNEC Hearings, Part 14-A, p. 7780 (adapted from TNEC questionnaire).

Table 8: Purchases of Crude Oil by Major Oil Companies (Excluding Imports) by years, 1929-1938

[In 42-gallon barrels]

Name of Company	1938	1935	1932	1929
Total	651,828,473	520,913,523	486,589,956	490,405,352
Atlantic Refining Co.	30,246,856	25,706,687	22,035,393	24,465,601
Cities Service Co.	20,711,380	22,709,671	17,249,454	11,890,509
Consolidated Oil Corp.	38,077,017	31,306,317	72,849,286	34,560,750
Continental Oil Company	15,777,504	16,086,003	16,388,068	--
Gulf Oil Corp. of Pennsylvania	42,217,337	24,280,503	16,384,010	18,523,811
Ohio Oil Co.	8,006,603	6,133,227	8,396,707	8,204,618
Phillips Petroleum Company	18,368,641	14,883,487	9,167,718	2,238,984
Pure Oil Co.	22,534,742	21,597,528	17,680,862	9,763,497
Shell Union Oil Corp.	49,518,914	45,012,529	35,739,389	36,459,878
Skelly Oil Co.	7,633,303	6,790,989	4,343,883	4,541,709
Socony-Vacuum Oil Co.	52,407,530	46,905,365	39,295,673	40,989,483
Standard Oil Co. (Indiana)	68,269,617	58,180,039	44,167,426	71,183,504
Standard Oil Co. (New Jersey)	106,793,455	72,176,685	86,799,179	105,300,438
Standard Oil Co. (Ohio)	19,896,595	13,369,651	9,859,947	9,792,043
Sun Oil Co.	24,112,083	20,250,620	14,157,845	11,436,328
Texas Corporation	66,946,126	54,133,222	31,826,228	34,984,622
Tide Water Assoc. Oil Co.	40,378,171	26,331,317	23,557,083	37,299,368
Union Oil Co. of California	19,932,599	15,059,683	16,691,805	28,770,209

Source: TNEC Hearings, Part 14-A, p. 7781 (adapted from TNEC questionnaire).

Table 9: Pipeline Mileage of the United States—Twenty Major Oil Companies and All Other Companies, January 1, 1938

	Crude Oil Mileage		Gasoline Mileage	
Name of Company	Total	Per cent	Total	Per cent
Consolidated Oil Corp.	13,383	11.6	124	1.9
Standard Oil Co. (N.J.)	11,220	9.8	544	8.4
Standard Oil Co. (Ind.)	8,382	7.3	40	.6
The Texas Corporation	7,998	6.9	258	4.0
Gulf Oil Corp. of Pa.	7,961	6.9	---	---
Socony-Vacuum Oil Co., Inc.	7,309	6.3	363	5.6
The Ohio Oil Co.	4,814	4.2	---	---
Shell Union Oil Co.	4,226	3.6	---	---
Tide Water Assoc. Oil Co.	3,856	3.3	13	.2
Cities Service Co.	2,314	2.0	186	2.9
Continental Oil Co.	1,682	1.5	629	9.7
Phillips Petroleum Co.	1,454	1.3	899	13.8
Atlantic Refining Co.	1,449	1.3	815	12.5
The Pure Oil Co.	1,357	1.1	345	5.3
Mid-Continent Pet. Corp.	1,212	1.0	404	6.2
Union Oil Co. of Calif.	911	.8	170	2.6
Standard Oil Co. (Ohio)	739	.7	158	2.4
Skelly Oil Co.	666	.6	302	4.6
Sun Oil Co.	646	.6	855	13.1
Standard Oil Co. of Calif.	560	.6	23	.4
Richfield Oil Corp.	407	.4	125	1.9
20 Major Oil Companies	82,546	71.8	6,253	96.1
All Other Companies	32,454	28.2	257	3.9
Total	115,000	100.0	6,510	100.0

Source: TNEC Hearings, Part 14-A, p. 7792 (adapted from TNEC questionnaire and I.C.C. Annual Reports of Pipeline Companies).

Table 10: Service Stations Owned or Controlled by Major Companies, 1929-1938

Name of Company	1929	1932	1935	1938
Atlantic Refining	394	580	261	131
Cities Service	1,031	2,869	2,317	2,515
Consolidated		14,244	9,172	9,611
Continental Oil	1,332	5,814	1,597	1,666
Gulf Oil	1,793	10,174	3,750	7,438
Ohio Oil	134	324	27	15
Phillips Petroleum	380	1,490	1,497	1,572
Pure Oil	464	952	579	36
Shell Union Oil	3,082	8,623	6,976	6,572
Skelly	285	388	538	630
Sacony-Vacuum	6,702	18,406	9,852	9,045
Standard Indiana	9,187	13,556	9,004	11,241
Standard New Jersey	156	17,012	7,981	417
Standard Ohio Oil	1,418	2,696	1,957	2,314
Sun Oil	345	474	677	682
Texas	5,571	23,459	13,143	9,607
Tidewater-Associated	880	1,233	1,794	2,166
Union Oil of California	550	915	4,201	4,053
Total	33,708	123,209	75,547	69,666

Source: Williamson et al., Energy, p. 707 (adapted from TNEC Hearings, Part 14-A, p. 7819).

Table 11: Engine Compression Ratios and Gasoline Octane Ratings, 1930-1941

Year	Engine compression ratios	Gasoline octane ratings*	
		Regular	Premium
1930		63.0	74.0
1931	5.23	63.0	75.0
1932	5.29	64.0	77.0
1933	5.57	68.0	77.0
1934	5.72	72.0	78.0
1935	5.98	72.0	78.0
1936	6.14	72.0	79.0
1937	6.25	73.3	81.0
1938	6.32	74.5	83.0
1939	6.32	74.5	83.0
1940	6.41	77.9	83.0
1941	6.60	80.4	85.3

*Research method.

Source: Williamson et al., Energy, p. 607 (adapted from John Lawrence Enos, Petroleum Progress and Profits, (Cambridge: MIT Press, 1962), p. 289).

Table 12: Domestic Refining Cracking Capacity, by Processes, 1929-1941

Year	Total capacity (thousands of barrels per day)	Percent of total						
		Burton process	Dubbs process	Tube and tank process	Holmes-Manley process	Cross process	Other thermal processes	Houdry (catalytic) process
1929	1,476	9.7	12.4	29.3		16.48	18.1	
1930	1,720	6.5	13.3	28.3	14.1	15.9	23.2	
1931	1,829	5.2	15.7	24.4	12.8	14.0	29.0	
1932	2,011	3.2	11.0	31.6	11.7	13.1	29.0	
1933	1,882	3.2	12.0	26.1	12.1	14.5	31.2	
1934	1,887	.8	12.0	26.7	13.0	13.2	33.9	
1935	2,153	.6	16.6	23.4	13.4	9.7	37.0	
1936	2,169	.5	17.6	21.5	12.7	8.7	38.7	
1937	2,195		19.3	21.0	13.0	7.1	40.0	.1
1938	2,348		20.7	19.8	12.5	7.5	39.7	.6
1939	2,138		21.5	19.4	11.7	7.9	39.2	1.1
1940	2,284		20.6	19.2	10.9	7.4	37.0	5.6
1941	2,352		21.4	18.2	10.2	8.1	35.7	6.6

Source: Williamson et al., Energy, p. 606 (adapted from Enos, Petroleum Progress, p. 287).

Table 13: Estimate of Houdry Process Royalities, 1936-1944

Year	Number	Capacity (000 barrels of charge per stream day)	Royalties (000 dollars)
1936	1	2.	270
1937	3	29.8	4,023
1938	--	--	--
1939	6	78.5	10,598
1940	4	50.	6,750
1941	1	16.6	2,241
1942	2	28.1	3,794
1943	3	37.	4,995
1944	4	47.5	6,412
Total	24	289.5	39,083

Source: Enos, Petroleum Progress, p. 315.

Table 14: Total Capitalization of the Sun Oil Company of New Jersey, 1918-1945

Year	Funded debt	Preferred stock outstanding	Common stock outstanding	Total surplus	Total capitalization
1918			$ 6,890,000		$ 6,890,000
1919	$ 6,000,000		6,890,000		12,890,000
1920	5,707,000		6,890,000	$22,356,310	34,953,310
1921	10,293,000		6,890,000	19,501,305	36,684,305
1922	13,617,000		30,520,000	865,209	45,002,209
1923	13,278,000		30,520,000	1,104,873	44,902,873
1924	14,445,000		30,520,000	1,722,653	46,687,653
1925	10,236,000		36,893,230	3,558,011	49,687,241
1926	9,836,000		39,202,026	3,539,566	52,577,592
1927	9,662,000		40,461,740	3,789,097	63,912,837
1928	8,992,000	$ 5,000,000	44,033,668	6,489,158	64,514,726
1929	8,695,000	5,000,000	47,917,187	9,136,519	70,748,706
1930	8,398,000	10,000,000	52,015,945	10,605,255	81,019,200
1931	12,131,000	10,000,000	52,010,380	11,502,220	85,643,600
1932	11,864,000	10,000,000	54,199,176	10,999,864	87,063,040
1933	10,541,000	10,000,000	59,006,338	10,778,750	90,326,088
1934	6,500,000	10,000,000	64,650,522	9,609,319	90,759,841
1935	6,232,000	10,000,000	69,493,800	9,512,102	95,237,902
1936	5,964,000	10,000,000	73,988,667	10,053,004	100,005,671
1937	9,000,000	10,000,000	79,873,770	10,953,754	109,827,524
1938	21,000,000	10,000,000	79,873,770	11,182,964	122,056,734
1939	21,000,000	10,000,000	80,024,678	15,091,224	126,115,902
1940	17,000,000	10,000,000	84,004,714	15,958,344	127,363,058
1941	20,000,000	9,319,700	97,878,416	13,604,990	140,803,106
1942	28,000,000*	9,319,700	97,878,416	18,263,495	153,461,611
1943	35,000,000*	9,319,700	97,878,416	27,062,503	169,260,619
1944	34,000,000*	9,319,700	107,623,564	27,344,493	178,287,757
1945	4,000,000	9,319,700	118,341,614	26,325,862	157,987,176

*Includes $8,000,000 of long term bank loans; 1943 includes $15,000,000 of bank loans; 1944, $14,000,000.

Source: John L. Kelsey, "A Financial History of the Sun Oil Company of New Jersey From its Inception through 1948," (M.B.A. Thesis, Wharton School, University of Pennsylvania, 1951), p. 60 (adapted from Poor's Manual of Investments-Industrials, 1917-1939 and Moody's Manual of Industrials, 1946, 1949).

Table 15: Gross Operating and Net Income of the Sun Oil company, 1915-1945

[In millions of dollars]

Year	Gross operating income	Net income
1915	10.9	1.8*
1916	19.5	6.1
1917	30.2	6.2
1918	31.2	5.7
1919	32.1	4.9
1920	52.7	10.1
1921	33.9	-.3
1922	44.3	1.6
1923	33.8	1.3
1924	--	2.0
1925	48.6	3.6
1926	52.2	3.3
1927	48.7	2.7
1928	64.2	5.0
1929	86.0	8.2
1930	98.3	7.7
1931	69.1	3.1
1932	67.1	4.2
1933	66.2	6.9
1934	82.6	6.6

(Cont.)

Table 15–Continued

Year	Gross operating income	Net income
1935	88.5	7.1
1936	105.4	7.5
1937	133.3	9.5
1938	115.0	3.0
1939	131.5	6.9
1940	147.7	7.9
1941	187.9	16.5
1942	310.7	8.6
1943	468.8	13.3
1944	600.8	13.3
1945	439.0	15.6

*Net income figures from 1915 through 1921 are before interest on funded debt and depletion charges, but after depreciation.

Source: Kelsey, "Financial History," pp. 25-26, 59 (adapted from Poor's Manual of Investments - Industrials, 1917-1939; Standard and Poor's Industrial Survey, 1949; and Moody's Investment Manual - Industrials, 1945, 1948).

Table 16: Shares of Common Stock Held by the 100 Largest Stockholders of the Major Oil Companies, December 31, 1938

Name of Company	Total number of common stock-holders	Total common shares outstanding	Shares held by 100 largest stockholders	Percent-age
Shell Union Oil Corporation	17,393	13,070,625	11,624,611	88.9
Sun Oil Co.	5,226	2,316,484	1,966,808	84.9
Skelly Oil Co.	3,152	995,349	817,245	82.1
Standard Oil Co. (Ohio)	3,532	753,740	521,166	69.1
Tide Water Assoc. Oil Co.	24,116	6,375,253	4,066,873	63.7
Gulf Oil. Corp. of Pennsylvania	15,135	13,751,846	7,430,934	54.0
Standard Oil Co. (New Jersey)	126,383	26,618,065	12,582,063	47.3
Ohio Oil Co.	31,287	6,563,377	2,955,244	45.0
Socony-Vacuum Oil Company	113,240	31,206,071	12,803,585	41.0
Continental Oil Company	29,969	4,738,593	1,688,030	35.6
Consolidated Oil Corp.	89,068	13,751,846	4,801,289	34.9
Standard Oil Co. (Indiana)	99,665	15,272,020	5,267,862	34.5
Pure Oil Co.	29,033	3,982,031	1,359,356	34.1
Phillips Petro-leum Co.	40,105	4,449,052	1,355,054	30.4
Union Oil Co. of California	26,524	4,666,270		28.1
Texas Corporation	86,380	10,876,882	2,605,090	24.0
Atlantic Refining Company	29,313	2,663,999	633,271	23.8
Cities Service Co.	466,658	3,704,067	776,599	21.0

Source: Cook, Control of the Petroleum Industry, p. 62 (adapted from TNEC questionnaire).

Table 17: Total Number of Shares of Sun Oil Company Common Stock Owned by the Pew Family, Broken Down for the Respective Owners, for the Years 1939 through 1945

Year	Number of Shares Owned by				Total
	Mary E. Pew	Mabel Pew Myrin	J. H. Pew	J. H. Pew, Jr.	
1939	297,491½	287,785½	303,616½	272,230½	1,161,124
1940	315,103	299,112	317,526	285,185	1,216,926
1941	328,324	316,975	335,122	300,128	1,280,349
1942	---	---	---	---	---
1943	---	---	---	---	---
1944	343,325	346,078	360,107	316,175	1,365,685
1945	365,069	379,110	394,661	343,041	1,481,881

Source: Kelsey, "Financial History," p. 104 (adapted from New York Stock Exchange Listings and Studley, Shupert and Co., "Analysis of the Oil Industry," [1948]).

Selected Bibliography

A. Manuscript Sources

Papers of The Sun Oil Company and Pew family. Accession #1317. Eleutherian Mills Historical Library, Greenville, Wilmington, Delaware. Accession #1317 has six subgroupings:
File 21-A: Administrative and Business Correspondence of the Sun Oil Company.
File 21-B: Joseph Newton Pew, Jr. Papers
J. Howard Pew Presidential Papers
File 71: J. Edgar Pew Production Files
Arthur E. Pew, Jr. Papers
Robert G. Dunlop Presidential Papers

Papers of Irénée duPont. File V.C. #25, Ethyl Corporation. Eleutherian Mills Historical Library, Greenville, Wilmington, Delaware.

Papers of Pierre S. duPont. Longwood Manuscripts, 10, File 771. Eleutherian Mills Historical Library, Greenville, Wilmington, Delaware.

Papers of Joseph N. Pew, Sr. Department of Records Management, Sun Oil Company Corporate Headquarters, Radnor, Pennsylvania. (This material will soon become a part of Accession #1317, Eleutherian Mills Historical Library).

Board Minutes of The Sun Oil Company of New Jersey. Office of the Assistant Secretary, Sun Oil Company Corporate Headquarters, Radnor, Pennsylvania.

Papers of Harold L. Ickes. Records Group 48, File 1-188, Department of the Interior, Office of the Secretary, Harold L. Ickes. National Archives, Washington, D.C.

B. Personal Interviews

Dunlop, Robert G. Retired President and Board Chairman, Sun Oil Company. St. Davids, Pennsylvania. Interviews: October 31, 1975; June 14, 1977.

Pew, Jno. G. "Jack." Retired Vice-President and Member of the Board, Sun Oil Company. Dallas, Texas. Interview: November 18, 1975.

Pew, R. Anderson "Andy." Secretary, Sun Oil Company. St. Davids, Pennsylvania. Interview: October 8, 1975.

Thayer, Clarence H. Retired Chief Engineer, Vice-President and Member of the Board, Sun Oil Company. Media, Pennsylvania. Interview: November 25, 1975.

C. Government Documents

Cook, Roy C. *Control of the Petroleum Industry by Major Oil Companies,* Temporary National Economic Committee Monograph No. 30. Washington, D.C.: U.S. Government Printing Office, 1941.

Farish, W.S., and Pew, J. Howard. *Review and Criticism on Behalf of Standard Oil Company (New Jersey) and Sun Oil Company of Monograph No. 39 With Rejoinder by Monograph Author,* Temporary National Economic Committee Monograph No. 39-A. Washington, D.C.: U.S. Government Printing Office, 1941.

Pennsylvania. *Meeting of Commission to Investigate the Oil Industry in the Commonwealth of Pennsylvania* (July 26-30, 1937). Harrisburg, 1937.

Pennsylvania. *Stenographic Report of Informal Conference by the Oil Industry Investigation Commission* (June 11, 1973). Harrisburg, 1937.

U.S. Congress, House: "Extension of remarks of Hon. Thomas F. Ford (California) on the production of synthetic rubber," July 21, 1942. 77th Congress, Second Session, *Congressional Record,* Vol. 88, Part 9, pp. A2885-6.

U.S. Congress, House. *Extract From Hearings Before a Special Committee on Un-American Activities on H.R. 282.* 75th Congress, Third Session. Washington, D.C.: U.S. Government Printing Office, 1938.

U.S. Congress, House. *Oil Marketing Divorcement. Hearings Before Subcommittee No. 3 of the Committee on the Judiciary on H.R. 2318.* 76th Congress, First Session. Washington, D.C.: U.S. Government Printing Office, 1939.

U.S. Congress, House. *Petroleum Investigation. Hearings Before a Subcommittee of the Committee on Interstate and Foreign Commerce on H. R. 441.* 73rd Congress, Recess. Washington, D.C.: U.S. Government Printing Office, 1934.

U.S. Congress, House. "Remarks of Hon. Leon H. Gavin (Pennsylvania) on food and oil production problems," March 30, 1943. 78th Congress, First Session, *Congressional Record,* Vol. 89, Part 10, pp. A1521-2.

U.S. Congress, House. *The Rubber Situation, Message From the President of the United States, September 10, 1942. House Document No. 836.* 77th Congress, Second Session. Washington, D.C.: U.S. Government Printing Office, 1942.

U.S. Congress, Senate. *Additional Report of the Special Committee Investigating the National Defense Program: Rubber. Senate Report 480, Part 7.* 77th Congress, Second Session. Washington, D.C.: U.S. Government Printing Office, 1942.

U.S. Congress, Senate. *Oil Concessions in Foreign Countries. Senate Document No. 97.* 68th Congress, First Session. Washington, D.C.: U.S. Government Printing Office, 1923.

U.S. Congress, Senate. *Petroleum Marketing Divorcement. Hearings Before Subcommittee of Senate Judiciary Committee on S. 2879 and S. 3752.* 75th Congress, Third Session. Washington, D.C.: U.S. Government Printing Office, 1938.

U.S. Congress, Temporary National Economic Committee. *Hearings Before the Temporary National Economic Committee. Part 14, Petroleum Industry.* Washington, D.C.: U.S. Government Printing Office, 1940.

U.S. Federal Oil Conservation Board. *Complete Record of Public Hearings Before the Federal Oil Conservation Board* (February 10-11, 1926). Washington, D.C.: U.S. Government Printing Office, 1926.

U.S. vs. *DuPont, General Motors et al.* Volume IV: U.S. District Court, DuPont's Post-Trial Briefs. Bound privately by Irénée duPont, 1954.

U.S. vs. *DuPont, General Motors et al.* Volume VI: U.S. District Court Proposed Findings of Fact and Conclusions of Law. Bound privately by Irénée duPont, 1954.

U.S. vs. *Standard Oil Company of New Jersey et al.* Supreme Court of the United States, Brief for Appellants, II. Washington, D.C.: U.S. Government Printing Office, 1909.

D. Secondary Sources: Books

Arnold, Thurman W. *The Folklore of Capitalism.* New Haven: Yale University Press, 1937.

______. The Symbols of Government. New Haven: Yale University Press, 1935.

Beard, Charles A. *President Roosevelt and the Coming of War, 1941.* New Haven: Yale University Press, 1948.

Berle, Adolf A., and Means, Gardiner C. The Modern Corporation and Private Property. New York: Macmillan, 1933.

Borkin, Joseph. *The Crime and Punishment of I.G. Farben.* New York: The Free Press, 1978.

Borth, Christy. *Pioneers of Plenty: The Story of Chemurgy.* New York: Bobbs-Merrill, 1939.

Brandes, Stuart D. *American Welfare Capitalism, 1880–1940.* Chicago: University of Chicago Press, 1970.

Calvert, Monte A. *The Mechanical Engineer in America, 1830–1910.* Baltimore: Johns Hopkins University Press, 1967.

Chandler, Alfred D., Jr. *Strategy and Structure: Chapters in the History of the American Industrial Enterprise.* Cambridge: M.I.T. Press, 1962.

______. *The Visible Hand: The Managerial Revolution in American Business.* Cambridge, Mass.: Harvard University Press, 1977.

Chandler, Alfred D., Jr., and Salsbury, Stephen. *Pierre S. duPont and the Making of the Modern Corporation.* New York: Harper and Row, 1971.

Cochran, Thomas C. *American Business in the Twentieth Century.* Cambridge: Harvard University Press, 1972.

Conkin, Paul. *The New Deal.* New York: Thomas Y. Crowell, 1967.

Culbert, David Holbrook. *News for Everyman: Radio and Foreign Affairs in Thirties America.* Westport, Conn.: Greenwood Press, 1976.

de Chazeau, Melvin G., and Kahn, Alfred E. *Integration and Competition in the Petroleum Industry.* New Haven: Yale University Press, 1959.

Dunlop, Robert G. *The Quiet Revolution.* Radnor, Pa.: Chilton, 1975.

Engler, Robert. *The Politics of Oil.* Chicago: Macmillan, 1961.

Enos, John Lawrence. *Petroleum Progress and Profits.* Cambridge: M.I.T. Press, 1962.

Fanning, Leonard M. *American Oil Operations Abroad.* New York: McGraw-Hill, 1947.

______. *The Story of the American Petroleum Institute.* New York: Leonard M. Fanning, 1959.

Feis, Herbert. *Petroleum and American Foreign Policy.* Stanford: Stanford University Press, 1944.

Filene, Edward A. *Speaking of Change.* Washington, D.C.: National Library Foundation, 1939.

Foster, George. "An Examination of Process Innovation in Petroleum Refining." In *Technology Transfer,* pp. 165–74, edited by Harold F. Davidson, Marvin J. Cetron, and Joel D. Goldhar. Leiden: Noordhoff International Publishing, 1974.

Frankel, P.H. *Essentials of Petroleum.* Reprint edition. New York: August M. Kelley, 1969.

Frey, John W., and Ide, H. Chandler. *A History of the Petroleum Administration for War.* Washington, D.C.: U.S. Government Printing Office, 1946.

Gardner, Lloyd C. *Economic Aspects of New Deal Diplomacy.* Madison: University of Wisconsin Press, 1967.

Gibb, George Sweet, and Knowlton, Evelyn H. *The Resurgent Years, 1911–1927,* Volume II, *History of Standard Oil (New Jersey).* New York: Harper & Brothers, 1956.

Giddens, Paul H. *Early Days of Oil: A Pictorial History of the Beginnings of the Industry in Pennsylvania.* Princeton: Princeton University Press, 1948.

______. *Standard Oil Company (Indiana), Oil Pioneer of the Middle West.* New York: Appleton-Century-Crofts, 1955.

Graham, Otis L., Jr., ed. *The New Deal, The Critical Issues.* Boston: Little Brown, 1971.

Hawley, Ellis W. *The New Deal and the Problem of Monopoly.* Princeton: Princeton University Press, 1966.

Hessen, Robert. *Steel Titan: The Life of Charles M. Schwab.* New York: Oxford University Press, 1975.

Hidy, Ralph W., and Hidy, Muriel E. *Pioneering in Big Business, 1882-1911,* Volume I, *History of Standard Oil Company (New Jersey).* New York: Harper & Brothers, 1955.

History of Petroleum Engineering. New York: American Petroleum Institute, 1961.

Hoover, Herbert. *The Memoirs of Herbert Hoover,* Volume II, *The Cabinet and the Presidency, 1920-1933.* New York: Macmillan, 1953.

Ickes, Harold L. *The Secret Diary of Harold L. Ickes,* Volume I, *The First Thousand Days, 1933-1936.* New York: Simon and Schuster, 1953.

______. *The Secret Diary of Harold L. Ickes,* Volume II, *The Inside Struggle, 1936-1939.* New York: Simon and Schuster, 1954.

______. *The Secret Diary of Harold L. Ickes,* Volume III, *The Lowering Clouds, 1939-1941.* New York: Simon and Schuster, 1954.

______. *Fightin' Oil.* New York: Alfred A. Knopf, 1943.

Ise, John. *The United States Oil Policy.* New Haven: Yale University Press, 1926.

James, Marquis. *The Texaco Story: The First 50 Years, 1902-1952.* New York: The Texas Company, 1953.

Johnson, Arthur M. *The Development of American Petroleum Pipelines: A Study in Private Enterprise and Public Policy, 1862-1906.* Ithaca: Cornell University Press, 1956.

______. *Petroleum Pipelines and Public Policy, 1906-1959.* Cambridge: Harvard University Press, 1967.

Johnson, Hugh S. *The Blue Eagle From Egg to Earth.* New York: Doubleday, 1935.

Jones, J. Edward. *And So—They Indicted Me!* New York: J. Edward Jones Publishing Corp., 1938.

Kelly, Patrick, and Kranzberg, Melvin, eds. *Technological Innovation: A Critical Review of Current Knowledge.* San Francisco: The San Francisco Press, 1978.

Kemnitzer, William J. *Rebirth of Monopoly.* New York: Harper & Brothers, 1938.

King, John O. *Joseph Stephen Cullinan.* Nashville: Vanderbilt University Press, 1970.

Koch, A.R. *The Financing of Large Corporations.* Washington, D.C.: National Bureau of Economic Research, 1943.

Kolko, Gabriel. *The Triumph of Conservatism.* New York: Glencoe, 1963.

______. *The Politics of War.* New York: Random House, 1968.

Kroos, Herman E. *Executive Opinion.* Garden City, N.Y.: Doubleday, 1970.

Larson, Henrietta M., and Porter, Kenneth Wiggins. *History of Humble Oil and Refining Company.* New York: Harper & Brothers, 1959.

Larson, Henrietta M.; Knowlton, Evelyn H.; and Popple, Charles S. *New Horizons, 1927-1950,* Volume III, *History of Standard Oil (New Jersey).* New York: Harper & Row, 1971.

Layton, Edwin T., Jr. *The Revolt of the Engineers: Social Responsibility and the American Engineering Profession.* Cleveland: The Press of Case Western University, 1971.

Leeston, Alfred M.; Crichton, John A.; and Jacobs, John C. *The Dynamic Natural Gas Industry.* Norman: University of Oklahoma Press, 1963.

Leuchtenburg, William E. *Franklin D. Roosevelt and the New Deal, 1932-1940.* New York: Harper & Row, 1963.

Lieuwen, Edwin. *Petroleum in Venezuela,* Volume 47, *University of California Publications in History.* Berkeley/Los Angles: University of California Press, 1954.

Lloyd, Henry Demerest. *Wealth Against Commonwealth.* New York: Harper & Brothers, 1894.

Mathews, John J. *Life and Death of an Oilman: The Career of E.W. Marland.* Norman: University of Oklahoma Press, 1951.

McKnight, David, Jr. *A Study of Patents on Petroleum Cracking.* Austin: University of Texas Press, 1938.

McLean, John G., and Haigh, Robert W. *The Growth of Integrated Oil Companies.* Boston: Harvard University Press, 1954.

The Mercantile, Manufacturing and Mining Interests of Pittsburgh, 1884. Pittsburgh: Chamber of Commerce, 1884.

Moody's Industrial Manual. New York: Moody's Investment Services, 1920–1975.

Myers, William Starr, and Newton, Walter H. *The Hoover Administration.* New York: Charles Scribner's Sons, 1936.

Nash, Gerald D. *United States Oil Policy, 1890–1964.* Pittsburgh: University of Pittsburgh Press, 1968.

Nelson, Daniel. *Managers and Workers.* Madison: University of Wisconsin Press, 1975.

Nevins, Allan. *Study in Power: John D. Rockefeller Industrialist and Philanthropist.* New York: Scribner's, 1953.

Noggle, Burl. *Teapot Dome: Oil and Politics in the 1920's.* Baton Rouge: Louisiana State University Press, 1962.

Notable Men of Pittsburgh and Vicinity. Pittsburgh: 1901.

Nourse, Edwin G., and Drury, Horace B. *Industrial Price Policies and Economic Progress.* Washington, D.C.: Brookings Institution, 1936.

O'Connor, Harvey. *The Empire of Oil.* New York: Monthly Review Press, 1955.

Oil for Victory. New York: Editors of *Look,* 1946.

Oil's First Century. Papers Given at the Centennial Seminar on the History of the Petroleum Industry. Cambridge: Harvard University Press, 1960.

Petroleum Facts and Figures. New York: American Petroleum Institute. First Ed., 1928; Second Ed., 1929; Third Ed., 1930; Fourth Ed., 1931; Fifth Ed., 1937; Sixth Ed., 1939; Seventh Ed., 1941; Eighth Ed., 1947; Centennial Ed., 1959.

Pettengill, Samuel B. *Hot Oil.* New York: Economic Forum Co., 1936.

Pittsburgh: Its Commerce and Industries, and the Natural Gas Interest. Pittsburgh: George B. Hill & Co., 1887.

Pogue, Joseph E. *The Economics of Petroleum.* New York, 1921.

Poor's Industrial Manual. New York: Poor's Publishing Company, 1900–1945.

Popple, Charles Sterling. *Standard Oil Company (New Jersey) in World War II.* New York: Standard Oil Company, 1952.

Porter, Glenn. *The Rise of Big Business, 1860–1910.* Arlington Heights, Illinois: AHM Publishing Corp., 1973.

Rauch, Basil. *Roosevelt: From Munich to Pearl Harbor.* New York, 1950.

Rister, Carl Coke. *Oil: Titan of the Southwest.* Norman: University of Oklahoma Press, 1949.

Rostow, Eugene V. *A National Policy for the Oil Industry.* New Haven: Yale University Press, 1948.

Rundell, Walter Jr. *Early Texas Oil. A Photographic History, 1866–1936.* College Station: Texas A&M University Press, 1977.

Sampson, Anthony. *The Seven Sisters.* New York: The Viking Press, 1975.

Schlesinger, Arthur M., Jr. *The Age of Roosevelt,* Volume II, *The Coming of the New Deal.* Boston: Houghton-Mifflin, 1959.

______. *The Age of Roosevelt,* Volume III, *The Politics of Upheaval.* Boston: Houghton-Mifflin, 1960.

Schroeder, Paul W. *The Axis Alliance and Japanese-American Relations, 1941.* Ithaca: Cornell University Press, 1958.

Scoville, John, and Sargent, Noel. *Fact and Fancy in the TNEC Monographs.* New York: National Association of Manufacturers, 1942.

Shwadran, Benjamin. *The Middle East, Oil and the Great Powers.* New York: Praeger, 1955.

Sobel, Robert. *The Age of Giant Corporations.* Westport, Conn.: Greenwood Press, 1972.
Spindletop, Where Oil Became an Industry. Beaumont, Texas: Spindletop 50th Anniversary Commission, 1951.
Stotz, Louis. *History of the Gas Industry.* New York: Stettiner Bros., 1938.
Tarbell, Ida M. *The History of the Standard Oil Company.* New York: Macmillan, 1904.
Thompson, Craig. *Since Spindletop: A Human Story of Gulf's First Half-Century.* Pittsburgh: Gulf Oil Corporation, 1951.
Translations of Captured German Scientific and Industrial Reports. F.L. Reports 99–117 (U.S. Government Technical Oil Mission, Reels 177–8). New York: Consultants Bureau, 1948.
Van Hise, Charles R. *Concentration and Control.* New York: Macmillan, 1941.
Van Winkle, Matthew. *Aviation Gasoline Manufacture.* New York: McGraw Hill, 1944.
Wall, Bennet H., and Gibb, George S. *Teagle of Jersey Standard.* New Orleans: Tulane University Press, 1974.
Watkins, Myron W. *Oil: Stabilization or Conservation.* New York: Harper & Brothers, 1937.
Whitby, G.S.; Davis, C.C.; and Dunbrook, R.F., eds. *Synthetic Rubber.* New York: Wiley, 1954.
White, Gerald T. *Formative Years in the Far West (Standard Oil of California).* New York: Appleton-Century-Crofts, 1962.
Whittlesey, Charles R. *National Interest and International Cartels.* New York: Macmillan, 1946.
Wilbur, Ray Lyman, and Hyde, Arthur M. *The Hoover Policies.* New York: Charles Scribner's Sons, 1937.
Wilkins, Mira. *The Emergence of Multinational Enterprise: American Business Abroad From the Colonial Era to 1914.* Cambridge, Mass.: Harvard University Press, 1970.
———. *The Maturing of Multinational Enterprise: American Business Abroad From 1914–1970.* Cambridge, Mass.: Harvard University Press, 1974.
Williamson, Harold F., and Daum, Arnold R. *The American Petroleum Industry,* Volume I, *The Age of Illumination, 1859–1899.* Evanston: Northwestern University Press, 1959.
Williamson, Harold F.; Andreano, Ralph L.; Daum, Arnold R.; and Klose, Gilbert C. *The American Petroleum Industry,* Volume II, *The Age of Energy, 1899–1959.* Evanston: Northwestern University Press, 1963.
Wilson, Joan Hoff. *Herbert Hoover: Forgotten Progressive.* Boston: Little Brown, 1975.
Wolbert, George S., Jr. *American Pipelines.* Norman: University of Oklahoma Press, 1951.
Wolfskill, George. *The Revolt of the Conservatives.* Cambridge, Mass.: Houghton-Mifflin, 1962.
Zimmerman, Erich. *Conservation in the Production of Petroleum.* New Haven: Yale University Press, 1957.
Zinn, Howard, ed. *New Deal Thought.* New York: Bobbs-Merrill, 1966.

E. Secondary Sources: Periodicals

"Acreage Proration Plan for East Texas Submitted to Sterling." *National Petroleum News* 23, 44 (November 4, 1931):23.
"Acreage Proration Plan Proposed to Check East Texas Drilling." *National Petroleum News* 23, 43 (October 28, 1931):43.
"Advocates of Price Fixing Control Code's Planning Committee." *National Petroleum News* 25, 36 (September 6, 1933):12.
"Alcogas." *Business Week* (February 8, 1933):9.
Ames, Judge C.B. "The Oil Situation." *National Industrial Conference Board Transcripts* 38:85–96.

"Another Oilman Goes East." *Business Week* (February 12, 1955):158-63.

"API—Past Living Presidents: Edwin B. Reeser, J. Edgar Pew." *National Petroleum News* 38, 44 (October 30, 1946):52-4.

Auerbach, Jerold. "New Deal, Old Deal, or Raw Deal: Some Thoughts on New Left Historiography." *The Journal of Southern History* 35 (February 1969):18-30.

Bernstein, Barton J. "Industrial Reconversion: The Protection of Oligopoly and Military Control of the Wartime Economy." *American Journal of Economics and Sociology* 26, 2 (April 1967):159-72.

______. "The Removal of War Production Board Controls on Business, 1944-46." *Business History Review* 39, 2 (Summer 1965):243-60.

"Big Oil Deal." *Toledo Evening Bee* (December 13, 1894).

"Bolt From the Sun." *Time* (October 24, 1932):49.

Boyd, T.A. "Gasoline Went to War and the Part Chemistry Had in it." *Record of Chemical Progress* (January/April 1947):11.

______. "Pathfinding in Fuels and Engines." *Society of Automotive Engineering (SAE) Quarterly Transactions* 4, 2 (April 1950):182-95.

Bridgeman, Oscar C. "Utilization of Ethanol-Gasoline Blends as Motor Fuels." *Industrial and Engineering Chemistry* 28:9 (September 1936):1102-12.

Burnham, John Chynoweth. "The Gasoline Tax and the Automobile Revolution." *Mississippi Valley Historical Review* 48, 3 (December 1960):435-59.

Butterworth, Irvin. "Natural Gas: What is it and can it Become a Competitor of Coal Gas as an Illuminating Agent?" *The American Gas Light Journal* 42, 8 (April 16, 1885):206.

Carll, Jno. F. "The Natural Gas Craze." *The American Gas Light Journal* 44, 6 (March 16, 1886);162.

Childs, Marquis W. "Pennsylvania's Boss Pew, A G.O.P. Power Out of Grim Hate of New Deal." *St. Louis Post Dispatch* (May 26, 1940):C-1.

"Code Fails Without Price Fixing Say Cost Recovery Advocates." *National Petroleum News* 25, 32 (August 9, 1933):14-5.

"The Country Turns Republican." *Life* (November 15, 1943):27-31.

"East Texas Acreage Plan Delayed by Numerous Protests." *National Petroleum News* 23, 45 (November 11, 1931):26.

Egloff, Gustav, and Morrell, J.C. "Alcohol-Gasoline as Motor Fuel." *Industrial and Engineering Chemistry* 28, 9 (September 1936):1080-8.

Ellsworth, Catherine C. "Integration Into Crude Oil Transportation in the 1930's, A Case Study: The Standard Oil Company (Ohio)." *Business History Review* 35 (Summer 1961):180-210.

Enos, John L. "A Measure of the Rate of Technological Progress in the Petroleum Refining Industry." *Journal of Industrial Economics* (June 1958):180-93.

"Farm-Brewed Fuel." *The Business Week* (March 15, 1933):14-5.

"The Fortune Director of the 500 Largest Industrial Corporations." *Fortune* (May 1975):208-35.

"Full Pay." *Time* (May 27, 1940):10, 12.

"40 Years of the 'Oil Age.'" *American Petroleum Institute Quarterly* (April 1944).

"Gas Light Companies in the United States." *The American Gas Light Journal* 4 (June 15, 1863):373.

"Gas Stock that Didn't Pay." *The American Gas Light Journal* 29, 2 (July 16, 1878):37.

Gressley, Gene M. "Thurman Arnold, Antitrust, and the New Deal." *Business History Review* 38, 2 (Summer 1964): 214-31.

Hagood, L.N. "Object is as Plain as the Nose on Your Face; Seek Oil Monopoly Through Governmental Control of the Industry; Will Governors of Oil States Stand For it?" *Inland Oil Index* (April 26, 1939):1.

Hallgreen, Mauritz A. "The N.R.A. Oil Trust." *The Nation* (March 7, 1934):272-3.

"Halt in Spread of Price Wars Causes Ickes to Postpone Price Fixing." *National Petroleum News* 25, 38 (September 20, 1933):11-5.

"Houdry Catalytic Operations at Marcus Hook." *World Petroleum*, Annual Refining Issue, 1943.

Ickes, Harold L. "After the Oil Deluge, What Price Gasoline?" *Saturday Evening Post* (February 16, 1935):39-42.

"Ickes Approves Revised Oil Agreements." *Oildom* 167, 15 (January 22, 1934):163-4.

"Ickes Considers Eastern Marketing Plan." *Oildom* 179, 13 (January 18, 1935):159, 161-3.

"Ickes Issues Statement on Agreements." *Oildom* 167, 16 (January 23, 1934):167-70.

"Important Uses of Natural Gas." *The American Gas Light Journal* 30, 17 (October 2, 1878):148.

"Inside Stories of Great Corporations, Sun Oil." *The Wall Street News* (December 7, 1925):3.

"Institute President Refutes Ickes' Charges in Magazine Article." *National Petroleum News* (February 13, 1935):17.

"Iowa Alcohol-Gasoline Proposal Tabled Temporarily as Idea Grips West." *National Petroleum News* 25, 9 (March 1, 1937):11-2.

"Its C.I.O. at Sun." *Business Week* (July 17, 1943):92.

Jensen, Michael C. "The Pews of Philadelphia." *The New York Times* (October 10, 1971):1, 9.

"General Johnson at Work on Revised Oil Code Following Stormy Hearings at Washington." *Oil and Gas Journal* 32, 12 (August 10, 1933):33.

"Johnson Prepares NRA Drive Wind-up." *The New York Times* (August 20, 1934).

"Last Chapter Finally Written in Madison's Anti Trust Case." *National Petroleum News* 33, 23 (June 4, 1941):5.

"Leaders Tell Oil Industry's Needs." *World Petroleum* (November 1932):458-9.

"Letter of Sun Oil President Probes Meaning of Vague Phrases in Anglo-American Pact." *National Petroleum News* (August 23, 1944):9-10, 14.

"Mamma Spank: 18 Major Oil Companies Indicted." *Time* (October 18, 1937):63.

Merrill, Harwood F. "Price Fixing: Right or Wrong?" *Forbes* (November 15, 1933):11-2, 36.

Marcosson, Issac F. "The Black Golconda." *Saturday Evening Post* (March 22, 1924):3-4, 161-6.

Meyer, Adolf. "The Combustion Gas Turbine: Its History, Development, and Prospects." *Institute of Mechanical Engineering Proceedings* 141 (1939):197-222.

Miner, H. Craig. "The Cherokee Oil and Gas Company, 1889-1902: Indian Sovereignty and Economy Change." *Business History Review* 46, 1 (Spring 1972):45-66.

"J.A. Moffett Resigning as Standard Official, Mentioned as Probable Oil Administrator." *Oil and Gas Journal* 32, 11 (August 3, 1933):9.

"Monsieur Houdry's Invention." *Fortune* (February, 1939):56-7, 127-32, 137-40.

"Mournful Talk on Postwar Jobs Plays Into Hands of 'Planners,' J. Howard Pew Warns." *National Petroleum News* (May 2, 1945):16-8.

"Muckraking by Earle." *Chester Times* (September 17, 1936):6.

"Murrayism: A Dramatization of Popular Unrest." *National Petroleum News* 23, 33 (August 19, 1931):19-20, 22.

"The Nation's Future Supply of Petroleum, Demand and Operating Conditions." *The Oil Age* 22, 9 (September 1925):1.

"No Provision for Estimating Crude Demand Will Stand in Final Texas Bill." *National Petroleum News* 23, 31 (August 5, 1931):19-20, 22.

Nordhauser, Norman. "Origins of Federal Oil Regulation in the 1920s." *Business History Review* 47, 1 (Spring 1973):53-71.

"Oil at the Capitols." *The Oil Weekly* (June 26, 1939):53-6, 65.

"Oil Code is Effective as Price Fixing Postponed." *National Petroleum News* 25, 36 (September 6, 1933):78.

"Oil Code Clause on Price Fixing Called Blunder." *New York Herald Tribune* (August 25, 1933):56.
"Oil Conservation Board Sets Feb. 10 and 11 for Hearings." *The Oil Age* 23, 2 (February 1926):85.
"Oil: Fifty Years of Sun Find Pew Family Still at the Helm." *Newsweek* (October 10, 1936):36-8.
"Oil Industry Can Promote Conservation Better Without National Interference." *The Oil Age* 23, 3 (March 1926):11.
"Oil Industry Will Wash its Linen on the Capitol's Doorstep." *National Petroleum News* 25, 27 (July 5, 1933): 11-2.
"Oil Men Will Plan Legal Curtailment." *New York Times* (December 10, 1926):23.
"Oil Producers Meet at Los Angeles." *New York Times* (January 20, 1926):8.
"Operators Split on Plan for Curtailment of East Texas at Commission Hearing." *National Petroleum News* 23, 21 (July 1, 1931):19-20.
Parton, Margaret. "Who Are America's 10 Richest Men" *Ladies Home Journal* (April 1957):72-3, 173-9.
Pew, J. Edgar. "Fifth Dimension in the Oil Industry." *Bulletin of the American Association of Petroleum Geologists* 25 (July 1941):1283-90.
______. "The Oil Industry; Its Relation to the Public." *The Oil Age* 22, 4 (April 1925):9.
"J. Edgar Pew." *National Petroleum News* (October 30, 1946):53-4.
"J. Edgar Pew, Sun Oil Company Dies." *National Petroleum News* (November 27, 1946):16.
Pew, J. Howard. "American Business in a Planned World." *National Industrial Conference Board Transcripts* 89, 62:33-49.
______. "Conserving Oil Reserves." *The Philadelphia Evening Bulletin* (November 20, 1936):10.
______. *The Economic Record* (October 1944):89.
______. "Oil Showing Strong at Monopoly Probe." *The Journal of Commerce* (November 15, 1939):1-A, 2-A.
______. "Preserving the Private Enterprise System." *Vital Speeches* (February 1, 1946):244.
______. "Price Control." *Vital Speeches* (May 1, 1948):440.
______. "A Squawk From the Petroleum Goose." *Nation's Business* (October 1932):20-1, 56.
"J. Howard Pew is Champion of Man's Rights and Freedom." *The Oil Daily* (November 19, 1969):p. 28.
"J. Howard and Arthur E. Pew Resign Sun Positions; Robert G. Dunlop Slated to Become President." *National Petroleum News* 39, 11 (March 12, 1947):19.
"J. Howard Pew is a Symbol of Men Who Have Made Oil Industry Great." *National Petroleum News* 39, 16 (April 16, 1947):23-4.
"J.H. Pew Hits Trend Toward Collectivism." *National Petroleum News* 37, 21 (May 23, 1945):16.
"J. Howard Pew Strikes at Anglo-U.S. Oil Pact as Forerunner of Cartels, Foe of Enterprise." *National Petroleum News* 37, 13 (March 28, 1945):16, 58.
"Mr. Pew at Valley Forge." *Time* (May 6, 1940):15-8.
"Pew Answers Charges Made by CIO Speaker." *National Petroleum News* 38, 27 (July 3, 1946):24.
"Pew, Arnold in Economy Talk at Haverford." *Chester Times* (January 18, 1945).
"Pew Denounces Oil Pact as Vicious Cartel." *National Petroleum News* 36, 43 (October 25, 1944):3, 14.
"Pew Hits at Rise in Gasoline Tax." *Wall Street Journal* (November 26, 1931):11.
"Pew Protests Federal Oil Price Fixing." *Oildom* (July 28, 1933):783.
"Pew to Make His Reply to O'Mahoney Committee." *The Tulsa Tribune* (June 11, 1939):1.
Pew, Joseph N., Jr. "Free Men Produce More Than Slaves." Excerpts from a letter published in *The Farm Journal* (March 1945). Philadelphia: Sun Oil Company, 1945.

"Pioneer in Petroleum Refining, J. Howard Pew Philadelphian of the Month." *Philadelphia Magazine* (July, 1947):22.

"Pittsburgh, Pennsylvania as Seen by an Englishman." *The American Gas Light Journal* 46, 3 (February 2, 1887):71.

Platt, Warren C. "Another Gag Rule by Secretary Ickes." *National Petroleum News* (October 25, 1933):5-6.

———. "Ickes Speaks As Would-Be Oil Dictator." *National Petroleum News* (February 13, 1935):13-6.

Potomachus. "Pew of Pennsylvania." *The New Republic* (May 8, 1944):625-7.

"Price Sections of Code Continue Recent Refinery Losses." *National Petroleum News* 25, 34 (August 23, 1933):17-8.

"Protection Control, Its Need, Legality, Technique Explained to A.P.I." *National Petroleum News* 23, 45 (November 11, 1931):23-5.

Pursell, Carroll W., Jr. "The Farm Chemurgic Council and the United States Department of Agriculture, 1935-1939." *Isis* 60, 3, 302 (Fall 1969):307-17.

"Rebuke by NLRB." *Business Week* (June 21, 1941):52.

"Report on Rubber." *Time* (July 20, 1942):18-9.

Rister, Carl Coke. "The Oilman's Frontier." *Mississippi Valley Historical Review* 37 (June 1950):3-16.

Robbins, L.H. "Ickes Defines the Task Ahead." *New York Times Magazine* (April 1, 1934):1-2, 22.

"Roosevelt Names Ickes to Administer the Provisions of the Oil Code." *National Petroleum News* 25, 35 (August 30, 1933):3-4, 6.

Rostow, Eugene V., and Sachs, Arthur S. "Entry into the Oil Refining Business: Vertical Integration Re-examined." *Yale Law Journal* (June/July 1952):856.

Rottenberg, Dan. "The Sun Gods." *Philadelphia Magazine* (September 1975):111-9, 180-97.

Schwab, Charles M. "What I've Learned About Business." *Nation's Business* (February 1930):198-9.

Shoemaker, Jane. "Sun Oil to Expand Overseas Operations." *Philadelphia Inquirer* (August 12, 1974):18-C.

"Short Sightedness of a Few Individuals Responsible for 10-Cent Crude." *National Petroleum News* 23, 28 (July 15, 1931):23.

Smith, George Otis. "The Petroleum Resources of the United States." National Industrial Conference Board Transcripts 6 (May 15, 1930):93-108.

Smith, Lawrence E. "'Poor Boy' Operators in East Texas, Promises Falling, Are on Way Out." *National Petroleum News* 23, 36 (July 1, 1931):25-7.

———. "Scramble for Crude as East Texas Output is Shut Off Under Martial Law." *National Petroleum News* 23, 33 (August 19, 1931):19-20, 22.

———. "Troops' Occupation of East Texas Fields Marked by Order and Efficiency." *National Petroleum News* 23, 34 (August 26, 1931):32-3.

Stafford, Roger B. "Oil Industry Leaders Split on Price Regulation Under N.R.A." *National Petroleum News* 25, 32 (August 9, 1933):7-8.

"Sterling Says He Will Veto Market Demand Provision." *National Petroleum News* 23, 31 (August 5, 1931):23.

"Stripper Well Operators Defend Definition." *The Independent Monthly* 5, 7 (November 1934):10, 14-16.

"Sun Oil." *Fortune* (February 1941):51-3, 112-9.

"Sun Oil Company." *Oil and Gas Journal* (February 1944):67.

"Sun Oil Company Bright Ray in Oil Trade." *Boston Financial News* (August 2, 1932):5.

"Sun Oil Employees Hand Testimonial to Pew, Directors." *Philadelphia Inquirer* (July 25, 1935):17.

"Sun Oil Planning to Buy Sunray DX." *The New York Times* (January 20, 1968):37.
"Sun Oil Split Into 4 Subsidiaries." *Philadelphia Inquirer* (August 7, 1975):4-B.
"Sun Shipbuilding." *Fortune* (February, 1941):54.
"Tanker Sea Power Essential Says Sun Oil Co. President." *National Petroleum News* 32, 36 (August 7, 1940);22.
Tedlow, Richard S. "The National Association of Manufacturers and Public Relations During the New Deal." *Business History Review* 50, 1 (Spring 1976):25–45.
"Texas Commission Meeting to Frame Oil Regulations." *National Petroleum News* 23, 34 (August 26, 1931):34.
Tunison, B.R. "The Future of Industrial Alcohols." *The Journal of Industrial and Engineering Chemistry* 12, 4 (April 1, 1920):370–6.
Wagner, Paul. "Court Decision Voiding Texas Proration Challenges Special Session." *National Petroleum News* 23, 30 (July 29, 1931):35–6.
"What Do They Mean: Monopoly?" *Fortune*, (March, 1938):120–8.
White, A.G. "Study of the Production and Consumption of Fuel Oil in This Country." *National Industrial Conference Board Transcripts* 3 (October 17, 1929):61-90.
Wik, Reynold Millard. "Henry Ford's Science and Technology for Rural America." *Technology and Culture* 3, 3 (Summer 1962):248–9.
Wilkins, Mira. "Multinational Oil Companies in South America in the 1920s." *Business History Review* 48, 3 (Fall 1974):414–46.
Wood, H.L. "Personnel of Petroleum." *The Oil Age* 22, 4 (April 1925):21.

F. Our Sun: Sun Oil Company Magazine

Note: *Our Sun* is numbered in two series. The First Series runs from Volume 1 (October 1923) through Volume 11 (July 1934). The Second Series runs from Volume 1 (November 1934) to the present.

"Advertising Reflects the Progress of Sunoco." 50th Anniversary Issue, 3, 1 (1936):36–9.
"Announce Stock Purchase Plan for Employees." 3, 8 (June 30, 1926):3.
Bauer, G.J. "*Our Sun*—25 Years." 13, 6 (Fall 1948):12.
Bergen, Daniel J. "Oklahoma." 13, 7 (July 1947):1–5.
"Beware! Little Red Riding Hood." 2, 1 (July 1935):2–4, 32.
"Billionth Gallon." 11, 1 (February 1945):15.
"Blue Sunoco 25th Anniversary, 1927-1952." 17, 2 (Spring 1952):29–33.
"Blue Sunoco Dynafuel." 11, 4 (November 1945):1–2.
Clayden, A. Ludlow. "Gasoline, What it Was, What it is, What it Will Be." 4, 5 (April/May 1927):5–7.
______. "Oil in the Internal Combustion Engine." 2, 7 (April 30, 1925):6–7.
"Code of Ethics for Petroleum Industry Now in Effect." 6, 6 (September/October 1929):5–7.
"Frank Cross, 1870-1946." 12, 4 (January 1947):9.
"The Drilling of Haymaker's Well." 50th Anniversary Issue, 3, 1 (1936):28–9, 47–8.
"Employees at Toledo Host to Two Sun Officers." 10, 4 (August/September 1944):2–7.
"The Employee Representation Plan." 10, 3 (August 1933):2.
"Facts About the Sun Oil Company." 1, 12 (September 1924):1–2.
"The First Fifty Years." 16, 4 (Autumn 1951):2–8.
"550 Mile Pipe Line For Gasoline to be Built by Company." 7, 2 (February 10, 1930):3.
"Golden Anniversary in London." 24, 2 (Spring 1959):23–4.
"History and Growth of Pipe Lines." 50th Anniversary Issue, 3, 1 (1936):16–8.
"Houdry Process." 5, 4 (January 1939):8–9.

"I Remember When. . . ." 16, 4 (Autumn 1951):19-25.
"Improvement Seen in Business." (*Sun Dial*) 1, 3 (January 1935):16.
"In the Oil Fields With Sun." 50th Anniversary Issue, 3, 1 (1936):30-4.
"Information and Your Job." 6, 3 (March 1929):3-6.
"Looking Back Over Fifty Years." 50 Anniversary Issue, 3, 1 (1936):2-7.
"A Miracle of Gasoline Chemistry." 6, 3 (November 1939):2-12.
"Mineraloelwerke Albrecht & Companie, A Progressive Sun Oil Affiliation." 3, 4 (January 30, 1926):3-7.
"Our Company—Its History, Scope and Facilities." 2, 12 (September 30, 1925):3-5.
"Our Company—Its History, Scope and Facilities." 3, 1 (October 31, 1925):2-3.
"Petroleum and Ships." 8, 1 (August/September 1941):2.
Pew, J. Edgar. "Condition of the Oil Business is Better Than For Several Years Past." 3, 5 (February 28, 1925):2.
———. "Savings and Re-Utilization of All Gas Proposed Over-Production Remedy." 4, 6 (June 30, 1927):16-7.
"J. Edgar Pew, 1870-1946." 12, 4 (January 1947):8.
Pew, J. Howard. "Government in Business." 2, 5 (April 1936):2-4, 26.
———. "Prop of Prosperity." 1, 1 (November 1934):1.
———. "3000 Years Ago They Tried Planned Economy." Address Delivered at the Institute of Public Affairs, University of Virginia. 2, 2 (September 1935):1-2.
———. "A Living Monument to the American System of Free Enterprise." A.P.I. Address. 5, 4 (January 1939):3-5, 40.
"J.N. Pew: A Biographical Sketch." 75th Anniversary Issue, 26, 3-4 (Summer/Autumn 1961):11-3.
Pew, J.N., Jr. "Faith in Free Enterprise Has Been Sun's Guiding Star for 75 Years." 75th Anniversary Issue, 26, 3-4 (Summer/Autumn 1961):2-3.
"Joseph N. Pew, Jr., Marks Half a Century of Service With Sun." 23, 3 (Summer 1958):12.
"Joseph N. Pew, Jr.: A Remembrance." 28, 3 (Summer 1963):3.
"Robert E. Pew Dies at Toledo." 2, 9 (June 30, 1925):2-3.
"Pioneer Venture." 16, 1 (Winter 1951):15-7.
"Pipelines Made Headlines." 9, 3 (March 1943):16.
"The Price Cutter." 1, 5 (February 1924):1.
"Prophet With Honor . . . At Home and Abroad." 12, 3 (October 1946):20-3.
"Recreation Has its Part in Company Activities." 50th Anniversary Issue, 3, 1 (1936):44-6.
"Rubber From Oil." 11, 3 (July/August 1945):12.
"The Scare of Scarcity." 2, 4 (February 1936):26-7.
"The Story of Sun." 75th Anniversary Issue, 26, 3-4 (Summer/Autumn 1961):17-33.
"Sunoco Gas Proves Itself in Special Test." 2, 8 (May 29, 1925):2.
"Sun Oil Company Extensions." 1, 4 (January 1924):1.
"Sun Oil Company Men Prominent in A.P.I. Work." 5, 1 (November 30, 1927):18.
"Sun Oil's Newest Plant for Aviation Fuel Dedicated." 10, 1 (December 1943):1-7.
"Sun Oil Now Building Plant for Butadiene Manufacture." 9, 2 (December 1942):33, 48.
"Sun Rise in the Southwest." 16, 1 (Winter 1951):18-22.
"Sun Ship's 30 Years." 12, 2 (July/August 1946):4.
"Sun Ship . . . The World's Largest Single Shipyard." 9, 3 (March 1943):13.
"Tankers Under the Sun Flag." 50th Anniversary Issue, 3, 1 (1936):20-3, 35.
"Ten Years of Progress." 14, 2 (Spring 1949):1-6, 32.
"The Toledo Refinery Looks Back Over 50 Years." 2, 3 (July/August 1945):1-9.
"True Blue." 14, 2 (Spring 1959):33.
"Turning Crude Into Sun Products." 50th Anniversary Issue, 3, 1 (1936):8-11, 19, 48.
"Twin Sun Leaders Mark 75th Milestone." 11, 4 (November 1945):14-5.

"Two Sun Pioneers." 75th Anniversary Issue, 26, 3-4 (Summer/Autumn 1961):14-6.
"You Are Asked to Take a Hand in the Battle of Marketers vs. Racketeers." 9, 3 (March 1932):3-5.
"A Warning to American Industry." 2, 3 (November 1935):10-1.
"With 'The Sun' in Venezuela." 3, 9 (July 31, 1926):3-7.

G. Pamphlets

Annual Report of the Houdry Process Corporation. Paulsboro, New Jersey/Wilmington, Delaware, 1933-1946.
Annual Report of the Sun Oil Company. Philadelphia: Sun Oil Company, 1925-1946.
Facts About Sun Oil Company. Philadelphia: Sun Oil Company, 1957.
Forty Years' Growth in the Oil Industry. Philadelphia: Sun Oil Company, 1926.
Fuel Oil. Volume 11 of a series of pamphlets issued monthly. Philadelphia: Sun Company, 1918.
Hydrolene. Volume 8 of a series of pamphlets issued monthly. Philadelphia: Sun Company, 1917.
Pew, J. Edgar. *United States Petroleum Resources.* Philadelphia: Sun Oil Company, 1945.
J. Edgar Pew, His Life and Times, 1870-1946. Philadelphia: Sun Oil Company, 1947.
Pew, J. Howard. *Governmental Planning and Control as Applied to Business and Industry.* Princeton: Guild of Brackett Lectures, 1938.
———. *The Oil Industry: A Living Monument to the American System of Free Enterprise.* Chicago: American Petroleum Institute Reprint, 1938.
———. *What Price Gasoline?* Reprint of Address Before the Meeting of the National Petroleum Association, 1937. Philadelphia: Sun Oil Company, 1937.
———. *Which Road to Take.* American Liberty League Publication No. 53, July 12, 1935.
The Story of Sun Oil Company. Philadelphia: Sun Oil Company, 1938.
Sun Oil: Building Flexibility for the Future. St. Davids, Pennsylvania: Sun Oil Company, 1975.
Tax Data. New York: American Petroleum Industries Committee, 1932.

H. Unpublished Manuscripts, Theses, and Dissertations

Bryant, Lynwood. "The Problem of Knock in Gasoline Engines." Unpublished Manuscript, 1972, in author's personal file.
Chandler, Alfred D. Jr. "The Standard Oil Company—Combination, Consolidation and Integration." Harvard Business School Case Study #9-362-001, BH 120. Cambridge: President and Fellows of Harvard College, 1973.
"Company's Beginnings—Southwestern Phase." Unpublished Manuscript, 1952, Folder: "Sun Oil History," Library File, Sun Oil Company Library, Marcus Hook, Pennsylvania.
Guider, Thomas C. "Sun Oil Company and the Butadiene Controversy." Unpublished Manuscript, 1973, Eleutherian Mills Historical Library.
"History of Sun Oil Company." Unpublished Manuscript, 1949, Folder: "Sun Oil History," Library File, Sun Oil Company Library, Marcus Hook, Pennsylvania.
"History of Sun Oil Company." Unpublished Manuscript, 1957(?), Folder: "Sun Oil History," Library File, Sun Oil Company Library, Marcus Hook, Pennsylvania.
Howard, Porter. "History of the Sun Oil Company." Unpublished Manuscript, 1951(?), Folder #1: "Sun Oil History," Library File, Sun Oil Company Library, Marcus Hook, Pennsylvania.

Kelsey, John L. "A Financial Study of the Sun Oil Company of New Jersey From Its Inception Through 1948." M.B.A. Thesis, Wharton School, University of Pennsylvania, 1951.

Longin, Thomas Charles. "The Search For Security: American Business Thought in the 1930's." Ph.D. Dissertation, University of Nebraska, 1970.

MacMurtrie, A.; Hughes, Ed; Marshall, John; Cross, Robert; Lampugh, George; and Herman, Murry. "Refinery Notes." Unpublished Manuscript, 1936, Accession #1317, Eleutherian Mills Historical Library, J.H. Pew Presidential, Box 58.

McFarland, John. "History of Sun Oil Company." Unpublished Manuscript, 1957, Folder #2: "Sun Oil History," Library File, Sun Oil Company Library, Marcus Hook, Pennsylvania.

"Milestones Along Sun's Fifty Year March." Unpublished Manuscript, 1936, Folder: "Sun Oil History," Library File, Sun Oil Company Library, Marcus Hook, Pennsylvania.

Nordhauser, Norman Emanuel. "The Quest for Stability: Domestic Oil Policy, 1919–1935." Ph.D. Dissertation, Stanford University, 1970.

"Our Sun, 65 Years." Unpublished Manuscript, 1950, Folder: "Sun Oil History," Library File, Sun Oil Company Library, Marcus Hook, Pennsylvania.

Pew, J. Edgar. "Notes Made by J. Edgar Pew." Unpublished Manuscript, 1942, Folder: "Sun Oil History," Library File, Sun Oil Company Library, Marcus Hook, Pennsylvania.

———. "Rough Notes by J. Edgar Pew on the Discovery of Spindletop." Unpublished Manuscript, 1947, Folder #2: "Sun Oil History," Library File, Sun Oil Company Library, Marcus Hook, Pennsylvania.

"Sun History." Unpublished Manuscript, 1944, Folder: "Sun Oil Chronology," Library File, Sun Oil Company Library, Marcus Hook, Pennsylvania.

"Sun Oil Company, Brief History 1933–1955." Unpublished Manuscript, 1961(?), Folder #1: "Sun Oil History," Library File, Sun Oil Company Library, Marcus Hook, Pennsylvania.

Sweeney, Matthew B. "Recollections of Experience With Sun Oil Company in Ohio." Unpublished Manuscript, 1936, Folder: "Sun Oil History, 1886–1911," Library File, Sun Oil Company Library, Marcus Hook, Pennsylvania.

Swope, Selden T. "The Credit Department of the Sun Oil Company." M.B.A. Thesis, Wharton School, University of Pennsylvania, 1936.

Tedlow, Richard S. "Keeping the Corporate Image: Public Relations and Business, 1900–1950." Ph.D. Dissertation, Columbia University, 1976.

Weston, W.W. "Sun Oil History." Unpublished Manuscript, 1943(?), Folder #1: "Sun Oil History," Library File, Sun Oil Company Library, Marcus Hook, Pennsylvania.

Index